Coping with Risk in Agriculture

3rd Edition

Applied Decision Analysis

Coping with Risk in Agriculture

3rd Edition

Applied Decision Analysis

J. Brian Hardaker
Emeritus Professor of Agricultural Economics, University of New England, Australia

Gudbrand Lien
Professor, Lillehammer University College; Senior Researcher, Norwegian Agricultural Economics Research Institute, Norway

Jock R. Anderson
Emeritus Professor of Agricultural Economics, University of New England, Australia; Consultant, Washington, DC, USA

Ruud B.M. Huirne
Professor of Cooperative Entrepreneurship, Wageningen University and Rabobank, The Netherlands

www.cabi.org

CABI is a trading name of CAB International

CABI
Nosworthy Way
Wallingford
Oxfordshire OX10 8DE
UK

Tel: +44 (0)1491 832111
Fax: +44 (0)1491 833508
E-mail: info@cabi.org
Website: www.cabi.org

CABI
38 Chauncy Street
Suite 1002
Boston, MA 02111
USA

Tel: +1 800 552 3083 (toll free)
E-mail: cabi-nao@cabi.org

A catalogue record for this book is available from the British Library, London, UK.

Library of Congress Cataloging-in-Publication Data

Hardaker, J. B., author.
Coping with risk in agriculture : applied decision analysis / J. Brian Hardaker, Gudbrand Lien, Jock R. Anderson, and Ruud B.M. Huirne. -- Third edition.
pages cm
Includes bibliographical references and index.
ISBN 978-1-78064-240-6 (pbk. : alk. paper) -- ISBN 978-1-78064-574-2 (hardback : alk. paper)
1. Farm management--Decision making. 2. Farm management--Risk assessment. I. Title.

S561.H2335 2015
630.68--dc23

2014035631

ISBN-13: 978 1 78064 240 6 (Pbk)
978 1 78064 574 2 (Hbk)

Commissioning editor: Claire Parfitt
Editorial assistant: Alexandra Lainsbury
Production editor: Lauren Povey

Typeset by SPi, Pondicherry, India.
Printed and bound by Gutenberg Press Ltd, Tarxien, Malta.

Contents

Preface to the Third Edition

The purpose of this edition of the book is the same as for earlier editions. Our aim is to support better agricultural decision making by explaining what can be done nowadays in risk analysis and risk management. As before, the intended readership includes senior undergraduate or graduate students of agricultural and rural resource management, agricultural research workers, people involved in advising farmers, such as extension workers, financial advisers and veterinarians, some farmers themselves, and policy makers.

Methods of risk analysis and management are evolving rapidly. In this third edition we have included some recent advances in both theory and methods of analysis. New material includes sections on state-contingent versus stochastic production functions and an introduction to the use of copulas for modelling stochastic dependency.

Improvements in available software continue to expand the scope to better represent and model real-world risky choices, and we have updated our advice concerning use of contemporary software.

We wish to thank Priscilla Sharland for her editorial input. Thanks are also due to Claire Parfitt, our commissioning editor, and to other staff at CABI, especially to Alexandra Lainsbury and Lauren Povey, for their professional work and encouragement in the production of this third edition.

Note

Dollar signs are used throughout to indicate money units but are not specific to any particular currency.

J. Brian Hardaker
Gudbrand Lien
Jock R. Anderson
Ruud B.M. Huirne

Preface to the Second Edition

Agricultural production is typically a risky business. For many decades governments around the world have intervened in order to try to help farmers cope more effectively with risk. Both national and international developments have led many countries to reorientate their agricultural policies towards deregulation and a more market-oriented approach. Much of the protection that many farmers had from the unpredictable volatility of markets may therefore be removed. In addition, such risky aspects as animal welfare, food safety and environmental impacts are being given increasing attention. Thus, it can be expected that in the future risk analysis and risk management in agriculture will receive increased attention.

The purpose of the first edition of the book published in 1997 was to contribute to improved agricultural decision making by explaining what can be done nowadays in risk analysis and risk management. The intended readership included senior undergraduate or graduate students of farm management, agricultural research workers, people involved in advising farmers, such as extension workers, financial advisers and veterinarians, some farmers themselves, and policy makers.

In this second edition, our goals and intended readerships are still the same. Since 1997 there have been some theoretical advances and some progress in developing new methods of analysis. In addition, since the last edition there has been progress in general understanding of possibilities and limitations of risk analysis in agriculture. All these changes provided the motivation for this revised and expanded edition of the book.

For this second edition the material covered has been somewhat restructured and most of the text has been revised. Some additional examples have been included to help clarify important concepts and we have striven to make the examples easier to understand. Topics that are more thoroughly dealt with in this second edition include:

- assessing and quantifying the degree of risk aversion of a decision maker;
- judging how important risk aversion is likely to be in particular circumstances;
- an improved approach for partially ordering risky stochastic distributions when the decision maker's attitude to risk is not fully known;
- stochastic simulation and its combination with optimization for the analysis of risky choice; and
- risk considerations in agricultural policy making.

We thank the School of Economics of the University of New England, the Farm Management Group at Wageningen University and Research Centre, the Agriculture and Rural Development Department of the World Bank, and the Norwegian Agricultural Economics Research Institute for general institutional and facility support, and the Research Council of Norway for financial and other support for Gudbrand Lien's work on the book. We are also grateful to Marcel van Asseldonk, Theo Hendriks, Steen Koekebakker and Ole Jakob Bergfjord for help on particular topics, and to Irene Sharpham for editorial assistance. Our special thanks go to Tim Hardwick, Rachel Robinson and other staff at CABI Publishing for their helpful cooperation.

J. Brian Hardaker
Ruud B.M. Huirne
Jock R. Anderson
Gudbrand Lien

Preface to the First Edition

The aim in this book is to introduce readers to:

- the nature of agricultural decision making under uncertainty;
- the concept of rational choice in a risky world and its foundations in theories of probability and risk preference;
- methods for the analysis of risky decisions that can be used in agriculture; and
- planning for risk management in agriculture.

The intended readership includes:

- people involved in advising farmers, such as extension workers, financial advisers and veterinarians;
- agricultural research workers;
- senior undergraduate or graduate students of farm management;
- some farmers themselves; and
- policy makers, for policy development itself and also to understand how farmers react to risk.

Risk adds a considerable degree of complexity to decision analysis. In explaining the methods that might be used we have striven to keep things as simple as possible. Although some use of mathematical notation has been unavoidable, the book is liberally seasoned with examples to explain the sometimes-difficult concepts in ways that people concerned with practical agriculture may understand. If the quality of agricultural decision making is improved by use of the methods outlined here, the book will have served its purpose.

Part of the motivation to write this book comes from the fact that two of the authors contributed to an earlier text on risk in agriculture:

> Anderson, J.R., Dillon, J.L. and Hardaker, J.B. (1977) *Agricultural Decision Analysis*. Iowa State University Press, Ames.

That book set the foundation for much work on risk analysis in agriculture that has followed. However, since 1977, several things have happened to make it timely to revisit the methods of risk analysis in agriculture. First, we have learnt more about what is possible and what is impossible among the procedures proposed in 1977 by Anderson *et al.* It therefore seems desirable to set down a revised set of guidelines for those brave souls who wish to venture into this field, without repeating the errors of the pioneers. Second, there has been a revolution in the world of computing that has brought to the desk-top or to the lap-top of many people working in agriculture, including many farmers, a computing capacity unimagined in 1977. Moreover, new software has emerged, meaning that analyses that would have taken many hours of tedious computer programming in those days are now done in seconds on the click of a mouse. This computing and software revolution should surely have revolutionized agricultural risk analysis. The fact that, by and large, it has not yet done so may indicate a need to spell out to potential users just what can be done today in risk analysis.

A number of people have contributed to the production of this book whose assistance we wish to acknowledge. First, we owe a great debt to Professor Aalt A. Dijkhuizen of the Department of Farm Management of the Wageningen Agricultural University. Without his enthusiasm, assistance and support, the book would not have been written. Erwin Nijmeijer and Marie-Ange Vaessen both helped with analytical work and the preparation of material for inclusion in the text. Shirley Hardaker helped with proof-reading, and Ernst van Cleef with preparing the graphical presentations. We are grateful to Tim Hardwick, Emma Critchley and Nigel Wynn of CAB International for their support and encouragement throughout, and to the CABI proof-reader Karin Fancett for guiding us through the complexities of preparing the camera-ready copy. A number of computer software houses made available products for use in the book. They are listed in the Appendix.

Thanks go to Wageningen Agricultural University, especially to the Department of Farm Management and its Head, Professor Jan A. Renkema, for providing hospitality and financial and other support for Brian Hardaker to work on the preparation of this book in The Netherlands. Thanks are also due to the University of New England, especially the Department of Agricultural and Resource Economics and its Head, the late Professor Ronald R. Piggott, for facilitating Ruud Huirne's work on the book in Armidale. The Australian Research Council and the Netherlands Organization for Scientific Research (NWO) contributed to the costs of the academic collaboration that went into the preparation of the book.

Finally, we wish to record our debt to [the late] Professor John L. Dillon for his pioneering efforts in agricultural decision analysis that got at least two of us into this business.

J. Brian Hardaker
Ruud B.M. Huirne
Jock R. Anderson

1 Introduction to Risk in Agriculture

Examples of Risky Decisions in Agriculture, and Their Implications

The development of agriculture in early times was partly a response to the riskiness of relying on hunting and gathering for food. Since then, farmers and others have tried to find ways to make farming itself less risky by achieving better control over the production processes. As in other areas of human concern, risk remains a seemingly inevitable feature of agriculture, as the examples below illustrate.

Example of institutional risk

A dairy farmer finds that the profitability of his herd is constrained by his milk quota. He now has the opportunity to buy additional quota, using a bank loan to finance the purchase. The farmer, however, has serious doubts about the profitability of this investment because he believes that milk quotas will be removed at some time in the future. Cancellation of quota would make the purchased quota valueless from that point in time. He also thinks it is likely that milk prices will drop significantly when the quotas go. His options, therefore, are to restrict milk output to the current quota or to invest to get the benefits from increased scale and intensity of milk production. But investing means that, if the quota was to be abolished, and prices fell, he might not be able to repay the loan.

Animal welfare example

Along with a group of other interested producers, a pig farmer is considering whether to switch to a method of production that takes more account of the welfare of the animals. In addition to her personal preferences, she believes that there is a potential premium to be earned from sale of pig meat produced in a way more in tune with good standards of animal welfare. To change to the new system, considerable investment would be needed to provide better and more spacious accommodation for the pigs. Production costs would also increase owing to the need to provide larger-size housing, straw for bedding and a wider range of types of feed. However, there is considerable uncertainty about whether the new product, pig meat produced in a more welfare-sensitive way, can be marketed successfully and whether consumers will pay the premium needed to cover the extra costs.

Conversion to organic farming example

Organic farming has become more popular in response to increased consumer demand for organic produce. In addition, in several countries, conversion grants and support schemes for organic farming have been introduced to encourage organic production. There are reasons to believe that conventional and organic farming systems respond differently to variations in weather, implying different impacts on farm income risk. For example, restrictions on pesticide and fertilizer use may give rise to different production risk in organic farming than in conventional farming. Prices for organic produce may fall as more producers switch to organic production, and smaller organic markets may mean greater price fluctuations. A farmer considering switching to organic production would need to take account of the changed exposure to risk.

Insurance example

A vegetable grower farms land close to a major river and is worried about the risk of a significant flood that would destroy her crops. Her insurance company offers flood insurance to cover the potential financial loss, but the annual premium is high because of the high risk of floods at that particular location. The grower is wondering what proportion of the risk she should insure, if any.

Flower growing example

A flower grower is concerned about energy use for glasshouse heating. His current heating system is obsolete and he is considering replacing it with a new one, either a conventional design or a low-energy system. The relative profitability of the two options depends on future energy prices. If future energy prices stay as they are or rise only moderately, the conventional system will be the most profitable choice. However, if future energy prices rise substantially, the low-energy system will be best, even though it is more expensive to install.

Potato marketing example

A potato farmer is about to harvest his crop. He has to decide whether to sell the potatoes now, at the current price, or to store them for sale later in the hope of a higher price. The first option gives him a sure return for his harvest. With the second option, however, he has to face storage costs and losses, yet the future price is uncertain and depends on the market situation later in the year. If the market is in short supply, prices will rise and he will make a good profit from storage. If supply is average, prices will not rise much and he may just break even. But an over-supplied market will mean a fall in prices, causing a significant loss from a decision to store rather than sell immediately. Yet, at the time the decision to sell or store has to be made, the farmer cannot be sure what the future supply situation will be since it will depend on yields of crops yet to be harvested, grown in other areas.

Farm purchase and financing example

A dairy farmer has sold her previous farm for urban development at a good price, leaving her with funds for re-investment. She is wondering whether to invest in stocks and bonds or to buy another dairy farm, which is the business she knows best. If she buys a new farm, she has to decide how large a unit to purchase. In order to buy a farm to run of a commercial size, she will have to borrow some of the funds needed for the investment. Since she believes that scale economies in dairy farming will become more important in the future, she has to decide how much she is prepared to borrow and on what basis. Options include a loan where the future interest rate varies with economic conditions or one where at least a part of the funds is at a fixed, but initially higher, interest rate.

Disease policy example

Foot-and-mouth disease poses serious risks to farming in many countries. Outbreaks of the disease require costly control measures, including the slaughter of all animals on infected farms and bans on the movement of animals in the vicinity. A basic policy choice is between routine vaccination or not. Routine vaccination is costly and not fully effective, so that some outbreaks could still occur. In addition, because vaccinated animals show positive for the disease in immunological tests, it is not possible to certify vaccinated stock as disease-free. However, without routine vaccination most animals would have no immunity. In the event of an outbreak, therefore, there is the potential for the disease to spread freely. In reaching a policy decision, in addition to ethical issues, policy makers evidently have to weigh the fairly certain costs and losses of vaccination against the less certain consequences of no vaccination. The latter choice would save vaccination costs and preserve access to export markets, but creates the risk that future outbreaks could be serious and costly.

The need to take account of risk in agriculture

As these examples show, risk can be important in agriculture. Indeed, risk and uncertainty are inescapable in all walks of life. Because every decision has its consequences in the future, we can seldom be absolutely sure what those consequences will be. Yet risk is not something to be too afraid of. It is often said that, in business, profit is the reward for bearing risk: no risk means no gain. The task rather is to manage risk effectively, within the capacity of the individual, business or group to withstand adverse outcomes.

Of course, farmers the world over have always understood the existence of risk and have adjusted to it in their own ways in running their farms. Yet, with a few notable exceptions, rather little practical use has been made of formal methods of risk analysis in agriculture. One reason may have been the effective elimination of at least some sources of risk provided by various government schemes to support the prices of farm products, such as the Common Agricultural Policy of the European Union (EU) and farm support programmes in the USA. The fact that prices of many agricultural products have been reasonably well assured in countries where such measures of protection have been in operation no doubt reduced the need to give a lot of attention to risk management. However, trade negotiations such as those at the World Trade Organization have led to changes in agricultural policies, with obligations on member countries to reduce levels of protection, especially via price supports. Moreover, many believe that these reforms are just first steps leading to further measures towards liberalization of international trade in farm products. The outlook, therefore, may be that many farmers will face greater exposure to competitive market forces and so will enjoy less predictable consequences than has been their experience.

A further reason for the relative neglect of risk is that the methods for the analysis of risky choice, although available for many years, are a good deal more complex than the more familiar forms of analysis under assumed certainty, such as the commonly used budgeting methods. The various methods that have been developed for analysing choices involving risk are collectively called *decision analysis*. Advances in computer software and hardware have made application of the methods of decision analysis simpler and quicker than in the past, bringing them within ready reach of farmers, farm advisers and agricultural policy analysts. It is therefore timely to examine the scope for wider application of decision analysis in agriculture.

Risk and Uncertainty

Some definitions

The terms 'risk' and 'uncertainty' can be defined in various ways. One common distinction is to suggest that risk is imperfect knowledge where the probabilities of the possible outcomes are known, and uncertainty exists when these probabilities are not known. But this is not a useful distinction, since cases where probabilities are objectively 'known' are the exception rather than the rule in decision making. Instead, in line with common usage, we define *uncertainty* as imperfect knowledge and *risk* as uncertain consequences, particularly possible exposure to unfavourable consequences. Risk is therefore not value-free, usually indicating an aversion for some of the possible consequences. To illustrate, someone might say that he or she is uncertain about what the weather will be like tomorrow – a value-free statement simply implying imperfect knowledge of the future. But the person might go on to mention that he or she is planning a picnic for the next day and there is a risk that it might rain, indicating lack of indifference as to which of the possible consequences actually eventuates.

To take a risk, then, is to expose oneself to a chance of loss or harm. For many day-to-day decisions, the risk is usually unimportant since the scope of possible loss is small or the probability of suffering that loss is judged to be low. Crossing a street carries with it the risk of death by being run over by a vehicle, yet few people would see that risk as sufficiently serious to prevent them making the crossing for quite trivial benefit, such as the pleasure of buying a newspaper or an ice cream. But, as the earlier examples show, for important business or personal decisions or for some government policy decisions, there is a good deal of uncertainty, and there are important differences between good and bad consequences. For these decisions, therefore, risk may be judged to be significant. In farming, many farm management decisions can be taken with no need to take explicit account of the risks involved. But some risky farm decisions will warrant giving more attention to the choice among the available alternatives.

Types and sources of risk in agriculture

Because agriculture is often carried out in the open air, and always entails the management of inherently variable living plants and animals, it is especially exposed to risk. *Production risks* come from the unpredictable nature of the weather and uncertainty about the performance of crops or livestock, for example, through the incidence of pests and diseases, or from many other unpredictable factors.

In addition, prices of farm inputs and outputs are seldom known for certain at the time that a farmer must make decisions about how much of which inputs to use or what and how much of various products to produce, so that *price* or *market risks* are often significant. Price risks include risks stemming from unpredictable currency exchange rates.

Governments are another source of risk for farmers. Changes in the rules that affect farm production can have far-reaching implications for profitability. For example, a change in the laws governing the disposal of animal manure may have significant impacts; so too may changes in income-tax provisions, or in the availability of various incentive payments. Horticultural producers may be badly affected by new restrictions on the use of pesticides, just as the owners of intensive pig or poultry operations may be affected by the introduction of restrictions on the use of drugs for disease prevention and treatment. Risks of these kinds may be called *institutional risks*. Institutional risks embody *political risks*, meaning the risk of unfavourable policy changes, and *sovereign risks*, meaning the risks caused by actions of foreign governments, such as a failure to honour a trade agreement. Also under this heading we might include *contractual risks*, meaning the risks inherent in the dealings between business partners and other trading organizations. For example, the unexpected breaking of agreements between participants in supply chains is an increasingly significant source of risk in modern agribusiness.

The people who operate the farm may themselves be a source of risk for the profitability and sustainability of the farm business. Major life crises, such as the death of the owner or the divorce of a couple owning a farm in partnership, may threaten the existence of the business. Prolonged illness of one of the principals may cause serious losses to production, or substantially increased costs. And carelessness by the farmer or farm workers, in handling livestock or using machinery for example, may similarly lead to significant losses or injuries. Such risks may be called *human* or *personal risks*.

The aggregate effect of production, market, institutional and personal risks comprise *business risks*. Business risks are the risks facing the firm independently of the way in which it is financed. Such risks comprise the aggregate effect of all the uncertainty influencing the profitability of the firm. Business risks affect measures of farm business performance such as the net cash flow generated or the net income earned.

In contrast to business risks, *financial risks* result from the method of financing the firm. The use of borrowed funds to provide some of the capital for the business means that a share of the operating profit must be allocated to meeting the interest charge on the debt capital before the owners of the equity capital can take their reward. Debt multiplies the business risk from the equity holders' viewpoint; an effect sometimes known as *leverage*. Moreover, the greater the proportion of debt capital to total capital, the higher the leverage, hence the greater the multiplicative factor applied to business risk. Only if the firm is 100% owner-financed is there no financial risk owing to leverage.

In addition to the financial risks associated with leverage, there are financial risks in using credit. The most significant of these are: (i) unexpected rises in interest rates on borrowed funds; (ii) the unanticipated calling-in of a loan by the lender; and (iii) the possible lack of availability of loan finance when required. Changes in the inflation rate can have positive or negative effects on both borrowers and lenders.

Impacts of risk

There are two reasons why risk in agriculture matters, as outlined in turn below.

Most people dislike risk

Most people are risk averse when faced with significantly risky incomes or wealth outcomes. A person who is risk averse will be willing to forgo some expected return for a reduction in risk, the rate of acceptable trade-off depending on how risk averse that individual is. Evidence of farmers' risk aversion is to be found in many of their actions, such as their willingness to buy certain kinds of insurance or their tendency to prefer farming systems that are more diversified than might seem best on profit grounds alone.

The existence of risk aversion means that analysis of risky choices in terms of their average or expected consequences will not always lead to the identification of the option that will be most preferred. Farmers, like most people, do not, and will not wish to, make choices based on what will pay best 'in the long haul' if that choice means exposing themselves to unacceptable chance of loss. Evidently, this aversion to risk has to be taken into account in developing and applying methods of decision analysis. We return to this matter in Chapter 5.

Downside risk

In the finance sector, downside risk is typically taken to mean an estimate of the potential that a security might decline in price if market conditions turn bad. Here we use the term rather differently and more generally to mean the risk that the payoff from a risky choice will be reduced if conditions are not as assumed. Downside risk, in our terms, can arise from two different causes that, in some situations, can operate in unison to magnify the risk that the consequences of some risky choice will be less than foreseen at the outset.

First, downside risk can occur when decisions are made under assumed certainty, based on some 'norm' or 'best estimate' of the consequences. The imprecision of such terms makes it hard to be sure what is implied, but evidently some single-valued measure of central tendency of some unspecified probability distribution of the payoff is implied – perhaps the mode (most likely) outcome. The risk here arises if the distribution of outcomes is negatively skewed so that the mode is above the mean (see Chapter 3, this volume).[1] Yet it is the latter that is the measure of central tendency that is more relevant in risky decision making. For a negatively skewed distribution of payoffs, there is a greater probability that the outcome will fall below the mode than above it, which is what we mean by downside risk.

Second, downside risk may also arise when a risky outcome depends on non-linear interactions between a number of uncertain quantities. The yield of a crop provides an obvious example. Yields depend on a large number of uncertainties, such as rainfall and temperature in each stage of the growing season. Large deviations of these uncertain variables in either direction from their expected values often have adverse effects – too much rain may be as bad as too little. While some smaller deviations may have beneficial effects, the 'law of diminishing returns' usually means that losses associated with adverse deviations from the mean level of, say, rainfall, are greater than the gains associated with favourable deviations of a similar magnitude. A natural definition of a 'normal' season is one in which all variables take values close to their expected values. The probability of a season defined in this way is necessarily small, however, and the non-linear interactions of the variables imply a high probability of yields less than those obtained in such a 'normal' season. Thus, the use of this notion of a normal year as the basis for analysis creates a situation in which downside risk is dominant. By contrast, downside risk is accommodated by the use of the mean of the series of observed yields (suitably adjusted for any trend), or of a well-considered estimate of the mean, as the basis for analysis.

Note that treating input variables as certain when they are not often leads to downside risk whether the distributions of the input variables are skewed or not. However, the downside risk is more severe when both causes are present. Assuming input variables take their modal or 'normal' values will commonly lead to greater downside risk than assuming they take their mean values.

We can illustrate the impact of these causes of downside risk with a simplified example relating to production of a crop the yield of which is assumed to be a quadratic function of growing-season rainfall:

$$y = 0.4\,x - 0.00075\,x^2 \tag{1.1}$$

where y is yield in tonnes per hectare and x is growing-season rainfall in millimetres. For this example, we assume that rainfall in the growing season is uncertain and can be represented by a relative frequency histogram based on historical information with a minimum of 100 mm, a maximum of 300 mm and a most likely value of 225 mm. The specified distribution has a mean value of 209 mm. Both the rainfall distribution and the crop yield function are illustrated in Fig. 1.1.

If we evaluate the crop yield on the assumption that the coming growing season will be 'typical', implying rainfall of 225 mm previously defined as 'most likely', the calculated yield is 52.0 t/ha. However, if we use instead the expected value (mean) of growing-season rainfall of 209 mm, the calculated yield is reduced to 50.8 t/ha. On the other hand, if we take full account of the whole distribution of rainfall, the expected yield[2] is 49.5 t/ha. The latter is the yield that a grower could 'expect' to get if he or she had no prior knowledge of what rainfall would be beyond the information represented in the specified relative frequency distribution. Such a state of knowledge would be typical of a grower at planting time. Thus, as shown in Table 1.1, we have two yield estimates under the unrealistic assumption of certain foreknowledge of rainfall, and one more satisfactory estimate obtained recognizing the entailed risk. The difference

[1] Assuming that the outcome is something desirable such that more is preferred to less, not something that is disliked with less being preferred to more.

[2] Calculated using stochastic simulation, outlined in Chapter 6, this volume.

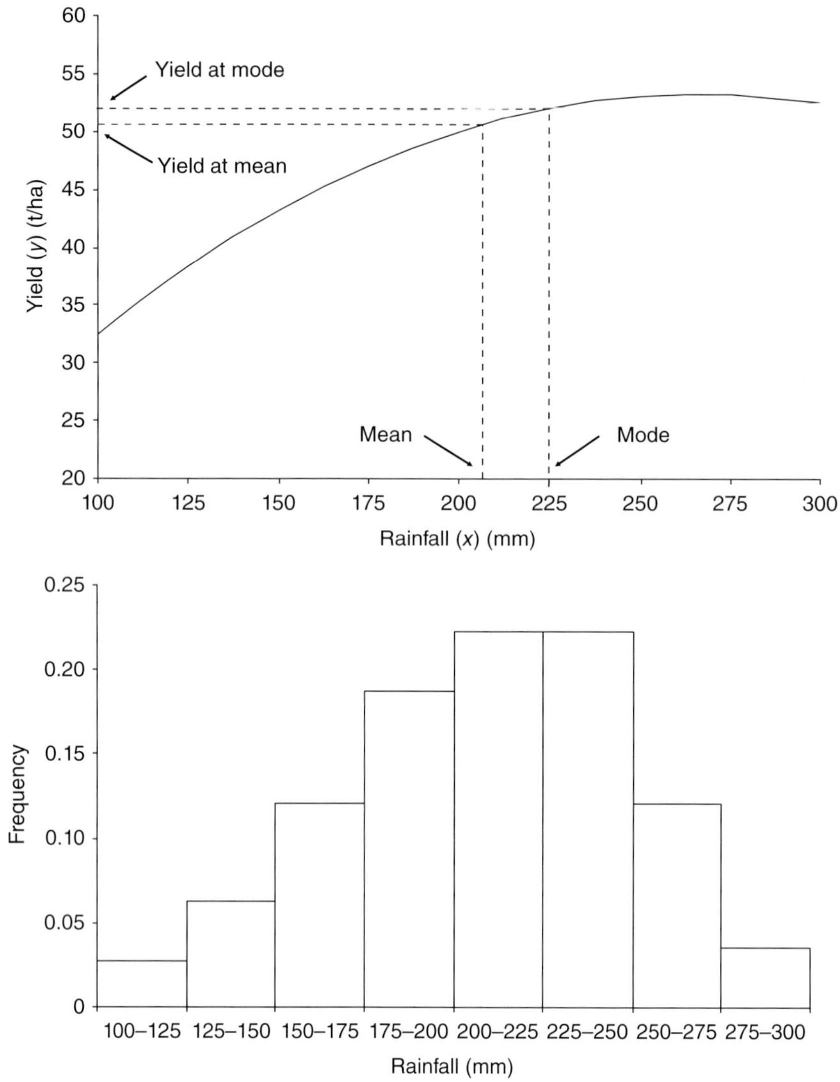

Fig. 1.1. Illustration of downside risk showing the assumed crop yield response to rainfall and the relative frequency distribution for growing-season rainfall.

Table 1.1. Effect of assumptions about rainfall on expected crop yield.

Assumption about rainfall	Calculated expected yield (t/ha)
Assumed certainty: Mode of 225 mm	52.0
Assumed certainty: Mean of 209 mm	50.8
Accounting for risk via actual distribution	49.5

between the third, more realistic yield estimate and the other two may be regarded as downside risk. Under the perhaps common assumption that the growing season rainfall will be 'normal' (i.e. the modal value) downside risk amounts to an over-estimate of expected yield of 52.0 – 49.5 = 2.5 t/ha. Even if the analysis under assumed certainty is based on the mean rainfall, the downside risk is still 50.8 – 49.5 = 1.3 t/ha.

In a similar vein, other uncertain factors besides rainfall will also generally not operate at their 'normal' values. There will be impacts of pests, diseases, frosts, strong winds, wildfires, foraging by uninvited animals, and so on, to be considered. The multiplicity of effects from many causes will tend to magnify downside risk to an extent that may make a decision-analytic assessment of the riskiness involved important and informative.

Some readers may believe that the difference in yield between assumed certainty and a risk analysis comes about mainly because of the somewhat skewed nature of the rainfall distribution. This is not the case since, for the data used, there is no difference in the calculated expected yield derived from a risk analysis whether that analysis is done using the given empirical distribution of rainfall or using a normal distribution with the same mean and standard deviation. However, in other cases, as noted above, skewness may contribute to downside risk.

Put most simply, downside risk occurs because nature tends to be unkind to human endeavours. In other words, bad outcomes are not fully offset by good ones. For example, catastrophes such as terrorist attacks or major natural disasters happen all too often, killing large numbers of people. Yet it is impossible to imagine any equivalently good outcomes that could compensate for the losses suffered. Were such an event to occur it would undoubtedly be called a miracle. When planning under uncertainty it is important to think about what can go wrong and what can go right, recognizing that the negatives typically outweigh the positives. Risk analysis is needed to account for this effect.

Who Needs to Think about Risk in Agriculture?

People and organizations who need to concern themselves with risk in agriculture include:

1. farmers;
2. farm advisers;
3. commercial firms selling to or buying from farmers;
4. agricultural research workers; and
5. policy makers and planners.

The need for farmers to plan for the risks they face hardly needs emphasis. The welfare of the farm family and the survival of the farm business may depend on how well farming risks are managed. In hard times, farm bankruptcy may occur, and many more farms survive when things go wrong only by 'belt tightening' and by such measures as finding off-farm employment. The ambition of many farmers to hand the farm on in a thriving condition to the next generation of the family can be frustrated if risk management is neglected.

Farm advisers also need to recognize that risk and risk aversion influence farmers' management decisions. Advice that overlooks the risks involved in changing from an existing production system to one that is supposedly 'superior' is bad, even negligent, advice. Advisers need to understand that the adoption of an untested 'improved' technology may entail a high degree of risk for the farmer, especially if the adoption of the technology requires a substantial capital investment. The perceived risk may be high because the farmer has had no first-hand experience of the new method. So, if the technology is 'lumpy', in the sense that it cannot be tried out on a pilot scale, there will be an 'adoption lag' until the farmer has accumulated sufficient evidence to make the perceived risk acceptable. Advisers may be able to speed the adoption process by measures that reduce the degree of risk perceived by the farmers, such as by supplying relevant technical and economic information, arranging field demonstrations, organizing visits to other farms where the technology is in use, and so on. More generally, it may be possible to present information and advice in ways that better portray the risks involved and that permit a farmer to decide more easily which choices best suit his or her particular circumstances and risk-bearing capacity.

Suppliers of inputs to farmers face similar challenges to those encountered by advisers. They need to recognize that the purchase of a new product may be a risky decision for farmers, especially where items involving a substantial investment are concerned. Suppliers can use the same sorts of methods as just mentioned to reduce the risks perceived by their intended clients. Leasing arrangements may be better than direct purchase for risk-averse farmer clients, so that those suppliers who can offer this option may have more success. Similarly, agricultural traders who buy from farmers may do well to consider the motivation of farmers to reduce price risks. Thus, some farmers will be willing to accept a lower price for their output if the buyer is prepared to offer a forward contract at an assured price. Both buyer and seller can benefit from such an arrangement.

Agricultural research workers, especially those working on the development of improved farming methods, may need to give more thought to risk aspects. For example, too often the results of trials are reported in terms of (differences between) treatment averages, with inadequate representation of the dispersion of results about the reported averages. Basing conclusions on 'statistically significant' differences in means between treatments fails to tell the whole story if there are differences in dispersion of outcomes. The variability of results across plots within a single trial is sometimes indicated, but seldom is information provided on the variability across different experimental sites or over a number of years. Yet the latter information may be vital for a proper assessment of relevant risks. Agricultural research that recognizes risk may entail more than simply better reporting of trial results. Thinking about farmers' production constraints and opportunities in a world in which risk is recognized will lead researchers to identify different research problems, or to address the problems they do identify in different, more complete ways. For example, both animal and crop breeders tend to base selection on performance under well-controlled conditions, such as good nutrition. This practice may mean the loss of genetic capacity in the animals or plants to thrive in less favourable conditions, so increasing the losses when things go wrong and making production more risky, for example, cereal varieties that lodge if the growing season is wet.

Agricultural policy makers and planners also need to account for risk and farmers' responses to it. Estimates of supply response obtained from models that ignore risk-averse behaviour by farmers can be significantly biased. In other words, models that include risk can provide better predictions of farmers' behaviour than those that do not. Similarly, policy making and planning are themselves risky activities, so that some of the methods of risk analysis discussed in the following chapters are relevant, although in a rather different context, as discussed in Chapter 13.

Policy makers and planners need to consider farmers' concerns and risk-averse behaviour when setting policies and programmes directly affecting (or causing) farming risk. For example, some time back, extensive beef producers in northern parts of Australia had a thriving market for live cattle in Indonesia. When a TV programme revealed significant cruelty in an Indonesian abattoir, the Australian authorities promptly banned the export trade. However justified that decision may have been, it caused massive problems for cattle producers in remote northern Australia, many of whom were simply unable to sell their cattle. Apart from significant losses for these producers, Indonesian consumers faced beef shortages, while drought and over-grazing on Australian cattle farms may well have led to many animals starving to death. It seems likely that the decision makers (DMs) had not foreseen the serious implication of their hasty intervention. Evidently, policy makers need to be aware of the consequences for farmers and others of risks created by their own unpredictable behaviour.

Risk Management and Decision Analysis

Many descriptions of the process of risk management view risk as rather like a disease that has to be treated. For example, the International Standard on Risk Management (ISO, 2009) has at its core the identification, analysis, evaluation and treatment of risks. While it can be helpful to look at risk in this way, this is not the approach we choose to follow in this book. Instead of treating risk management as something that is separate from general management of an organization, we see a need to account for risk as an integral part of all management decision making. We take this view because just about every decision has its consequences in the future, and we can never be certain about what the future may bring. So most if not all management decisions create some risk exposure. Making risk management a separate process ignores this reality. Moreover, economics teaches that profit is the reward for risk taking – no risk means no gain. So what is needed is a process to balance risk against possible rewards. Separating out the treatment of risk may obscure the need to get the balance right.

Obviously, some decisions are more risky than others, and those for which the range of possible consequences is narrow, with little or no chance of a really bad result, can be handled easily with a bit of common sense. But there are also other decisions for which the range of possible consequence is wide, perhaps with a non-trivial chance of bad outcomes. For these decisions much more careful consideration will certainly be warranted. However, dealing with such risky choice is not easy – there may be many options to choose between and the consequences of each may depend on many uncertain factors. Procedures aimed at getting to grips with just such important but difficult risky choices fall under the general heading of *decision analysis* and it is these procedures that form the main part of this book.

Decision analysis may be defined as the philosophy, theory, methods and practices necessary to systematically address important risky decisions. In this book we focus on an approach to decision analysis founded on the theory of *subjective expected utility*, as described in the next chapter. Decision analysis includes methods and tools for identifying, representing and assessing important aspects of a risky decision, leading to a recommended course of action consistent with careful consideration of the possible consequences of the alternative choices, the associated probabilities of those consequences, and the relative preference for possible outcomes. In other words, it is a prescriptive theory of choice.

When Formal Decision Analysis Is Needed

Formal analysis of risky choice in agriculture obviously has costs, including the costs of the time to do the required thinking and figuring. Not all decisions will warrant this input of effort. There are at least two cases where formal analysis may be thought worthwhile.

The first is for repeated risky decisions for which a sensible strategy might be devised that could be applied time and again. The benefit from better individual decisions may be small, but the accumulated benefit over many decisions may justify the initial and continuing investment of time and effort in analysis. An example is the development of a strategy for the treatment of mastitis in dairy cows. This is a commonly occurring disease that, if untreated, can damage the animal and can lead to penalties for the farmer when milk delivered to the dairy factory is found to contain high bacterial loads. Treatment with antibiotic has costs, including the need to divert milk from cows being treated from that being sold. Because the disease may occur repeatedly, and on-farm tests for the presence of the disease are not wholly reliable, the design and implementation of a good strategy for managing mastitis makes sense.

The case of decision analysis for many repeated decisions may be extended to advisory situations where an analyst may be justified in putting considerable effort into devising good risk-management procedures for some aspect of farming that could be adopted by many farmers. The choice of a planting strategy for potatoes in frost-prone areas might be such an example. It may be known that yields are generally improved by early planting, but early-planted crops are obviously more vulnerable to late spring frosts. The adviser may be able to use weather records in a decision analysis to develop recommendations to guide local potato growers.

The second case for formal analysis arises when a decision is very important, in the sense that there is a considerable gap between the best and the worst outcomes, one big enough to have a significant impact on wealth. A case in point is a major investment decision. A farmer may be deciding whether it would be worthwhile to invest in the purchase of another farm, in order to expand the size of the business with the goal of attaining long-run viability in the face of an expected cost-price squeeze. Yet the investment may require considerable borrowing that could lead to bankruptcy if things go wrong.

The fact that a decision or a set of decisions is sufficiently important to justify efforts to reach a better choice is not the only consideration in deciding whether to undertake a formal analysis. Many real choices in agriculture are complex. Some common characteristics of complex decision problems are:

1. The available information about the problem is incomplete.
2. The problem involves multiple and conflicting objectives.
3. More than one person may be involved in the choice or may be affected by the consequences.
4. Several complex decision problems may be linked.
5. The environment in which the decision problems arise may be dynamic and turbulent.
6. The resolution of the problem may involve costly commitments that may be wholly or largely irreversible.

The psychological response of people to such complexity varies and may be more or less rational. Some simply defer choice, even when to do so is to court disaster. For example, in hard financial times when debt servicing is becoming a problem, it has been observed that some farmers may cut off communication with their bankers at the very time when they should be discussing with them how to resolve the emerging crisis. Other more or less rational responses include:

1. simplifying the complexity by searching for a course of action that is 'good enough', rather than 'best' (an inevitable procedure in all decisions, though the question of what opportunities have been forgone may or may not get considered);
2. avoiding uncertainty or taking steps to reduce it (perhaps rationally or perhaps at considerable cost);
3. concentrating on incremental measures rather than on those involving fundamental changes (perhaps thereby closing off risky but potentially rewarding options); and

4. seeking to reduce conflict of interest or perceptual differences through discussion among concerned people (often wisely, but maybe simply to defer a hard decision by 'forming a committee' or seeking to shift the burden of choice to the shoulders of others).

The fundamental question in considering the role of formal methods of decision analysis is whether, in a particular case, the need to sweep aside much of the complexity of the real decision problem will leave a representation of the problem that is:

1. sufficiently simple to be capable of systematic analysis, yet
2. sufficiently like the real situation that an analysis will aid choice.

Obviously, it is our contention that the answer to these questions will be in the affirmative sufficiently often to make it worthwhile for agriculturists and others to familiarize themselves with the formal methods of decision analysis set out in the following chapters. Whether we are right will be determined by the reactions of agricultural DMs to these ideas.

Outline and General Approach of the Book

In the next chapter, the basic approach to the analysis of risky choice is described. The central idea underlying the methods described is to break any risky decision down into separate assessments of:

1. the nature of the uncertainty affecting the outcomes of alternative actions, and hence the riskiness of the possible consequences of those actions; and
2. the preferences of the DM for the various consequences that might arise from his or her choice.

Having dissected the decision problem in this way, the two sets of judgements are then integrated in some suitable analytical framework to work out what choice would be best.

The measurement of uncertainty using subjective probabilities is the topic of Chapter 3, and ways of making better probability judgements are discussed in Chapter 4. The assessment of preferences for consequences, and hence of risk attitudes, is introduced in Chapter 5. In this chapter we also consider options when the elicitation of an individual's utility function proves too difficult. Methods of integrating beliefs in the form of probabilities and preferences expressed as utilities to reach a choice are described in Chapter 6.

Most of the rest of the book is devoted to an elaboration of these ideas. Chapter 7 deals with situations where, for whatever reason, the risk attitude of the DM cannot be accurately assessed. Chapter 8 deals with the state-contingent approach to risky choice while Chapter 9 describes methods of whole-farm planning accounting for risk, based mainly on extensions of the familiar 'programming' approach to farm planning. In Chapter 10, the methods of preference assessment introduced in Chapter 5 are extended to cases where consequences have more than one attribute that must be taken into account. Time is an important dimension to many risky farm decision problems; methods of analysis that give explicit consideration to the time dimension are introduced in Chapter 11. Some of the strategies that farmers use, or might consider using, to deal with risk are described and analysed in Chapter 12, while aspects of agricultural risk affecting policy making and planning are the topic of the final chapter, Chapter 13. In the Appendix the sources of the main special-purpose software used in the book are given.

Throughout, the aim is to make the explanations as simple as possible to appeal to a wide cross-section of readers. Often, the language of mathematics is used as the most convenient way of expressing some concept in an efficient and unambiguous way. However, for those not comfortable with the mathematical approach, other ways of explaining the same ideas have been provided, usually via a worked if simplified example. Although the use of mainly very simple examples may give the impression that decision analysis can only be applied to simple problems, this is certainly not so. The methods illustrated can also be applied to realistic decisions, but at the cost of extra complexity that would cloud rather than clarify the exposition of the underlying concepts.

Similarly, it is difficult in a text of this kind to avoid being overly prescriptive in describing the methods that can be applied. It would make the book too long and dull to discuss the pros and cons of every aspect of the methods described. Yet, as a study of the relevant literature will show, there are often 'more ways than one of skinning a cat' in decision analysis. Methods other than those described here may better suit particular situations, and practitioners should not be afraid to try out alternative approaches or to develop new ones to suit particular circumstances and needs.

An analogy can be drawn between the way a particular decision problem is conceived and analysed, and the way a landscape artist sees and represents a particular view. A good painting is not necessarily a close representation of the reality; rather it is a representation that best helps a viewer of the painting to come to a better appreciation of the scene depicted. Similarly, the best analysis of some decision problem is one that truly helps the DM to make a good choice. Such analysis generally will not be a close representation of all the detail and complexity of the real situation. Indeed, too much detail in the analysis may obscure rather than clarify the choice for the DM.

There are many constraints on what is possible to do in decision analysis. Lack of time, information, access to computer facilities and software, and limited access to the DM are just some of the factors that may limit the applicability of the methods described here. In these circumstances, novice analysts may become discouraged. They should not be. Decision analysis is the art of the possible! It is plausible to presume that an analysis that has elements of the approach is likely to be better than one that does not. To pretend that risk does not exist when it obviously does, just because accounting for risk is more difficult, is mistaken, even dishonest. The truth is that risk, along with death, is one of the few certainties of the human condition. So what is at stake is not the need to recognize risk – it is the question of how the risk is to be accounted for in reaching a decision. This question is pursued from several analytical standpoints in the subsequent chapters.

Selected Additional Reading

The approach we follow in this book, which departs from earlier traditions such as established by Knight (1933), was first popularized in US business schools such as at Harvard University (e.g. Raiffa, 1968; Schlaifer, 1969). Subsequently, a vast literature on decision analysis has evolved dealing with both the general approach and the applications to specific fields or topics. Examples include Clemen (1996) and McNeil *et al.* (2005), while Meyer (2003) provides an overview of the economics of risk.

Significant early works dealing with application to agriculture include Halter and Dean (1971), Dillon (1971), Anderson *et al.* (1977), Barry (1984) and Huirne *et al.* (1997). More recent texts relating to agriculture include Harwood *et al.* (1999), OECD (2009) and Moss (2010). Just (2003), Chavas *et al.* (2010) and Hardaker and Lien (2010) are useful expositions of the future opportunities and challenges for risk research in agriculture.

References

Anderson, J.R., Dillon, J.L. and Hardaker, J.B. (1977) *Agricultural Decision Analysis*. Iowa State University Press, Ames, Iowa.

Barry, P.J. (ed.) (1984) *Risk Management in Agriculture*. Iowa State University Press, Ames, Iowa.

Chavas, J.-P., Chambers, R.G. and Pope, R.D. (2010) Production economics and farm management: a century of contributions. *American Journal of Agricultural Economics* 92, 356–375.

Clemen, R.T. (1996) *Making Hard Decisions: an Introduction to Decision Analysis*, 2nd edn. Duxbury Press, Belmont, California.

Dillon, J.L. (1971) An expository review of Bernoullian decision theory: is utility futility? *Review of Marketing and Agricultural Economics* 39, 3–80.

Halter, A.N. and Dean, G.W. (1971) *Decisions under Uncertainty with Research Applications*. South-Western Press, Cincinnati, Ohio.

Hardaker, J.B. and Lien, G. (2010) Probabilities for decision analysis in agriculture and rural resource economics: the need for a paradigm change. *Agricultural Systems* 103, 345–350.

Harwood, J., Heifner, R., Coble, K., Perry, J. and Agapi, S. (1999) *Managing Risk in Farming: Concepts, Research, and Analysis*. Agricultural Economic Report No. 774. Market and Trade Economics Division and Resource Economics Division, Economic Research Service, United States Department of Agriculture, Washington DC.

Huirne, R.B.M., Hardaker, J.B. and Dijkhuizen, A.A. (eds) (1997) *Risk Management Strategies in Agriculture: State of the Art and Future Perspectives*. Mansholt Studies 7. Wageningen Agricultural University, Wageningen, The Netherlands.

International Organization for Standardization (ISO) (2009) ISO 31000:2009, *Risk Management – Principles and Guidelines*. ISO, Geneva, Switzerland.

Just, R.E. (2003) Risk research in agricultural economics: opportunities and challenges for the next twenty-five years. *Agricultural Systems* 75, 123–159.

Knight, F.H. (1933) *Risk, Uncertainty and Profit*. Houghton Mifflin, Boston, Massachusetts.

McNeil, A.J., Frey, R. and Embrechts, P. (2005) *Quantitative Risk Management: Concepts, Techniques and Tools*. Princeton University Press, Princeton, New Jersey.

Meyer, D.J. (ed.) (2003) *The Economics of Risk*. W.E. Upjohn Institute for Employment Research, Kalamazoo, Michigan.

Moss, C.B. (2010) *Risk, Uncertainty and the Agricultural Firm*. World Scientific Publishing, Singapore.

Organisation for Economic Co-operation and Development (OECD) (2009) *Managing Risk in Agriculture: a Holistic Approach*. OECD Publishing, Paris.

Raiffa, H. (1968, reissued in 1997) *Decision Analysis: Introductory Readings on Choices under Uncertainty*. McGraw Hill, New York.

Schlaifer, R. (1969) *Analysis of Decisions under Uncertainty*. McGraw-Hill, New York.

2 Decision Analysis: Outline and Basic Assumptions

Basic Concepts

As explained in Chapter 1, decision analysis is the name given to the family of methods used to rationalize and assist choice in an uncertain world. In this chapter we focus on the concepts and methods of decision analysis. Figure 2.1 provides an outline of the typical steps in decision analysis of a risky choice. These stages are discussed in turn below.

Establish the context

This first step is concerned with setting the scene and identifying the parameters within which risky choice is to be analysed. Particularly in a large organization, it may be important to take note of the level in the organizational structure at which the choice will be made. For example, different sorts of decisions may be made at different levels, perhaps with important strategic issues decided upon at board level, with key tactical decisions made by senior management and with a range of more routine choices made at the operational level. Identifying the level may lead to identifying the decision maker (DM) or makers – a critical need for proper conduct of the steps to follow. Similarly, it will be important to identify the stakeholders – those who will be affected by the outcomes of the decision. For a family farm, the principal stakeholders are farm family members who typically will be concerned with their standard of living and the continued survival of the family business. Other stakeholders include staff, the local community, buyers of the produce of the farm, etc.

Also important in setting the context is the relationship between the person doing the analysis and the DM(s), when these are not the same. Decision analysis, as propounded here, is fundamentally a *personal* approach to rationalizing risky choice, yet in practice the analyst and the DM(s) are often not the same person. In decision support in agriculture, the analyst may be an extension or research worker, seeking to assist a large number of farmers, creating obvious difficulties in trying to 'personalize' the analysis. We return to this issue in later chapters. For the moment, it is sufficient to note the need to be clear about the roles of the different people or groups involved in the analysis and in making the final decision, as well as establishing the relationships between those involved.

Identify the important risky decision

As noted in Chapter 1, most standard approaches to risk management start with the identification and ranking of risks. That is good sense but does not cover all situations. It will usually also be important

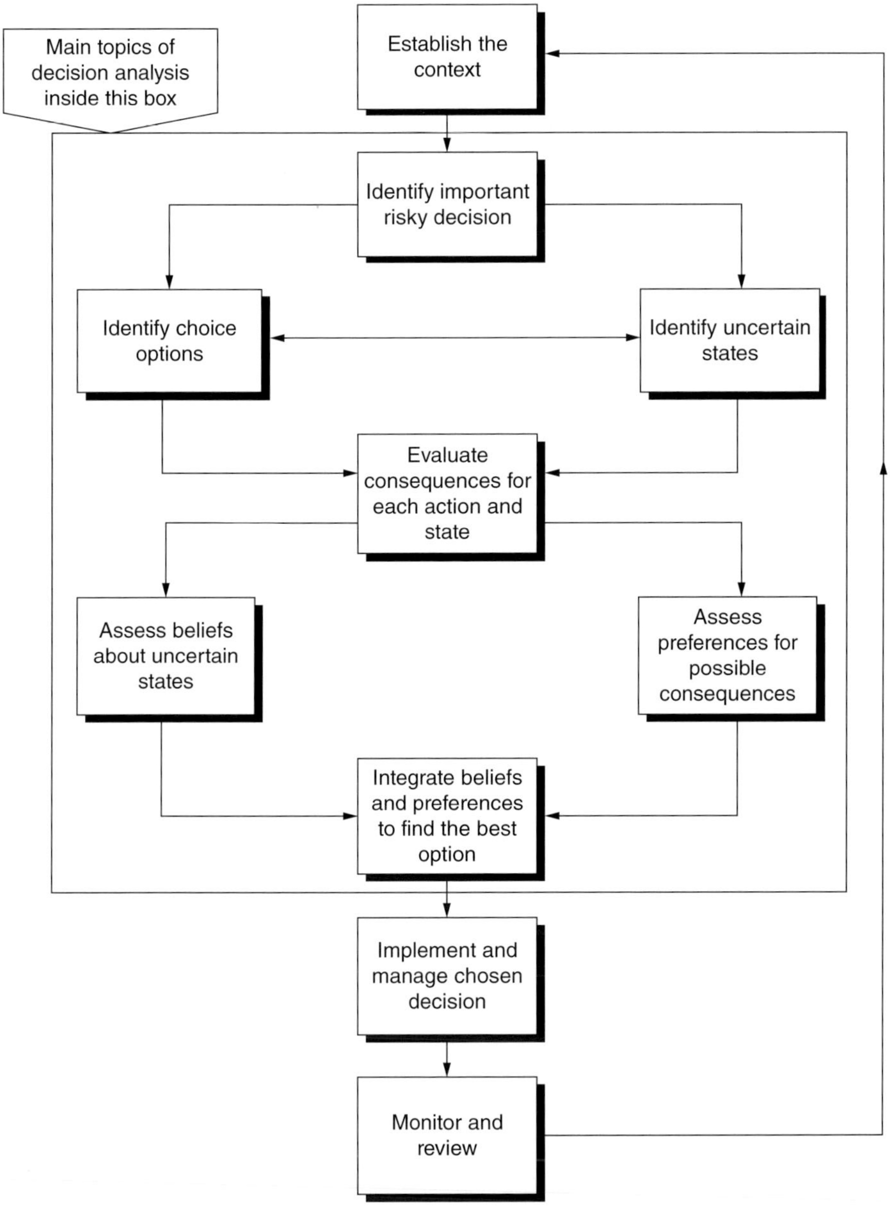

Fig. 2.1. Steps in decision analysis for risky choice.

to consider any new opportunities. Farmers and other managers need to constantly strive to do things better, yet change is inherently risky. So the risky decision for analysis may be either how to deal with threats to the business (or other organization), or whether to take up some new but risky investment opportunities that improve business performance.

If the concern is with risks to the business as it presently exists, the logical starting point is the identification and assessment of the various risks. It is usual to consider: (i) what can go wrong; (ii) how likely that eventuality is; and (iii) how serious the consequences might be. These steps lead logically to some ranking of the risks and hence to the identification of one or more 'important risky decisions' (Fig. 2.1), namely serious risks that may either be 'treated', or that can be borne. If some treatment is judged necessary, there may be more than one method of treating a particular risk. Therefore there may be choices to be made about whether to treat a given risk and, if so, how. Typical examples of such decision problems relate to what risks should be insured, and to what extent. We give such an example later in this chapter.

New risky opportunities may arise because of changes to the general environment of the organization. New technologies may become available that require considerable capital to implement. Changes in markets may signal a need to switch to different forms or methods of production, again often requiring capital to be invested. It may simply be necessary to expand the scale of operations to remain competitive. Large investments, especially if partly or wholly funded by borrowed funds, are inherently risky, suggesting a need for careful consideration and review before a firm decision to go ahead is made.

Identify choice options and uncertain states

Once the important risky decision is identified, the next stage, as shown in Fig. 2.1, is to consider and identify both the choice options or actions and the important uncertain states than constitute the sources of risk for each possible choice option. These two processes will generally occur more or less concurrently since there will nearly always be important interactions between them. Deciding which uncertainties and hence which states will be important may depend on which choice options are being considered, and the choice options considered to be worth investigation may partly depend on perceptions about what are the major sources of uncertainty impinging on the future of the business or other organization.

This stage of the analysis may require considerable investigation and thought. Whether dealing with a new initiative or searching for how best to treat a threat, the range of options may be considerable, and it may be far from clear what possible uncertain events may affect the outcome of different actions. For some decisions the implications may stretch well into the future, perhaps imposing a need to set a time horizon for the analysis. Generally it will be important to identify the time sequence of actions, unfolding uncertain events and possible reactions to those events, as discussed further in Chapter 11, this volume.

Evaluate consequences

This step involves evaluating the consequences of each possible action given each possible state. Obviously, there could be a large number, perhaps an infinite number, of choice options. The same goes for the number of uncertain events that could affect the consequences of those options. For some forms of analysis described later, such as decision trees and payoff tables, judgement is usually needed to cull the scope of the analysis heavily to what really matters. Other forms of analysis, such as mathematical programming or stochastic simulation, described in later chapters, can allow large numbers of choice options or states to be evaluated. State-contingent analysis, also considered in a later chapter, is usually confined to just a few states while generally allowing for continuous decision variables.

While the consequences may take different forms, in this book we focus mainly on business decisions in which consequences are expressed in monetary terms, such as of wealth (equity), income, or gains and losses. The implication is that non-monetary consequences can either be valued in money terms, or are left as 'below the line' items, to be considered informally after the main analysis is done. However, as explained in Chapter 10, this volume, decision analysis can be applied to problems with consequences expressed in terms of multiple attributes, if that is considered to be necessary.

Assess beliefs and preferences

These assessments form the key stage in decision analysis. We assume that important risky decisions are best dealt with by breaking the choice down into separate judgements about the uncertainty that affects the possible consequences of the decision, and about the preferences of the DM for different consequences. This is done for each choice option and then these two parts of the analysis are reunited to find what is the 'best' choice for a 'rational' DM.

A number of things need to be understood about this approach. First, in a risky world (i.e. in the real world), there is no way of knowing ahead of time what will be the 'correct' choice to make – that can be known only after the uncertainty surrounding the decision has unfolded, and seldom even then. In most cases, we eventually experience the consequences of choices we have made in the past, but rarely can we know what would have been the consequences of some other choice. So, decision analysis is based on the proposition that a 'good' decision is one that is consistent with the carefully considered beliefs of a person making the decision about the uncertainty surrounding that decision, and with that person's preferences for the different possible consequences. A 'good' decision certainly does not, and cannot, guarantee a good outcome.

Second, the approach suggested here is based on an assumption about how most people will wish to approach the task of making important risky decisions. Decision analysis, as described here, is a *prescriptive model of choice*. This prescriptive choice model is a logical derivation from some *axioms* or supposedly self-evident truths about how a rational person would wish to act in making important risky decisions. These axioms are given later in this chapter.

Third, and related to the previous two points, not everyone will want to address every important decision in this way. Some may be uncomfortable with the 'clinical' exposure of their beliefs and preferences. They may be discomforted by the emphasis on subjective beliefs and preferences and may look for some more reliable basis for their risky choices. Unfortunately, they are unlikely to find one and may turn to more conservative choice models such as some of the so-called '*safety first*' rules that give priority to minimizing the chances of specified bad outcomes. These conservative choice rules are intrinsically arbitrary but are appreciated by some pragmatic analysts.

Fourth, many people, especially those with a scientific background, find the emphasis on subjectivity in decision analysis disturbing if not unacceptable. The response to them is in two parts. First, the aim in decision analysis is to make the analysis as 'objective' as possible (or at least, as objective as is worthwhile in the context of the particular decision). Second, and notwithstanding the first point, 'objectivity in science is a myth, in life an impossibility, and in decision making an irrelevance' (Anderson *et al.*, 1977, p. 18). The history of science shows time and again that many supposed objective truths have been found to be anything but true. In a world full of uncertainty, and when making choices that will be affected by events in the future, the hard reality is that seldom if ever is it possible to eliminate a degree of subjectivity. It may be possible to be objective about the past and the present, but the future is inevitably wholly imaginary.

Integrate beliefs and preferences to find the best option

The integration stage flows from and is linked to the previous steps of the analysis. For individual decisions, it is necessary for the DM to consider, for each choice option, her probabilities of different outcomes and her relative preferences for these outcomes. With a single DM, who is presumed to bear the consequences of what is eventually decided, it makes sense to use that person's subjective beliefs about the likelihood of occurrences of different possible outcomes and that person's relative preferences for different consequences. We explain in subsequent chapters how beliefs can be calibrated via subjective probabilities and how preferences for consequences can be encoded via a utility function. So we ask readers to suspend any doubts they may have about these ideas until they have read the explanations and rationalizations to follow. If these two ideas are accepted, it is then possible to calculate the probability-weighted average of the utility of each action, called the *expected utility* of that action. The axioms given later in this chapter confirm that this expected utility is the relevant criterion for rational choice (i.e. the utility of any choice option is its expected utility and alternatives can be ranked by their expected utilities). The action with the highest utility is the best.

Some decisions, particularly but not solely public policy choices, have consequences that affect other people than the DM (or the group of DMs). As noted above, decisions involving many people add an extra level of complexity. We discuss group decision making in Chapter 5 and policy choices in Chapter 13.

Implement and manage

Implementing a decision simply means doing what has been decided upon. Implementation may be a simple matter in small organizations when the person making the risky decision is also the manager. However, implementation may be much more difficult in large organizations in which management responsibility is delegated. Especially in such cases, but also generally, active management is likely to be needed to make sure that the implementation is going ahead as planned.

Monitor and review

Once a decision or decision strategy has been decided upon and implemented, it should, of course, be maintained. Moreover, because all such choices are made with imperfect information, it is likely that some decisions will turn out to be unsatisfactory. Monitoring and review are therefore necessary to establish whether the plans decided upon are working and to identify aspects where further decisions need to be made. If adjustments are needed, the steps in risk management may have to be revisited to deal with the problem appropriately.

In a risky world, to paraphrase Robert Burns, 'the best laid plans of mice and men often go awry'. Even carefully considered decisions may turn out poorly. Monitoring and review processes are essential parts of the process of learning about the choice that was made, so that remedial action can be taken if necessary, better plans for the future can be devised and put into operation, or better methods of analysis can be adopted.

A Simple Risky Decision Problem

To illustrate the application of the basic decision analytic approach, we take as an example a choice faced by a dairy farmer in Europe who is worried about the risk of another outbreak of foot-and-mouth disease (FMD). If an outbreak should occur, under the current national control strategy, if an infection is discovered on a farm, government policy is to slaughter the herd and destroy the carcasses. The same measures are applied even if there is a serious suspicion of infection, for instance if it is established that there has been contact with another infected herd.

For those farms in the affected area for which there is no suspicion of infection, bans are imposed on the movement of animals, along with a range of other compulsory preventative measures. No compensation is paid to these farmers.

On farms where the animals are slaughtered, the government compensates the farmers for the value of the animals lost. But there is no government compensation for business interruption losses. Yet, after the stock are slaughtered, the farm might have to remain without livestock for 6 months or more and it would take a long time once the outbreak is controlled for herd numbers and productivity to be built back to the previous level.

There is an insurance product on the market that provides full cover for the additional costs if the farm is affected by regulations and bans imposed as a result of a nearby disease outbreak. Also insured are consequential losses if the farmer's animals have to be slaughtered. However, understandably, the premium is rather high and our farmer is wondering whether to insure. She is thinking seriously about insurance because she has recently invested heavily in new dairy buildings and equipment, funded by a bank loan. A catastrophe such as an FMD outbreak could seriously affect the viability of the business.

Key information about the farmer's decision problem is set out in Table 2.1. Her financial position is shown and the insurable losses are indicated. In practice, the losses are uncertain, depending mainly on the duration of the outbreak. However, for the purposes of this simplified example, this additional uncertainty is ignored and emphasis is placed on the main source of uncertainty – the possibility that there will be a disease outbreak in the area. The premium to indemnify this farmer against the possible losses is also shown in the table. What should the farmer do, insure or not insure?

Decision tree representation

The simple decision problem described above can be represented using a decision tree. This is a diagram that shows the decisions and events of the problem and their chronological relationships. The time

Table 2.1. Information about a dairy farmer's decision on insurance against foot-and-mouth disease (FMD).

Financial position of the farmer:	
Total assets	$1,000,000
Total debts	$500,000
Net assets	$500,000
Insurable losses if there is an outbreak:	
Government-imposed bans only	$10,000
Losses following animal slaughter	$200,000
Cost of protection:	
Insurance premium	$7,245

scale runs from left to right in a decision tree, usually with the first choice represented at the extreme left, and the eventual consequences shown at the extreme right.

Decision problems are shown with two different kinds of forks: one kind representing decisions and the other representing sources of uncertainty. *Decision forks*, conventionally drawn with a small square at the *node*, have branches 'sprouting' from the *decision node* representing alternative choices. *Event forks*, drawn with a small circle at the node, have branches sprouting from the *chance node* representing alternative events or states. The two kinds of fork are very different because the DM has the power to decide which branch of a decision fork is to be selected, but the branch of an event fork that applies is determined by chance or 'luck'.

The decision tree for the disease insurance example is shown in Fig. 2.2. In this case there is only one decision fork representing the choice between insuring or not. There are, however, two event forks or 'states of nature' that can occur for each decision. First, there may or may not be an FMD outbreak during the coming 12 months in the area where the dairy farm is located. Second, if such an outbreak does occur, the government may decide to: (i) impose bans; or (ii) slaughter and destroy the cows and young stock.

The values given at the right-hand end of each act–event sequence are the payoffs or consequences for the farmer. In this case, these numbers are expressed in terms of terminal wealth. To calculate these numbers the losses along each act–event sequence were first added up and then resulting (negative) totals were added to the farmer's initial net asset position shown in Table 2.1. This was done because, as we shall explain in Chapter 5, it is likely that the farmer would be better able to appreciate the significance of possible consequences expressed in terms of (positive) wealth rather than as (negative) losses.

Inspection of the tree shows that the consequences are identical after a decision to insure, regardless of the state of nature. That, after all, is the purpose of full-cover insurance. We can therefore simplify the tree by omitting the event forks after a decision to insure. The result is shown in Fig. 2.3.

Observe that this simple example is truly a decision problem because neither option is clearly superior to the other. If a decision is made to insure and there is no disease outbreak, the premium has been paid to no advantage. On the other hand, not insuring is naturally considerably inferior if the disease does strike the area, and especially so if the farmer suffers the misfortune of having to have her cattle slaughtered.

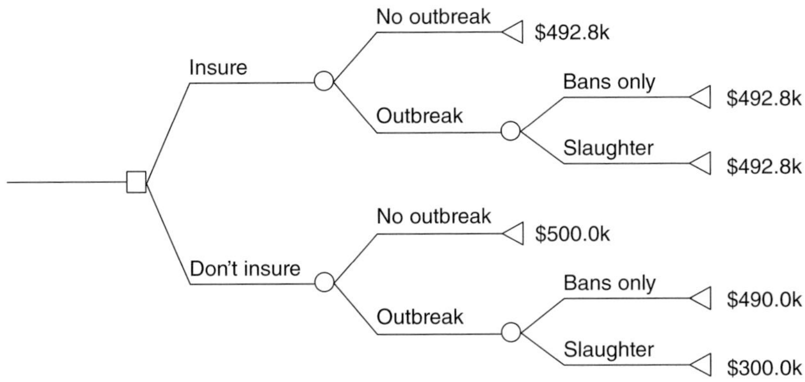

Fig. 2.2. A decision tree for the disease insurance example.

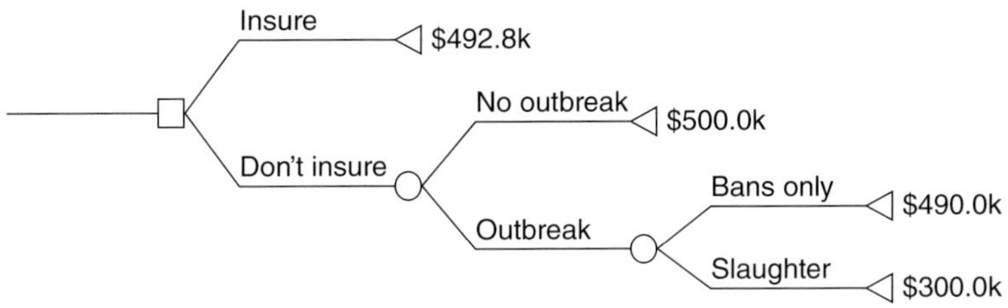

Fig. 2.3. Simplified decision tree for the disease insurance example.

Analysing the decision tree

The first step in analysing a decision tree is to find a way of dealing with the risk represented by the event forks in the tree. To do this fully in the spirit of the decision analysis exposited in this book, we ultimately shall need to know the farmer's subjective probabilities for the uncertain events, and her utility function for wealth. Since we deal with the elicitation of probabilities and utility functions in subsequent chapters, we can conveniently side-step these issues for the moment by introducing the concept of a certainty equivalent.

Certainty equivalents

Consider the event fork in the decision tree in Fig. 2.3 that follows the decision not to insure and the event that there is a disease outbreak.

This fork, reproduced in Fig. 2.4, represents a source of risk for the dairy farmer in that she faces uncertain consequences or payoffs. A decision with risky consequences is called a *risky prospect*. For every DM faced with a decision with risky payoffs, there is a sum of money 'for sure' that would make that person indifferent between facing the risk or accepting the sure sum. This sure sum is the lowest sure price for which the DM would be willing to sell a desirable risky prospect, or the highest sure payment the DM would make to avoid an undesirable risky prospect. This sure sum is called the *certainty equivalent* (CE) of that DM for that risky prospect. In general, CEs will vary between people, even for the same risky prospect, because people seldom have identical attitudes to risk (utility functions) and may also hold different views about the chances of better or worse outcomes occurring (subjective probabilities).

In this case we can ask the dairy farmer to consider a situation in which there has been a disease outbreak with no insurance. She now faces the risky prospect of having net assets of either \$490,000 or \$300,000, depending on whether she has to cope only with government bans or whether her cattle have to be slaughtered. To find her CE for this prospect, we use a progressive questioning procedure, as illustrated below.

We ask the farmer to express a preference between the risky prospect or the sure sum in a *payoff table*, as shown in Table 2.2.

This table is formed with the choice options forming the main columns and the possible events forming the rows, and with the payoffs for each action corresponding to each event shown in the body of the table. Note that the 'sure sum' is the wealth position the farmer could reach regardless of how the disease outbreak affects her farm. We replace the \$$x$ in the payoff table with a sequence of trial values, with a view to zeroing in on the CE.

Suppose we were to set the sure sum x = \$300,000. It is then obvious that any DM would opt for the risky prospect since the worst that can happen in each case is a terminal wealth position of \$300,000, while the risky prospect offers some chance of ending up with as much as \$490,000. Now suppose we set x = \$490,000. The reverse situation now prevails in that any DM will opt for the sure sum because it offers the same as the best outcome for the risky prospect, but with no chance that the outcome will be as low as \$300,000. Evidently, there is a reversal of preference between values of x = \$300,000 and x = \$490,000, and the purpose of the questioning is to find the point at which that reversal takes place.

We might set a trial value for x of \$400,000. In making the choice now, the farmer will obviously have to consider both her assessment of the chances that her cows and young stock will or will not have to be slaughtered, and her attitude towards the possible terminal wealth positions in each case. Suppose that she opts in this case for the risky prospect. Then we can substitute a larger trial CE value for the \$400,000 and again ask for her preference. But if she opts for the sure sum when x = \$400,000, we need to reduce the trial value to, say, \$380,000. By a series of such questions we can zero in on a figure at which the farmer is just indifferent between the two, and then we have established her CE for the risky prospect.

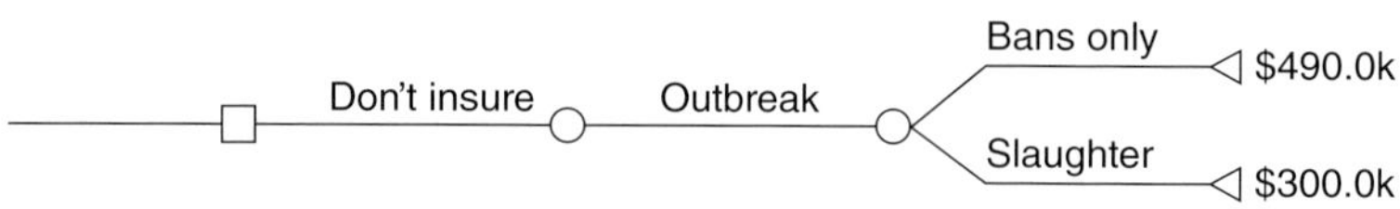

Fig. 2.4. Risky prospect faced if no insurance is bought and a disease outbreak occurs.

Table 2.2. Payoff table for dairy farmer's event fork if insurance is purchased.

Outbreak entails	Risky prospect	Sure sum
Bans only	\$490,000	\$$x$
Slaughter	\$300,000	\$$x$

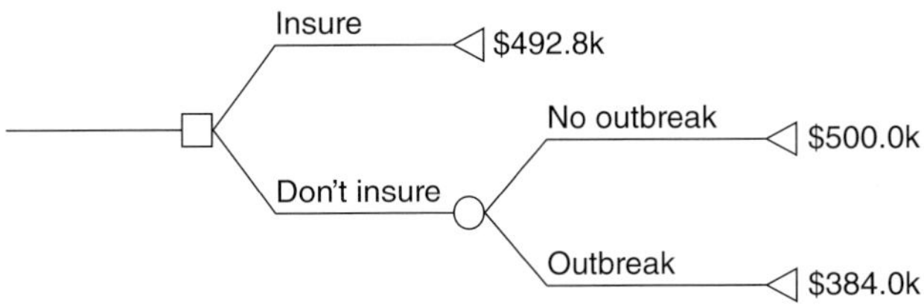

Fig. 2.5. Decision tree for the disease insurance example with terminal chance node replaced by its certainty equivalent (CE).

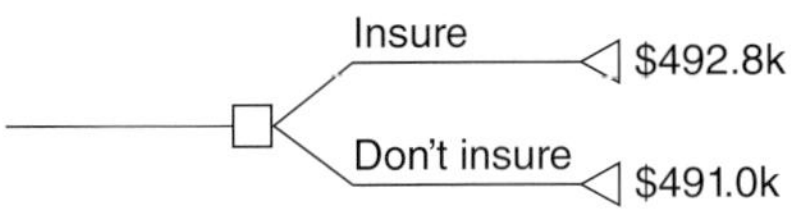

Fig. 2.6. Decision tree for the disease insurance example pruned back to the initial choice to insure or not.

A CE established in this way may be assumed to encapsulate both the farmer's attitude to risk, as encoded in her (as yet unknown) utility function and her (also not yet explicitly assessed) subjective probabilities for the uncertain government-imposed control measures.

Suppose that the DM assigns a CE of $384,000 to the event fork in the decision tree in Fig. 2.3. That means that we can replace the event fork reflecting uncertainty about the disease impact with a sure payment of $384,000, as shown in Fig. 2.5. Making such a replacement is legitimate because we have established that a sure sum of $384,000 is equivalent to the risky prospect in the judgement of the dairy farmer.

We have made some progress, but we still have not completed the analysis. It is still unclear which is the best option. To go further we need to replace the remaining event fork by its CE. We repeat the process, therefore, of asking the dairy farmer to consider trading this risky prospect for a sure sum. This time she will have to consider how she feels about terminal wealth positions of $500,000 versus $384,000, as well as about the chances that there will be a disease outbreak in the area during the currency of the policy, should she buy it. Note that, although the consequences of an outbreak, should it occur, are uncertain, for the elicitation of this CE, they can be treated as certain, since we have established the equivalent sure value of the next event fork for this farmer.

Suppose the farmer, after due consideration, decides that her CE for the remaining event fork in Fig. 2.5 is a net asset position of $491,000. Again, we can prune the decision tree back further by replacing the event fork with this CE, with the result shown in Fig. 2.6.

Now the choice is straightforward. The farmer faces a simple option between a sure payoff of $492,800 and a risky prospect worth the equivalent of $491,000 for sure. True, the difference is small, though it may seem more significant if we note that the CEs differ by about $1800 in favour of insuring. Evidently, our dairy farmer should buy the FMD insurance policy.

Steps in decision tree analysis

The method used to solve this very simple decision tree can be extended to more complex decision problems. For problems that can be adequately represented with a single objective of money payoffs, the steps in the analysis are as follows:

1. Calculate money payoffs for each terminal node of the decision tree by summing along the relevant decision–event sequence.
2. Working back from the terminal branches, resolve the tree by replacing each event fork by the corresponding CE, and resolving decision forks by selecting the branch with the highest CE.
3. Mark off dominated decisions at each stage so that the optimal path through the tree is apparent.

There are some obvious limitations to this approach. It may well prove difficult to obtain a large number of CEs for the resolution of complex trees, especially if some decision forks have many branches. In general, it is likely to prove easier and more effective to elicit the DM's utility function to encapsulate his or her attitude to risk, which can then be applied to the evaluation of every event fork. Also, there will be merit in explicitly identifying the probabilities relevant to each event fork. Armed with a utility function and the relevant probabilities, the tree can be evaluated in terms of expected utility which, as we show in Chapter 5, is identical to maximizing the CE. Methods of eliciting probabilities and utility functions are addressed in the following chapters.

Components of decision analysis and its axiomatic foundation

It was pointed out in the context of the example above that the specification of a CE would require the DM to consider both her subjective probabilities associated with risky prospects and her preferences for the possible consequences, encoded as utility values. Bearing these additional aspects in mind and on the basis of the disease insurance example, we can list six components of a risky decision problem:

1. *Decisions*, between which the DM must choose, denoted by a_j. Note that we use a range of other common terms for decisions, such as choices, options, alternatives, actions, or simply acts. They all have the same meaning. Because the outcomes of most decisions are risky, they are sometimes called risky prospects.
2. *Events* or uncertain states of nature over which the DM has no control, denoted by S_i.
3. *Probabilities* measuring the DM's beliefs about the chances of occurrence of uncertain events, denoted by $P(S_i)$. This topic is addressed in detail in Chapters 3 and 4, this volume.
4. *Consequences* or outcomes, sometimes called payoffs, that indicate what might happen given that a particular act or sequence of acts is chosen, and that a particular event or sequence of events occurs. The consequences may be expressed in terms of a single attribute, such as terminal equity, net income or as some partial measure of net benefit. Or consequences may be expressed in terms of two or more attributes, say net income, debt level and hours worked on the farm. In either case, consequences from taking the *j*-th act given the *i*-th state of nature occurs are denoted by X_{ij}.
5. The DM's *preferences for possible consequences*, explicitly measured as utilities $U(a_j)$ for the *j*-th act, where $U(.)$ denotes the DM's *utility function*. By the application of the subjective expected utility (SEU) hypothesis, outlined in the axioms below, $U(a_j) = \Sigma_i U(a_j|S_i)P(S_i)$ (i.e. the utility of each action is its expected utility). Note that $a_j|S_i$ should be read as a_j 'given' S_i. Hence the utility of the *j*-th action is calculated as the utility value of the consequences for that action given that the *i*-th state eventuates, $u_{ij} = U(a_j|S_i)$, these values weighted by the corresponding subjective probabilities $P(S_i)$. Equivalently, preferences may be captured by the corresponding CEs such that $U(CE_j) = U(a_j)$, where CE_j is the sum of money for sure that

makes the DM indifferent between it and the risky prospect a_j. The derivation of CEs is addressed further in Chapter 5, this volume.

6. A *choice criterion* or *objective function* – taken to be choice of the act with the highest (expected) utility, which, as we shall show in Chapter 5, is exactly equivalent to maximizing CE.

The decision analysis approach outlined in Fig. 2.1 is founded on a set of axioms. Axioms are propositions that are sufficiently self-evident that they can be regarded as widely accepted truths. Anderson *et al.* (1977, pp. 66–69) provide one of many versions of the axioms that underlie decision analysis. Most of the alternative sets of axioms vary only in minor ways from those given below.

These axioms relate to choices among risky prospects, each of which is characterized by a probability distribution of outcomes (including a probability of one for a single outcome in some cases). As explained above, the consequences of the *j*-th risky prospect a_j are uncertain and depend on the state of nature S_i that eventuates, and the DM has no control over states of nature. The axioms are:

1. *Ordering.* Faced with two risky prospects, a_1 and a_2, a DM either prefers one to the other or is indifferent between them.

2. *Transitivity.* Given three risky prospects, a_1, a_2 and a_3, such that the DM prefers a_1 to a_2 (or is indifferent between them) and also prefers a_2 to a_3 (or is indifferent between them), then the DM will prefer a_1 to a_3 (or be indifferent between them).

3. *Continuity.* If a DM prefers a_1 to a_2 and a_2 to a_3, then there exists a subjective probability $P(a_1)$, not zero or one, that makes the DM indifferent between a_2 and a lottery yielding a_1 with probability $P(a_1)$ and a_3 with probability $1 - P(a_1)$.

4. *Independence.* If the DM prefers a_1 to a_2 and a_3 is any other risky prospect, the DM will prefer a lottery yielding a_1 and a_3 as outcomes to a lottery yielding a_2 and a_3 when $P(a_1) = P(a_2)$.

These four axioms are sufficient to deduce what is called *the subjective expected utility hypothesis*, originally proposed by Daniel Bernoulli in 1738 and re-invented by von Neumann and Morgenstern in 1947. The hypothesis states that, for a DM who accepts these axioms, there exists a *utility function U* that associates a single utility value $U(a_j)$ with any risky prospect a_j. Moreover, the function has the following remarkable properties:

1. If a_1 is preferred to a_2, then $U(a_1) > U(a_2)$ and vice versa. In other words, utility values can be used to rank risky prospects and to identify the one with the highest utility as the most preferred.

2. The utility of a risky prospect is its expected utility, i.e. $U(a_j) = E[U(a_j)]$ where:

$$U(a_j) = \Sigma_i U(a_j|S_i)P(S_i) \tag{2.1}$$

for a discrete distribution of outcomes. For a continuous probability distribution for S, $f(S)$, this expression becomes:

$$U(a_j) = \int U(a_j|S) f(S)\, dS \tag{2.2}$$

The implication of Eqns 2.1 and 2.2 is that higher order moments of utility such as variance do not enter into the assessment of risky prospects.

3. The function U is defined only up to a *positive linear transformation*. In other words, it is like temperature, with an arbitrary origin and scale. This property limits how utility measures can be used and interpreted, as amplified in Chapter 5. For example, it makes no sense to say that one risky prospect yields 20% more utility than another, since a shift in the origin or scale will change the proportional difference.

Although it may not be obvious, the axioms imply a unified theory of utility to measure preference, and of *subjective probability* to measure degree of belief. The proof is not given here but may be found in Savage (1954). It is surely amazing that such a simple and reasonable set of axioms could lead to such powerful implications. Other axiomatic formulations (e.g. Quiggin, 1993) have led to other powerful and more general theories. However, the consensus seems to be that these other generalized utility theories are more relevant in modelling behaviour rather than for prescriptive use, which is the focus of this book (Edwards, 1992, Chapters 1–3). This issue is revisited in Chapter 5, this volume.

Other Assumptions

Before we can put this theory to work, we need to specify some additional assumptions upon which we rely. The first relates to what has been called *asset integration*. We assume that a rational DM faced with an important risky choice will regard the possible losses and gains associated with that decision in terms of their impact on wealth. Depending on the context of the decision, this may mean the personal wealth of the DM for a private decision, or it may mean the net assets of the business or other organization. We justify this assumption purely in terms of its good sense – losses and gains really are just changes in wealth. Oddly, however, empirical evidence suggests that people will often act otherwise, especially when the magnitudes of the losses and gains are small. Often people will accept unfair bets, meaning a risky prospect with a negative expected value such as a lottery ticket. Yet in other situations, even the same person may refuse a risky prospect with much better than fair odds, implying a very strong aversion to losses. Such inconsistent behaviour is quite human and is not unreasonable for prospects with small stakes. However, it can be shown that, over many decisions, such behaviour is likely to lead to significant actual losses or forgone gains. Fortunately, behaviour of this kind seldom extends to prospects with bigger stakes, which are our main concern in decision analysis. In particular, the problem often seems to evaporate if DMs are encouraged to think about the impacts of alternative outcomes of risky choice on wealth, so this is the approach we mainly adopt in this book.

The second assumption we need to make relates to the notion of subjective probabilities. For those trained to believe that probabilities can come only from relative frequency information, the notion of subjective probabilities usually makes them uncomfortable. We shall return to this issue in Chapters 3 and 4, but for now we need to note that we assume that a rational DM will aim for consistency in his or her network of beliefs. Specifically, that means forming judgements about probabilities that are consistent with any relevant evidence. Clearly, therefore, in situations of abundant relevant data, it is likely that there will be little or no difference between the probabilities chosen according to the subjective probability view and those chosen according to the relative frequency view. Yet decision analysis can still be undertaken when relevant frequency data are sparse or absent if subjective probabilities are accepted, while those who reject such probabilities are powerless to make a considered choice.

Rational Choice under Risk

As noted above, *rational choice* under risk may be defined as choice consistent with the DM's beliefs about the chances of occurrence of alternative uncertain consequences and with his or her relative preferences for those consequences. The DM's beliefs are reflected in subjective probabilities assigned to uncertain events, while preferences for consequences are captured in the way risky payoffs, with their associated probabilities, are converted to a CE that can be used as the criterion for choice. As noted, we have also extended our concept of rational choice to include asset integration when assessing preferences for consequences. Moreover, we have asserted that a rational DM will strive to make probability judgements consistent with other relevant beliefs.

The measurement of uncertainty using probabilities is the topic of Chapter 3, with some refinements discussed in Chapter 4, while the elicitation of preferences is the topic of Chapter 5.

Selected Additional Reading

References cited in the chapter offer more information on the axioms and assumptions underlying the approach to prescriptive decision analysis presented here. There has been lively debate about the legitimacy of the approach. We revisit some of these issues in subsequent chapters.

The use of decision trees to structure and analyse risky choice was popularized by Raiffa (1968). Clemen (1996, Chapters 1–3) provides a good introduction to decision analysis, including the basics of decision trees.

For a more recent and more wide-ranging treatment of decision analysis, see Goodwin and Wright (2009). The edited collection by Edwards *et al.* (2007), dealing with the foundations and applications of decisions analysis, includes chapters by some distinguished authors who tell the story of the development of decision analysis and important advances made.

References

Anderson, J.R., Dillon, J.L. and Hardaker, J.B. (1977) *Agricultural Decision Analysis*. Iowa State University Press, Ames, Iowa.

Bernoulli, D. (1738) *Specimen Theoriae Novae de Mensura Sortis*. St Petersburg. Translated by Somer, L. (1954) Exposition of a new theory on the measurement of risk. Econometrica 22, 23–36.

Clemen, R.T. (1996) *Making Hard Decisions: an Introduction to Decision Analysis*, 2nd edn. Duxbury Press, Belmont, California.

Edwards, W. (ed.) (1992) *Utilities Theories: Measurements and Applications*. Kluwer, Boston, Massachusetts.

Edwards, W., Miles, R.F. Jr and von Winterfeldt, D. (eds) (2007) *Advances in Decision Analysis: from Foundations to Applications*. Cambridge University Press, New York.

Goodwin, P. and Wright, G. (2009) *Decision Analysis for Management Judgment*, 4th edn. Wiley, Chichester, UK.

Quiggin, J. (1993) *Generalized Expected Utility Theory: the Rank Dependent Model*. Kluwer, Boston, Massachusetts.

Raiffa, H. (1968 reissued in 1997) *Decision Analysis: Introductory Readings on Choices under Uncertainty*. McGraw Hill, New York.

Savage, L.J. (1954) *Foundations of Statistics*. Wiley, New York.

von Neumann, J. and Morgenstern, O. (1947) *Theory of Games and Economic Behavior*, 2nd edn. Chapter 3 and Appendix. Princeton University Press, Princeton, New Jersey.

3 Probabilities for Decision Analysis

Probabilities to Measure Beliefs

In the previous chapter we explained the components of a decision problem. One of the major components is the existence of uncertain *events*, also called *states of nature*, over which the DM has effectively no control. *Probability distributions* are usually used in decision analysis to specify those uncertainties.

The SEU model introduced in Chapter 2 implies the use of subjective probabilities to measure uncertainty. Because the notion of subjective probabilities may be unfamiliar to some readers, we first explain it. Then we provide some discussion of the thorny problem of bias in subjective assessments of probability, leading to a discussion of methods of eliciting and describing probability distributions. In the final section we explain Monte Carlo sampling from probability distributions, which is a frequently used method in decision analysis.

Different notions of probability

There are different ways of thinking about probability. Most people have been brought up in the frequentist school of thought. According to this view, a probability is defined as a relative frequency ratio based on a large number of cases – strictly, an infinite number. Thus, the probability of a flood in a particular area may be found from a sufficiently long historical trace of river heights. The data would be used to calculate the frequency of occurrence of a river height sufficient to overflow the banks.

A little thought shows that this is not a very practical definition for decision making. Seldom will there be sufficient relevant data from which to calculate probabilities pertinent to the choice problem. For example, consequences of a particular decision may depend critically on the price of grain at some future date. The DM may believe that this price will be largely affected by the outcome of ongoing international trade talks. Clearly, while historical data on grain prices have some relevance, such data alone cannot reflect the uncertainty about the specific current and future situations. Even in the above example of the chance of flooding, historical data will not tell the whole story if climatic change is believed to have affected river flows, or if the hydrological characteristics of the river have been changed by engineering works or land-use changes.

Such problems, which cannot readily be handled in the frequentist school, are conveniently dealt with by taking a *subjectivist* view of probability. According to this point of view, a probability is defined as the *degree of belief* an individual has in a particular proposition. Thus, for instance, a DM contemplating cutting a field of grass for hay might assign a probability of 0.2 to the proposition that there will be rain over the next 3 days.

In summary, probabilities provide the means to measure uncertainty. In decision analysis they are viewed as subjective statements of the degree of belief that a person has in a given proposition. The view of a probability as a degree of belief can be given a formal basis by considering the notion of a reference lottery.

Subjective probabilities and the reference lottery

Suppose we offer you a bet that there are more dairy cows in Denmark than in The Netherlands. We assume that you don't know for sure whether this is the case and you are not allowed to check the statistical information before taking on the bet. You win \$100 if there were more cows in Denmark than in The Netherlands at the most recent agricultural census date, and nothing otherwise. Even though you are uncertain about the relative cow numbers in the two countries, we would expect that you would be willing to play since you stand to lose nothing by doing so, and you could win \$100.

Now we offer you another bet with the same payoffs based on a single random spin of a perfectly balanced wheel, such as that illustrated in Fig. 3.1. After the wheel is spun, the pointer may be in the shaded segment or the white segment. The relative sizes of the two segments can be adjusted before each spin without affecting the balance of the wheel. This second bet is called the *reference lottery* for the first bet. You win \$100 if the pointer is in the shaded zone and zero if it is in the white segment. Suppose we adjust the size of the shaded and unshaded areas on the wheel until the two are exactly equal. You must now make a choice between the two bets, one based on the relative numbers of cows in Denmark and The Netherlands, and the alternative of a 50:50 chance based on a spin of the wheel. Which will you prefer, the cows or the wheel?

To make a choice between the two bets, you need to think about your probability that there are more cows in Denmark than in The Netherlands. Note there is no 'objective' or 'logical' probability for this event, since it is either true or false. Nor are the numbers of cows in the two countries at the last census date subject to random variation. Yet the proposition about the relative sizes of the two populations is still uncertain for most people and that uncertainty can be measured as a probability. If you think there is more than a 50:50 chance that there were more cows in Denmark, you will bet on that, otherwise you will prefer to bet on the wheel.

We now vary the size of each of the two segments on the wheel, and continue to ask you to choose between the two bets until we find a proportion of shaded to white areas that make you just indifferent between the two risky prospects. Suppose it is 30% shaded and 70% white. Then your probability that there were more cows in Denmark than in The Netherlands at the last census date is 30/100 = 0.3.

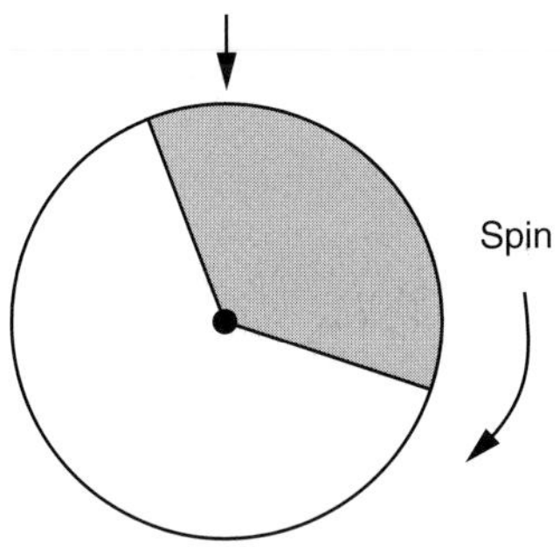

Fig. 3.1. A probability wheel for a reference lottery.

The concept of a reference lottery such as that just illustrated can be used as the basis for a definition of probability. Such a definition implies two rules of probability:

1. All probabilities are numbers between zero and 1.0.
2. The sum of the probabilities across all the possible outcomes for a particular uncertain prospect must be 1.0.

If probabilities are needed for only three or four alternative uncertain events, the probability wheel can be modified to show an appropriate number of segments, the relative angle of each of which can be varied. Computer versions of such a probability wheel are available on the Internet – search for 'spinning wheel probability'.

It is important to strive for consistency in probability judgements – with real feelings of uncertainty and with available evidence (including any relevant relative frequency data if they exist). Rational people will strive for consistency in their whole network of beliefs and so will try to form probability judgements using all relevant information. The aim is to make probabilities as 'objective' as possible, while recognizing that the process of collecting and processing information to refine probabilities has costs that must be matched against possible benefits from better (i.e. better informed) decisions. Ways to improve the quality of probability judgements and how probabilities should be updated in light of new evidence are discussed in Chapter 4, this volume. Later in this chapter we deal with the way in which probabilities should be updated in the light of new evidence.

Whose probabilities?

If all probabilities are subjective, they are also personal; two people can reasonably assign different probabilities to the same uncertain event. Such differences can help explain differences in behaviour. But whose probabilities should be used to analyse a risky decision problem? If there is only one chief DM, the answer is obvious. It is the DM's probabilities that are appropriate since the DM is the one who must take responsibility for the choice made and, in most cases, must bear all or most of the consequences. Of course, when DMs are not very well informed about the uncertainty at stake, they might decide to consult somebody more expert, perhaps being willing to adopt that person's probabilities as their own.

When there is more than one DM, there are difficulties in modelling risky choice. A group such as a family, a village, a government or nation, or an organization such as a business, may face collective risky choices. How can differences in opinion about the chances of occurrence of important events bearing on the decision be resolved? This tricky issue of group decision making is deferred to Chapter 5, this volume.

Quite often, a decision analyst working in agriculture or resource economics will be required to formulate some probabilities for someone else, such as an individual client, a group of clients or a policy maker. Much published academic work on risk in agriculture is of this nature. Yet there is the obvious problem that probabilities for decision analysis are essentially 'personal', and in the typical situations just described the analyst cannot know the beliefs of each actual or potential client. It would be desirable to have a widely accepted 'code of conduct' for such cases but, as far as we know, no such code exists. Hence, what follows might be seen as first steps towards the development of such a code. For further discussion, see Hardaker and Lien (2005, 2010).

First, it seems obvious that what we choose to call 'public' probabilities should be formulated thoughtfully and with care. Second, the process should be documented and disclosed (or available for disclosure) to clients. Sources of information or frequency data should be listed and any manipulation of data, such as detrending, should be described. If experts are consulted, their qualifications for being judged as bringing specific knowledge to the task at hand should be indicated. Any training in probability assessment the experts underwent should be noted, and such training might be regarded as a necessary part of good practice. The way probabilities were elicited from the experts and then combined to reach the final assessments should also be explained. Any steps taken to minimize bias, or to account for it, as discussed below, should be described. Finally, the limitations thought to apply to the final assessments should be described. For example, to which areas, types of farms, scales of operation are the assessed probabilities thought to apply, and which not.

Bias and Ways to Deal with It

Studies by experimental psychologists and other behavioural scientists have led to the discovery and naming of very many sources of bias in human assessments of risk in general and of probabilities in particular. For a readable overview of the many problems and issues, see Gardener (2008). Gilovich *et al.* (2002) describe some of the relevant research.

It is clear that most people are not particularly good at thinking sensibly about risk. For example, *availability bias* means that many of us judge how likely a bad event is by how easy it is to recall instances. The wide exposure in the media of catastrophes of one sort or another can lead people to believe that the world is an increasingly dangerous place when, for most of us, the reverse is true. Yet it seems we have short memories, so that, as time passes following some adverse event, such as a drought or a wildfire, memory of that event becomes less 'available' and there is a tendency to assign a lower probability or importance to such a thing happening again. Many people stopped travelling by air after the September-11 terrorist attacks, opting for the much more risky mode of long-distance road travel. Yet air passenger numbers quite quickly returned to former levels as those dreadful events faded somewhat in memory.

A related form of bias is *neglect of priors* whereby a few instances of some event may be deemed to indicate that there is a high risk when the reality is the opposite. For example, a single case of mad cow disease in a country may lead to closure of major markets for producers, even though the prior probability that the disease is at all widespread may be very low indeed.

There is evidence of a phenomenon called *anchoring*. So, for example, in assessing a probability for some important uncertain variable, focusing first on a measure of central tendency may cause the assessor to 'anchor' on this measure and to assign lower probability to extreme values than would be the case otherwise. Anchoring may also contribute to *conservatism*, meaning that initial assessments of a probability are revised too little in the light of new information. And there is other evidence that the way questions about uncertain events are asked can lead to different answers – a phenomenon called *framing bias*.

Many of the cognitive failures in probability judgements lead assessors to believe that they know more than they do. The result has been called *catastrophic overconfidence*, meaning the widespread tendency for people to assign far too high a probability to the chance that some forecast they make will come true. In practice, that often means that the risk of a bad outcome is seriously under-estimated, with potentially damaging consequences.

While there is a large literature on these and other forms of bias that can distort probability assessments and hence (or also) risky decision making, there is less published evidence on what can be done to avoid such bias. But some things are known. Most obviously, using any relevant data, even if sparse, is likely to be beneficial. Moreover, if relevant data are lacking, it surely makes sense to look first at the scope for collecting relevant data before undertaking a risk analysis using only potentially biased and misleading subjective assessments. While a decision to collect more data is not likely to be costless, and may or may not be worthwhile, the methods of decision analysis can be used to evaluate, *ex ante*, whether such additional investigation would be worthwhile.

Similarly, it is surely obvious that thoughtful assessments of probabilities that draw on any relevant non-quantitative evidence are needed. For example, it may be possible to use information about the nature of the processes being considered to identify, say, upper or lower bounds, and perhaps a modal value, for some uncertain quantity to provide a useful starting point for assessment of the full distribution. In the case of crop yields, for example, it is clear that negative yields are impossible and that yields below some threshold would not be worth harvesting. At the other end of the scale, there may be information about the highest attainable yields from experience or experimental work.

There is also good evidence that training in probability assessment can help reduce bias. Potential assessors are presented with almanac-based questions and are asked to provide probability distributions for uncertain quantities, or to give confidence levels that their binary (yes-no) answers are correct. The answers can then be calibrated using *proper scoring rules*. These are rules that reward or penalize assessors (usually with 'points' rather than cash) according to how well or poorly the probabilities they assign conform to their true uncertainty about the actual value under consideration. Typically, such training will lead to a reduction in overconfidence of assessors. Although applied far too rarely, such training using calibration would appear to be vital for any serious risk analysis.

There are many different proper scoring rules and several ways they can be applied. To illustrate the general principle, consider the proposition that the average yield of potatoes in France is higher than that in England. (If you know nothing about potato production in these two countries, substitute any form of production and any two locations for which you are uncertain about the actual yields but for which it would be possible to get reliable data later to find whether the statement is true or not.) Think about your personal probability that the proposition is true. If you really have no idea, you would logically assign a probability of 0.5 to the proposition being true (and so the same to it being false). If you were certain that it was true, you would assign a probability of 1.0. Similarly, if you were sure it was false, your probability would be zero. If you thought it more likely than not that it was true, you might assign p = 0.7 or 0.8. A proper scoring rule assigns points to all such probability values according the actual truth or falsehood of the proposition – which has to be known by the person doing the training. To illustrate, Fig. 3.2 shows

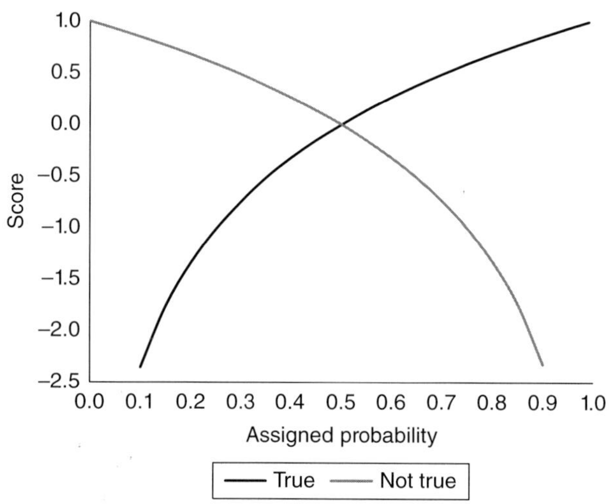

Fig. 3.2. A scaled logarithmic scoring rule of a probability assessment for a binary proposition.

scaled scores based on a logarithmic scoring rule. So a probability of 0.5 scores zero points whether the proposition is actually true or not – compare the black and grey curves. However, if the proposition is true and you assigned $p = 1.0$ to that statement, you would score 1.0, as indicated by the black curve. On the other hand, if you said $p = 1.0$ and the proposition was actually false, you would be penalized with a large negative score, plotted on the grey curve but with a value way off the scale of the figure. Other scores for other values of p can be seen in the figure for the two possibilities that the proposition is true or false.

As this simple example shows, the scoring rule rewards you for assigning a probability that truly represents your uncertainty about the proposition. Over a reasonable number of such almanac-type propositions, some of which would be chosen to be true and some false, arranged at random, your aim would be to maximize your total score (or minimize your total penalty). The calibration process with scoring might be done in competition with others, or it could be set up so that you would aim to improve your scores over a number of sessions. The evidence is that even a modest amount of such training can considerably improve the quality of subjective probability assessments.

Assessors can also learn from calibration 'on the job' through comparisons of actual outcomes versus the assessments they made. For example, when probabilities of rain tomorrow are given daily, it is evident that, following issued forecasts of a probability of rain of, say, 0.3, it should then have rained on close to 30% of occasions. Such calibration of assessment will reveal any consistent bias that assessors can correct. While rather few probability assessment tasks allow such ready calibration, occasional feedback and review may still be educational.

It may also be possible to improve probability assessments by considering the range of events and outcomes that are likely to impinge on the uncertain quantity of direct interest, then to think about the chances of those events and outcomes happening and to assess how they might individually and collectively affect the uncertain quantity of interest for a risk analysis. Such approaches are commonly used in engineering to assess the risk of failure of a system by considering the reliability of its component parts. So, for example, in considering what might be the future prices of some commodity, the analyst might consider factors causing shifts in supply and demand, how likely and how large these shifts might be, and how they might affect the future price. The analysis might be largely intuitive or might be more formal, using available data and a structured model. Either way, by 'dissecting' the assessment in this way, new sources of uncertainty may be revealed, forcing the assessor to recognize the inappropriateness of a too-confident direct assessment. Modelling methods along these lines may include influence diagrams, Bayesian networks and adaptations of Brunswik's probabilistic functionalism (Brunswik, 1955, originally published in German in 1934) – all considered to be beyond the scope of this book.

Eliciting and Describing Probability Distributions

Elicitation

Direct elicitation for simple discrete events

Eliciting or specifying probabilities for discrete states with few possibilities is usually a not-too-difficult task. The reference lottery concept can be used to help attain consistency with both inner feelings of uncertainty and the laws of probability. For example, a farmer may be concerned about the number of breakdowns of the harvester during the coming harvesting season. Each breakdown could be expensive in terms of repair costs and could result in a serious delay in harvesting, affecting the quality of the harvested crop. The most likely number of breakdowns for a machine in good condition might be zero, but (for the purposes of this example) it is also possible to have one, two or a maximum of three breakdowns.

Subjective assessments can be made, drawing on any relevant information, such as the reliability of the machine in the previous season and the condition it is known to be in now. But these assessments need to be expressed as proper probabilities. To this end, the assessor is first asked:

1. What is the probability that the number of breakdowns during the next harvest will be fewer than two versus two or more?

These two probabilities can be calibrated against a reference lottery exactly as illustrated above. Suppose the answer is:

Fewer than two breakdowns: probability = 0.91
Two or more breakdowns: probability = 0.09

(Note that the use of a reference lottery has ensured that these two probabilities add up to 1.0, as required by the logic of probability theory.)

Then, two more questions need to be addressed, again calibrating the answers with reference lotteries:

2. Given that the number of breakdowns of the harvester during the coming season will be fewer than two, what are the probabilities that it will break down exactly once, or not at all in the coming harvest season?
3. Given that the number of breakdowns during the coming season will be two or more, what are the probabilities that there will be two breakdowns or three breakdowns this coming harvest season?

Suppose the answers to these two questions are as follows:

Given number of breakdowns is fewer than two:
Zero breakdowns next harvest: probability = 0.76
One breakdown next harvest: probability = 0.24
Given number of breakdowns is two or more:
Two breakdowns next harvest: probability = 0.78
Three breakdowns next harvest: probability = 0.22

The full probability distribution can now be derived, as shown in the *probability tree* in Fig. 3.3. As the figure illustrates, a probability tree is a decision tree with only event forks, no decision forks. Finding the terminal probabilities in such a tree is simply a matter of multiplying the probabilities along the branches leading to each end point. The results are shown at the right of the tree.

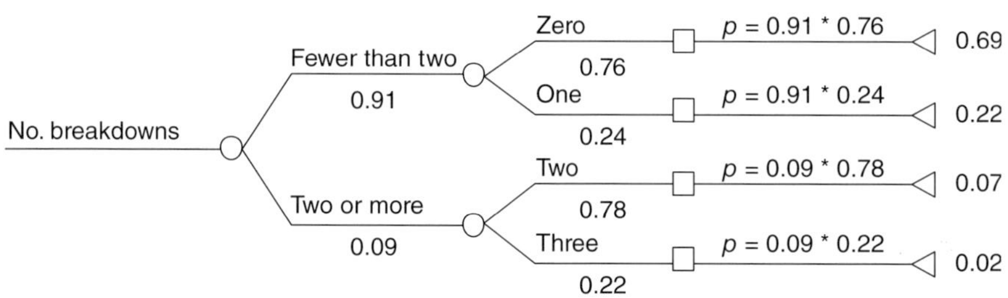

Fig. 3.3. Probability tree to describe harvester breakdowns.

Extrapolating from this example, it is clear that it would be too tedious to use a reference lottery for every probability assessment. Nor is it necessary. Rather the notion of such a lottery can be used to get assessors to think clearly about what they are being asked to judge. It helps assessors to gain a clearer understanding of the nature of (subjective) probability so that they can then make judgements that are consistent with their real feelings of uncertainty, as well as with the rules of probability mentioned above.

Visual impact methods

In cases where the uncertain event of interest can have several outcomes, *visual impact methods* can be useful ways of conceptualizing probability judgements.

As an alternative to the probability wheel, a tabular approach can be used in which counters are used to represent relative probabilities rather than segments of the wheel. The method generally works well so long as the number of possible outcomes is not too great – ideally only about five or six, though most people can readily handle a few more. A table is prepared with the possible alternative outcomes listed and with associated spaces provided for counters – ordinary matchsticks are quite suitable. Sufficient counters – perhaps 25 or so – are then given to the assessor to allocate across the spaces in the table according to the degree of belief that the actual value will be as specified for that row of the table. So, for example, if an assessor thinks that one outcome is twice as likely as another, twice as many counters should be allocated to the space for the first outcome as to that for the second. The assessor should be told that it is not necessary to allocate all the counters, and also that more are available if required. When the assessor is fully satisfied with the allocation of counters, probabilities can be calculated according to observed cell frequencies.

The above example of the number of breakdowns for the harvester might therefore have been approached using a layout as illustrated below. If the allocation of counters was more or less consistent with the answers given above using the reference lottery, the outcome might be as illustrated in Fig. 3.4.

Note that the resultant probabilities are a little different from those given before for the same uncertain events. Such differences are to be expected, if only as a result of the rounding error from using a mere 25 counters. In principle, greater accuracy could be obtained by using more counters. In practice, however, the precision of the whole process is limited by the capacity of assessors to express their feelings of uncertainty in quantitative terms. Given the difficulty most people have in this regard, it is pointless to aim for an unattainable degree of precision. Moreover, in many real decision situations, the final choice will not be affected by minor variations in the probabilities, but in the event that it is, experience suggests that the difference in consequences will seldom matter much.

Number of breakdowns	Counters	Count	Implied probability
Zero	••••• ••••• ••••• ••	17	17/25 = 0.68
One	•••••	5	5/25 = 0.20
Two	••	2	2/25 = 0.08
Three	•	1	1/25 = 0.04

Fig. 3.4. Visual impact method for assessing the probabilities of breakdown of a harvester.

The method extends readily to the case of a single continuous uncertain variable by dividing the range into a convenient number of intervals – say five to seven – then proceeding as before. For example, a particular analysis might depend on the annual average milk production per cow on a dairy farm. The farmer (or other expert) is first asked to think about the possible range of this uncertain quantity. First, the highest possible value is sought on the basis that the DM would be very surprised indeed if the actual herd-level yield ever turned out to be more than this value.[1] Then a lower limit is found by asking for a yield such that the farmer would again be very surprised if the actual herd average were less than this. Suppose that the upper limit is judged to be 7400 kg, and that the lower limit is 6700 kg. This range can be divided into seven equal intervals as follows. As the total range is 7400–6700 = 700 kg, intervals of 700/7 = 100 kg milk are a convenient choice. The first interval ranges from 6700 to 6799 kg, the second from 6800 to 6899, and so on. (Of course, we 'cheated' in this hypothetical example to make the intervals work out neatly. In practice it may be best to use equal 'rounded' intervals for most of the identified range, leaving the 'leftovers' for the highest and lowest intervals. Obviously, the differences in interval width should be pointed out to the assessor.)

Next we set up a table, as for the example of the number of breakdowns above, but this time with seven rows corresponding to the seven intervals in milk yield just derived, as illustrated in Fig. 3.5. Again, the farmer is given a suitable number of matchsticks or other counters and is asked to allocate them to the rows in the table to reflect personal judgements of the relative chances that the actual yield will fall in each interval. Usually there is no difficulty in getting people to approach this judgemental task in a thoughtful way, especially if they have been introduced to the notion of subjective probability using the reference lottery analogy.

Assume that the farmer's allocation is as shown in Fig. 3.5. Then by counting matchsticks, a discrete approximation to the required continuous distribution has been obtained, as illustrated in Fig. 3.6. The continuous distribution can be approximated from this discrete distribution by a method to be explained a little later. First, it is necessary to provide some explanation about graphical representations of continuous distributions.

Range in milk yield (kg per cow/year)	Probability weight (counters)	Count	Probability[a]
6700–6799	•	1	1/26 = 0.038
6800–6899	•••	3	3/26 = 0.115
6900–6999	••••• •••	8	8/26 = 0.308
7000–7099	••••• ••	7	7/26 = 0.269
7100–7199	••••	4	4/26 = 0.154
7200–7299	••	2	2/26 = 0.077
7300–7400	•	1	1/26 = 0.038
	Totals	26	1.000

[a]Probabilities in this column do not sum exactly to 1.0 due to rounding.

Fig. 3.5. Milk yield probability distribution assessed by the visual impact method.

[1] The exact meaning of 'very surprised' depends on how many counters are used to derive the distribution. If there are about 25, then, recognizing the rounding up or down implicit in the approach, the probability of an outcome above the upper bound (and similarly of one below the lower bound) is less than about 1 in 50, i.e. < 0.02.

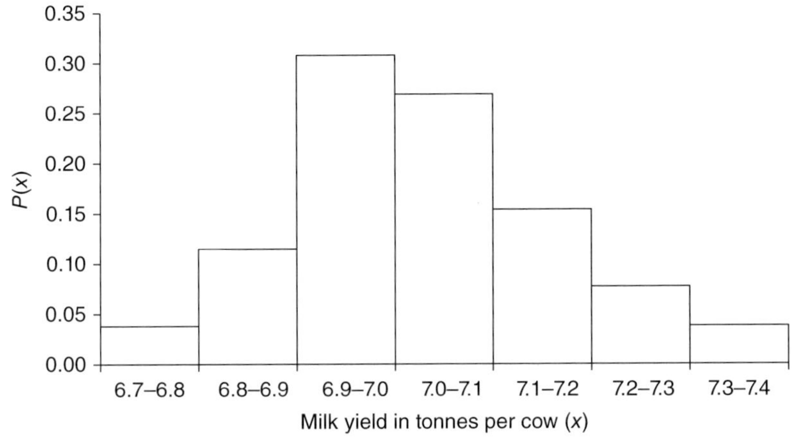

Fig. 3.6. Milk yield per cow approximated as a discrete probability distribution $P(x)$ based on the visual impact method of assessment.

Graphical representation of probability distributions of continuous variables

Two ways of graphing distributions

Many people are used to thinking about probability distributions of continuous uncertain quantities such as dairy cow milk yields or growing-season rainfall in terms of *probability density functions* (PDFs). A PDF will often (but by no means always) be shaped like a bell, with a central peak indicating the most likely value or *mode* of the uncertain quantity and with low-probability 'tails' on either side of the peak stretching out to the upper and lower extremes. Other PDFs may be shaped like a reversed letter J, may have two or more peaks, or may be of many other different forms.

An example of a PDF is given in the upper part of Fig. 3.7. The PDF for an uncertain quantity x is conventionally denoted by $f(x)$. In this case, $f(x)$ is a typical bell-shaped distribution ranging from a minimum value of a to a maximum of b and with a mode at m. Like every PDF, it has the property that the area under the whole curve is 1.0. The PDF can be 'read' by noting that the area under the curve between two points on the horizontal x-axis measures the probability that the value of the uncertain quantity x will lie in this range. For example, the shaded area A in Fig. 3.7 shows the probability that x will lie in the range of a (the minimum value of x) to X. We could use one of a number of ways to estimate this area relative to the total area under the curve, and hence find the probability of interest. One way is by counting squares when the curve is plotted on graph paper. Another is to cut out and weigh the two parts of the plotted graph. Clearly, such methods are not very convenient.

There are some problems associated with the use of PDFs in decision analysis, partly because they are usually difficult to draw. In particular, it may be challenging to make sure that the area under the curve really is equal to 1.0, as the rules of probability require. More convenient in many situations are the *cumulative distribution functions* (CDFs). The CDF for an uncertain quantity x is conventionally denoted by $F(x)$.

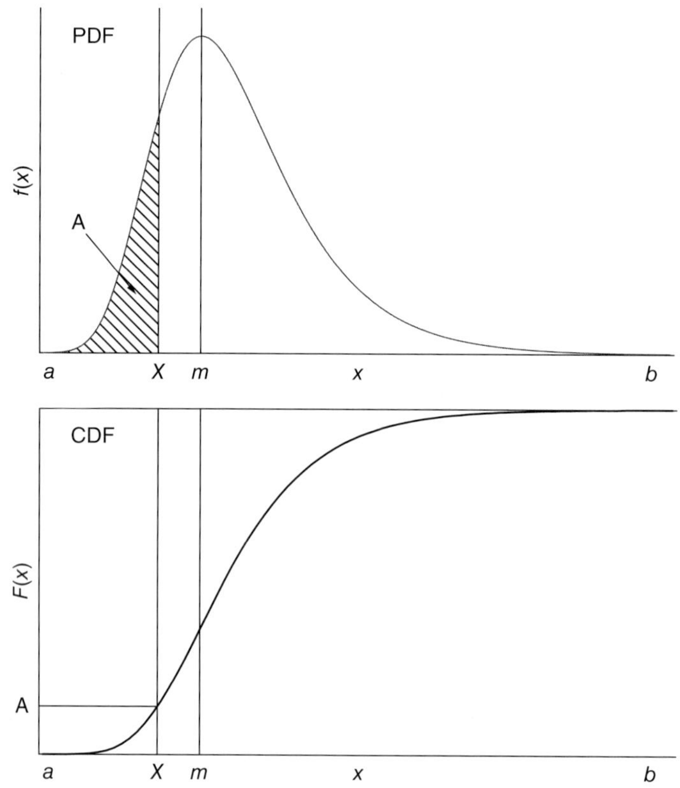

Fig. 3.7. The relationship between a probability density function (PDF) and a cumulative distribution function (CDF).

Formally, the CDF of an uncertain quantity X is the function given by:

$$F(x) = P(X < x) \quad (3.1)$$

where the right-hand side represents the probability that the X takes on a value less than or equal to x.

The relationship between a PDF and a CDF can be illustrated by reference to Fig. 3.7. Suppose we were to take the value X and move it in small steps along the x-axis from the minimum to the maximum value of x, at each point assessing the area to the left of the vertical line. This series of areas will start at zero, when X is set at the minimum point a, and will progressively rise to be 1.0 at the maximum point b. In the tails of the PDF the cumulated area under the curve will increase only slowly as X is moved to the right, whereas it will increase most rapidly in the region of the mode m. The resulting curve is drawn in the lower part of Fig. 3.7.

As illustrated, and for the reasons just explained, the CDF is S-shaped for a typical bell-shaped PDF, with the point of inflection on the CDF corresponding to the mode of the PDF. Of course, CDFs for PDFs that are not of the typical bell shape will themselves be of correspondingly different shapes.

CDFs and fractiles

From the observed frequencies of the elicited distribution found in Fig. 3.6 above, points on the CDF can be calculated as the cumulative probability to that value. The full CDF can then be smoothed through these points. The curve may be drawn by eye or by using some suitable curve-fitting device or software (e.g. kernel smoothing (Silverman, 1998) or splines (Ahlberg *et al.*, 1967)). Often, a hand-drawn curve will serve well since the process allows the analyst to incorporate judgements about additional features of the distribution, such as the locations of the upper and lower limits and the point of inflection (= the mode). However, if using software, it is important to judge the goodness of fit by inspecting the location of the curve relative to the plotted points since there is more to a fitted CDF than may be indicated by typical statistics of goodness of fit, such as R^2. For example, in addition to getting a good fit to the plotted points, it is necessary to ensure that the curve meets the horizontal lines at 0.0 and 1.0 on the cumulative probability axis. For some decision analyses, where extreme outcomes may bring catastrophic consequences, it may be necessary to take special care in locating the tails of the distribution, usually the lower tail.

Once the CDF is drawn, probabilities for particular intervals can then be read from the curve. Again, this can be done by eye or from the formula of the curve if this is to hand, such as having been generated by curve-fitting software.

To illustrate, the elicited data points for the forecast annual milk yield per cow can be represented as cumulative probabilities as shown in Table 3.1.

These data are plotted as a CDF in Fig. 3.8 and a curve has been smoothed through the data points (including the points for the specified minimum and maximum values corresponding to cumulative probabilities of 0.0 and 1.0, respectively).

From the CDF in the figure, the full distribution can be quantified by reading off as many *fractile* values as may be needed. A 0.*b* fractile, denoted $f_{0.b}$ (read as 'point *b* fractile'), is that value of the uncertain variable x for which $P(x < f_{0.b}) = 0.b$. For example, the 0.5 fractile, $f_{0.5}$, can be read from a graph as about 7020 kg milk per cow.

Table 3.1. Conversion of probabilities for intervals of milk yield to cumulative probabilities.

Range in milk yield (kg per cow/year)	Probability	Cumulative probability[a]
Less than 6700	0.000	0.000
6700–6799	0.038	0.038
6800–6899	0.115	0.154
6900–6999	0.308	0.462
7000–7099	0.269	0.731
7100–7199	0.154	0.884
7200–7299	0.077	0.962
7300–7400	0.038	1.000

[a]Probabilities do not reconcile exactly because of rounding.

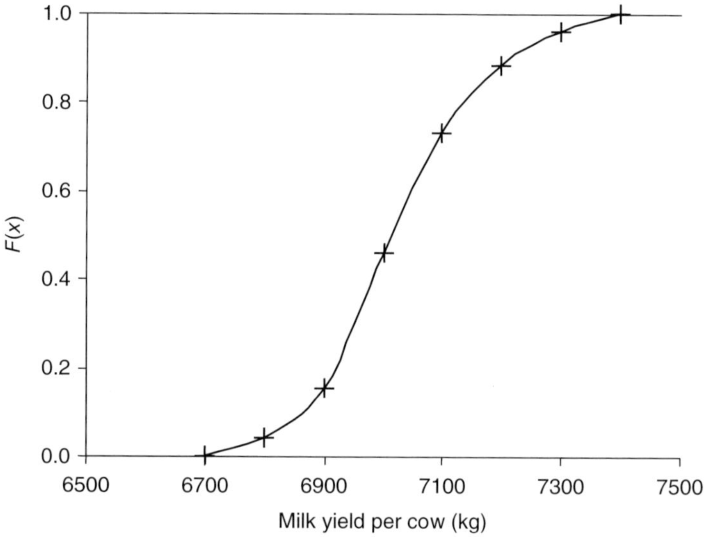

Fig. 3.8. CDF for the elicited probability distribution for milk yield per cow.

Elicitation of continuous distributions using fractiles

Fractiles can be used to provide an alternative to the visual impact method for eliciting subjective probability distributions for continuous uncertain quantities. Most obviously, respondents can be asked for the $f_{0.0}$ and $f_{1.0}$ fractiles that define the range of the distribution. Next, the $f_{0.5}$, also known as the *median*, can be elicited by asking for the value of the uncertain quantity such that it is equally likely that the true value will lie above this value as below it. These three fractiles give three points on the CDF which, for some applications, might be judged to be 'enough'. However, a sparse number of points can easily be augmented by obtaining further fractiles. Thus, the $f_{0.1}$ and $f_{0.9}$ values can be found by seeking values such that there is a 10% chance that the actual value will lie, respectively, below or above the specified values. Some decision analysts (e.g. Raiffa, 1968; Anderson *et al.*, 1977, p. 24) suggest that the best approach is to work with 'equally likely' probabilities since 0.5:0.5 probabilities are better understood than any other values. They therefore propose a series of questions that progressively 'split the difference' on the CDF. Thus, for example, having found that $f_{0.0}$ = 11.0 and $f_{0.5}$ = 13.0, the next question might be: 'Given that the value of this uncertain quantity is less than 13.0, for what value would it be equally likely that the true value would be below this value or between this value and 13.0?' The answer to this question, of course, gives the $f_{0.25}$ value. By such steps it is possible to elicit progressively the 0.25 and 0.75 fractiles, and then the 0.125, 0.375, 0.625 and 0.875 fractiles, and so on. The process of 'splitting the difference' can be continued until enough points have been obtained to draw the CDF with the required degree of confidence. This procedure is sometimes known as the *judgemental fractile method.*

Description of distributions by moments

For some types of analysis it is necessary or useful to calculate some statistics that describe certain features of a distribution. For example, the *mean* or average of a distribution is a good measure of central tendency; it can be thought of as the 'centre of gravity' of the distribution. Dispersion from the mean is typically described by calculating the *variance*, often expressed as its positive square root, the *standard deviation*. When comparing the relative variability of two or more distributions it is often convenient to use the *coefficient of variation*, which may be calculated by dividing the standard deviation by the mean. The coefficient of variation indicates the dispersion of the distribution relative to the mean.

The mean and variance are technically called two of the *moments* of the distribution. A moment is the probability-weighted average of deviations from a specified point, with the deviations raised to some power. In what follows we use the notation $M_k'[x]$ to indicate the k-th moment about the origin and drop the prime to indicate the k-th moment about the mean.

The mean is the first moment about the origin, which, for a discrete distribution, is given by the formula:

$$M_1'[x] = \Sigma_i P(x_i)\,(x_i - 0)^1 = \Sigma_i P(x_i)\,x_i = E[x] = E \qquad (3.2)$$

where $P(x_i)$ is the probability of x_i and $E[.]$ is the expectation operator. For convenience, we may denote the expected value or mean of x, $E[x]$, as simply E.

Higher order moments are usually calculated as deviations from the mean. The first moment about the mean is always zero; the second is the variance:

$$M_2[x] = \Sigma_i P(x_i)\,(x_i - E)^2 = V[x] = V \qquad (3.3)$$

and measures dispersion. As this formula shows, the variance is the probability-weighted average of deviations from the mean, all squared. Because the square of a negative number is positive, variance is a measure of absolute deviation from the mean. Again, for convenience, we denote the variance of x, $V[x]$, as simply V.

Variance is usually more conveniently calculated by the equivalent formula:

$$V[x] = M_2'[x] - M_1'[x]^2 = E[x^2] - E^2 \qquad (3.4)$$

i.e. the probability-weighted average of the squared values of x less the square of the mean.

The third moment about the mean, $M_3[x]$, is one convenient measure of skewness. A positively skewed distribution has a longer tail to the right of the mode than to the left, and vice versa for a negatively skewed distribution. Often skewness is reported in terms of the *coefficient of skewness*, one definition of which is the third moment divided by the standard deviation raised to the third power, i.e. $M_3[x]/SD^3$. Higher moments can also be calculated, though have less familiar interpretations.

These calculations can be illustrated drawing on the example of the discrete distribution for the number of breakdowns of a harvester, the elicitation of which was explained above. The probabilities from the tree in Fig. 3.3 are more conveniently summarized in Table 3.2.

Then the first three moments of the distribution can be calculated applying the above formulae, as follows:

$$M_1' = E = 0.69\,(0 - 0)^1 + 0.22\,(1 - 0)^1 + 0.07\,(2 - 0)^1 + 0.02\,(3 - 0)^1 = 0.42 \qquad (3.5)$$

Table 3.2. Probabilities of harvester breakdowns.

Number of breakdowns	Probability
0	0.69
1	0.22
2	0.07
3	0.02

$$M_2' = V = 0.69\ (0 - 0.42)^2 + 0.22\ (1 - 0.42)^2 + 0.07\ (2 - 0.42)^2 + 0.02\ (3 - 0.42)^2 = 0.50 \qquad (3.6)$$

Or equivalently:

$$V = [0.69\ (0)^2 + 0.22\ (1)^2 + 0.07\ (2)^2 + 0.02\ (3)^2] - (0.42)^2 = 0.50 \qquad (3.7)$$

and

$$M_3 = 0.69\ (0 - 0.42)^3 + 0.22\ (1 - 0.42)^3 + 0.07\ (2 - 0.42)^3 + 0.02\ (3 - 0.42)^3 = 0.61 \qquad (3.8)$$

These calculations show that the expected number of breakdowns is $M_1' = E = 0.42$, with variance $M_2 = V = 0.50$. Hence, the standard deviation is $(0.50)^{0.5} = 0.71$ breakdowns (the positive square root of the variance). The third moment M_3 is a positive number, which means that the number of breakdowns of the harvester is positively skewed, i.e. it has a longer tail to the right, as is obvious from inspecting the elicited distribution. The same effect is shown by the coefficient of skewness, which is $0.61/(0.71)^3 = 1.70$ (M_3 divided by the third power of the standard deviation).

The formulae for the moments given above have been provided for discrete distributions of the uncertain quantity x. Similar formulae apply for continuous distributions but involve some possibly unfriendly integration signs, and so are not given. In any case, except for certain distributions of standard form, for which the required integrals have been worked out, it is common in decision analysis to approximate continuous distributions as equivalent discrete ones to calculate moments. The approximations can be made as precise as required by narrowing the intervals for the approximation as necessary.

To calculate moments from an elicited CDF, the elicited curve may be treated as a series of k joined linear segments. Such a CDF can be mathematically described by the co-ordinates of the end points of the segments beginning at the zero fractile, x_1 at cumulative probability $F_1 = 0.0$, through to the co-ordinate pairs (x_i, F_i) to the final fractile x_{k+1} at cumulative probability $F_{k+1} = 1.0$.

Formulae for the mean and variance of such a distribution are:

$$E[x] = 0.5\ \Sigma_i(F_{i+1} - F_i)\ (x_{i+1} + x_i) \qquad (3.9)$$

$$V[x] = \Sigma_i(F_{i+1} - F_i)[0.5(x_{i+1} + x_i)]^2 - (E[x])^2 \qquad (3.10)$$

The calculation for the distribution of milk yield per cow has been done using a spreadsheet, as illustrated in Fig. 3.9. The calculation is done with milk yield expressed in tonnes rather than kilograms to avoid having to deal with very large numbers. The expected value or mean milk yield per cow is found to be 7.03 t or 7030 kg. The variance is also calculated in the spreadsheet, using the formula given above. The result of the calculation is a variance of 0.0212 t^2, corresponding to a standard deviation of 145 kg per cow.

Note that, except in a few special cases, such as the normal distribution, which are fully characterized by only the mean and variance, summarizing any probability distribution using only a few moments implies some loss of information. In particular, the variance (or other statistics derived from it such as the standard deviation) by itself, tells little about the degree of risk, since it indicates nothing about the location of the distribution, nor about its skewness.

Eliciting distributions with three parameters

In some situations it may be desirable to use a more approximate but less time-consuming and demanding method of eliciting or assessing probability distributions. Such is often the case when seeking probability

	A	B	C	D	E	F
1	**Calculation of statistics of distribution from 0.1 fractiles**					
2						
3	i	F_i=B(k)	x_i=C(k)	=B(k+1)-B(k)	=C(k+1)+C(k)	=[0.5(E(k)]^2
4	1	0.0	6.700	0.1	13.570	46.036
5	2	0.1	6.870	0.1	13.790	47.541
6	3	0.2	6.920	0.1	13.870	48.094
7	4	0.3	6.950	0.1	13.935	48.546
8	5	0.4	6.985	0.1	14.000	49.000
9	6	0.5	7.015	0.1	14.065	49.456
10	7	0.6	7.050	0.1	14.140	49.985
11	8	0.7	7.090	0.1	14.235	50.659
12	9	0.8	7.145	0.1	14.370	51.624
13	10	0.9	7.225	0.1	14.625	53.473
14	11	1.0	7.400			
15						
16		E[x] =	7.030	=0.5*SUMPRODUCT(D4:D13,E4:E13)		
17		V[x] =	0.021	=SUMPRODUCT(D4:D13,F4:F14)		
18		SD[x] =	0.1432	=SQRT(C17)		

Fig. 3.9. Excel worksheet to calculate statistics of milk yield per cow from the smoothed CDF (milk yield per cow expressed in tonnes).

assessments from experts, especially if several such experts are to be consulted or if distributions are required for many uncertain quantities. In such cases it can be reasonable to elicit judgements about each distribution in terms of just three parameters – the minimum, maximum and most likely (modal) values, say a, b and m. With just these three values it is readily possible to fit one of two convenient distributions: the triangular distribution or the PERT distribution.

The triangular distribution

The formula for the PDF of the triangular distribution is:

$$f(x) = 2\,(x - a)/(b - a)\,(m - a),\; x \le m$$
$$f(x) = 2\,(b - x)/(b - a)\,(b - m),\; x > m \qquad (3.11)$$

and the formula for the CDF is:

$$F(x) = (x - a)^2/(b - a)\,(m - a),\; x \le m$$
$$F(x) = 1 - (b - x)^2/(b - a)\,(b - m),\; x > m \qquad (3.12)$$

The first two moments of the triangular distribution can be found using the formulae:

$$E[x] = (a + m + b)/3 \qquad (3.13)$$
$$V[x] = [(b - a)^2 + (m - a)\,(m - b)]/18 \qquad (3.14)$$

For example, a dairy farmer is planning the establishment of a new herd and knows that profitability is strongly affected by the average calving interval. Because the new herd does not yet exist, this is uncertain. However, on the basis of previous experience, the farmer is able to assign a value to the lowest possible herd average interval (a = 11 months), as well as to the highest possible value (b = 16 months). The farmer believes that the most likely or modal value will be about 12.5 months (m = 12.5). The distribution is then as shown in Fig. 3.10.

There are at least three ways in which this elicited distribution might be used in an analysis of the dairy farmer's decision option. First, the distribution, defined by the values of a, b and m, might be incorporated into a stochastic budgeting analysis or stochastic simulation, as illustrated in Chapter 6, this volume. Second, the graph of the CDF, or the formula for the CDF given above, could be used to obtain fractile values for use in decision analysis. Third, the dairy farmer might do the analysis of alternative options using the following moments of the distribution:

$$M_1 = E = (a + m + b)/3 = (11 + 12.5 + 16)/3 = 13.17 \text{ months} \qquad (3.15)$$

$$M_2 = V = [(b - a)^2 + (m - a)(m - b)]/18 = [(16 - 11)^2 + (12.5 - 11)(12.5 - 16)]/18 = 1.10 \text{ months}^2 \qquad (3.16)$$

The PERT distribution

Unlike the triangular distribution, the PERT distribution uses the same three parameters to create a smooth curve that assessors may judge to better represent their opinions.

The PERT distribution gets its name from the Program Evaluation and Review Technique (PERT), in which this distribution is generally used to represent the uncertainty in completion times for tasks in project implementation.

The PERT distribution is a special case of the beta distribution. Typically, sampling from the beta distribution requires minimum and maximum values and two shape parameters, v and w. In the PERT distribution the values of a, b and m are used to generate the shape parameters v and w. An additional scale parameter λ is needed, the default value of which is 4, usually regarded as arbitrarily predetermined.

In the PERT distribution, the mean is calculated as:

$$E[x] = E = (a + b + \lambda m)/(\lambda + 2) \qquad (3.17)$$

It is then possible to use this value for the mean to calculate the shape parameters:

$$v = \{(E - a)(2m - a - b)\}/\{(m - E)(b - a)\} \qquad (3.18)$$

$$w = \{v(b - E)\}/(E - a) \qquad (3.19)$$

Calculating the variance of the PERT distribution is not simple but it is possible to use Monte Carlo sampling from the distribution, as explained below, to generate enough values to calculate any required moments. Hence the distribution could be used in the same ways as suggested above for the triangular except that the difficulty in finding the second moment excludes its use directly in E,V analysis. For more information, see Vose (2008).

Applying a PERT distribution to the same parameters as for the triangular distribution gives the PDF and CDF shown in Fig. 3.11. These graphs of the two functions may be compared with those for the triangular distribution in Fig. 3.10.

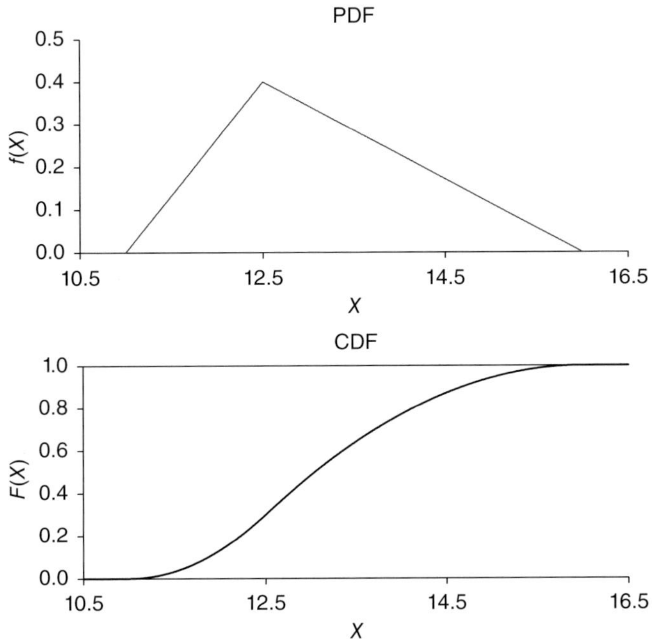

Fig. 3.10. The PDF and the CDF of a triangular distribution for calving interval.

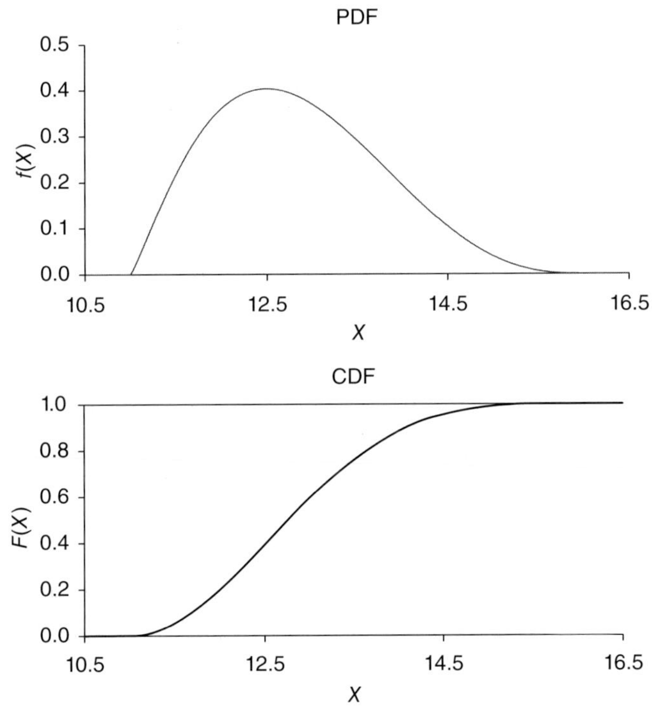

Fig. 3.11. The PDF and the CDF of a PERT distribution for calving interval.

Summary

The triangular distribution has the obvious advantage of simplicity and ease of use compared with the PERT. On the other hand, the latter avoids the unrealistic shape of the PDF of the triangular distribution. Generally a PERT distribution assigns more probability to outcomes near the mode and less to the tails than does a triangular distribution with the same parameters, which may be a factor to consider when choosing between them.

Monte Carlo sampling

Monte Carlo sampling is widely used in risk analysis. Models can be built to calculate the consequences of various actions in a risky world by representing the key uncertainties bearing upon the choice and the outcomes as probability distributions. Such a model can then be evaluated repeatedly using random drawings from the specified probability distributions to find the distribution of consequences.

The CDF is the key to understanding Monte Carlo sampling. As explained above, a cumulative distribution function $F(x)$ is a function that gives the probability P that the variable X will be less than or equal to x, i.e.:

$$F(x) = P(X \leq x) \tag{3.20}$$

where $F(x)$ ranges from zero to one.

As an example, in Fig. 3.12 the CDF shows a cumulative probability $F(x)$ of about 0.62 on the vertical axis for getting a value of the random variable X less than or equal to x on the horizontal axis.

In Monte Carlo sampling the inverse function is used, namely the value of x implied by a given value of $F(x)$. The inverse function can be written as:

$$x = G(F\{x\}) \tag{3.21}$$

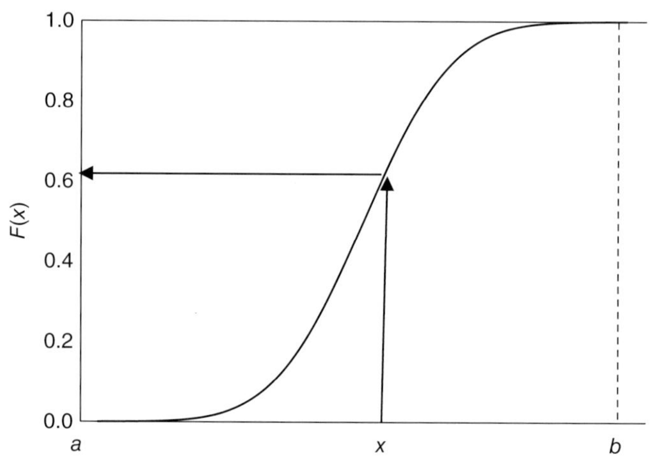

Fig. 3.12. The cumulative distribution function (a = minimum distribution value, b = maximum distribution value).

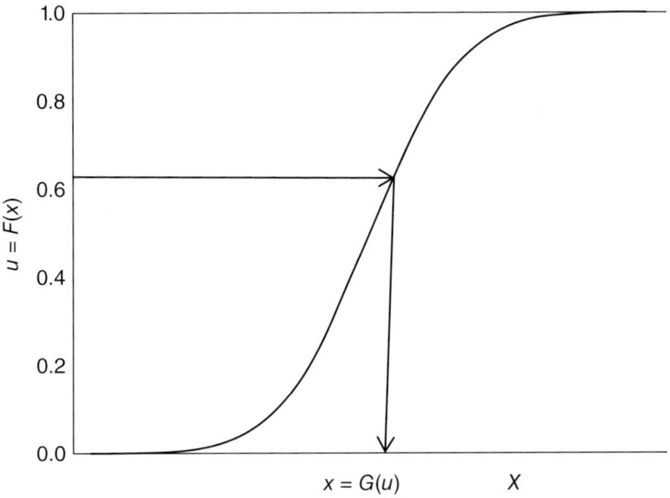

Fig. 3.13. The principle of Monte Carlo sampling using the inverse CDF.

This inverse function can be used to generate values of x on the horizontal axis with the frequency that, for a large sample, will represent the original distribution. Such a sample is generated by repeatedly selecting uniformly distributed values u between zero and one (which means that every value of u from zero to one is equally likely), then by setting $F(x)$ equal to each such value of u in Eqn 3.21 to find the corresponding x value, i.e. for each sampled value of u:

$$x = G(u) \tag{3.22}$$

This procedure is illustrated in Fig. 3.13 for a single variate. Samples for x, of course, are more likely to be drawn in areas of the distribution with higher probabilities of occurrence (where the CDF is steepest). With enough iterations, Monte Carlo sampling will re-create the distribution, meaning that, as full convergence is approached, additional samples do not markedly change the shape or statistics of the sampled distribution. One measure of convergence (among several) is that the mean changes by less than a specified threshold percentage as the number of iterations increases.

For highly skewed or long-tailed distributions we need a large number of samples for convergence, or we could use some form of stratified sampling techniques (e.g. Latin hypercube sampling).

The way we simulate values of a single random variable, illustrated above, is also almost the same when we simulate several uncertain quantities. The difference is that, in the multivariate case, we need account for their co-dependence (if any). This topic is dealt with in Chapter 4, this volume.

Selected Additional Reading

A comprehensive formal treatment of the foundations of subjective probability is to be found in Savage (1954). Somewhat easier introductions are to be found in Lindley (1985, Chapters 2 and 3), Raiffa (1997, Chapter 5), and Clemen and Reilly (2014). Winkler (2003) gives a good introduction to subjective probabilities, their assessment and to Bayesian statistics. In a collection of papers edited by Wright and Ayton (1994), the first five chapters provide an introduction to statistical and philosophical views of subjective probability. Vick (2002) covers both foundations and applications of subjective probabilities, but in an engineering context.

Spetzler and Staël von Holstein (1975) is an often-cited basic reference on probability assessment. A very readable history of the development of ideas in risk assessment is to be found in Bernstein (1996), although he largely overlooks the subjectivist viewpoint. Hubbard (2009) provides a critique of the way much risk management is done, urging, among other things, the greater use of scoring rules to train probability assessors.

For a pioneering treatment of the mathematics of probability and statistics for decision analysis, see Schlaifer (1959), expanded into a heavy-duty treatment of the topic in Pratt *et al.* (1995). See also Smith (1988) for a brief treatment of decision analysis in a subjectivist Bayesian framework. Anderson *et al.* (1977), especially Chapters 2 and 3, cover the topic in an agricultural context.

Applications of the subjective probabilities to agricultural decision problems are surprisingly few, presumably because most agricultural economists have been brought up in the frequentist school and so are uncomfortable with the use of subjective assessments. One exception is by Smith and Mandac (1995). For one attempt to remedy the dearth of such studies in agricultural and resource economics, see Hardaker and Lien (2010).

References

Ahlberg, J.H., Nilson, E.N. and Walsh, J.L. (1967) *The Theory of Splines and their Applications.* Academic Press, New York.

Anderson, J.R., Dillon, J.L. and Hardaker, J.B. (1977) *Agricultural Decision Analysis.* Iowa State University Press, Ames, Iowa.

Bernstein, P.L. (1996) *Against the Gods: the Remarkable Story of Risk.* Wiley, New York.

Brunswik, E. (1955) Representative design and probabilistic theory in a functional psychology. *Psychological Review* 62, 193–217.

Clemen, R.T. and Reilly, T. (2014) *Making Hard Decisions with Decision Tools*, 3rd edn. South-Western, Mason, Ohio.

Gardener, D. (2008) *Risk: the Science and Politics of Fear.* Scribe, Melbourne, Australia.

Gilovich, T., Griffin, D. and Kahneman, D. (eds) (2002) *Heuristics and Biases: the Psychology of Intuitive Judgment.* Cambridge University Press, Cambridge.

Hardaker, J.B. and Lien, G. (2005) Towards some principles of good practice for decision analysis in agriculture. Norwegian Agricultural Economics Research Institute. WP 2005:1. Available at: http://www.nilf.no/publikasjoner/Notater/2005/N200501Hele.pdf (accessed 8 September 2014).

Hardaker, J.B. and Lien, G. (2010) Probabilities for decision analysis in agriculture and rural resource economics: the need for a paradigm change. *Agricultural Systems* 103, 345–350.

Hubbard, D.W. (2009) *The Failure of Risk Management: Why It's Broken and How to Fix It.* Wiley, Hoboken, New Jersey.

Lindley, D.V. (1985) *Making Decisions*, 2nd edn. Wiley, Chichester, UK.

Pratt, J.W., Raiffa, H. and Schlaifer, R. (1995) *Introduction to Statistical Decision Theory*. MIT Press, Cambridge Massachusetts.

Raiffa, H. (1968, reissued in 1997) *Decision Analysis: Introductory Readings on Choices under Uncertainty*. McGraw Hill, New York.

Savage, L.J. (1954) *Foundations of Statistics*. Wiley, New York.

Schlaifer, R. (1959) *Probability and Statistics for Business Decisions*. McGraw-Hill, New York.

Silverman, B.W. (1998) *Density Estimation for Statistics and Data Analysis*. Chapman & Hall, London.

Smith, J. and Mandac, A.M. (1995) Subjective vs objective yield distributions as measures of production risk. *American Journal of Agricultural Economics* 77, 152–161.

Smith, J.Q. (1988) *Decision Analysis: a Bayesian Approach*. Chapman & Hall, London.

Spetzler, C.S. and Staël von Holstein, C.-A.S. (1975) Probability encoding in decision analysis. *Management Science* 22, 340–352.

Vick, S.G. (2002) *Degrees of Belief: Subjective Probability and Engineering Judgment*. ASCE Press, Reston, Virginia.

Vose, D. (2008) *Risk Analysis: a Quantitative Guide*, 2nd edn. Wiley, Chichester, UK.

Winkler, R.L. (2003) *An Introduction to Bayesian Inference and Decision*, 2nd edn. Probabilistic Publishing, Gainesville, Florida.

Wright, G. and Ayton, P. (eds) (1994) *Subjective Probability*. Wiley, Chichester, UK.

4 More about Probabilities for Decision Analysis

Having introduced what might be called the basics of the principles and practice of using probability to measure beliefs for decision analysis, in this chapter we address some issues relating to the application of those principles, covering the derivation of probability distributions in situations where there are abundant, sparse or no data available for the task. We then introduce the Bayesian approach to updating prior probability judgements in light of new information. Finally, we address the thorny issue of how to account for situations where a number of uncertain quantities impinge on the outcomes of some risky choice and these quantities are correlated in some way.

The Relevance of Data in Probability Assessment

Making the best use of abundant data

Although we have argued in the previous chapter that all probabilities for decision analysis are necessarily subjective, for those occasions when there are abundant, reliable and relevant data that are pertinent to some uncertain quantity of interest, any sensible person will want to base probability judgements on such information. Probabilities in such cases may be viewed as 'public' because many people can be expected to share almost the same probabilities – at least once the information has been brought to their notice. Note, however, that everyone may not agree on the relevance of a given set of data. For example, farmers will often not share the confidence of a research agronomist that data from trials on the research station can be replicated on their own farms – scepticism that, unfortunately, is too often well founded.

There are various ways of summarizing such abundant data. Three, which are not mutually exclusive, are:

1. The data can be treated as a sample, and estimates of moments of the distribution can be calculated.
2. The data can be arranged as empirical CDFs and allowed to 'tell their own story'.
3. If appropriate, some appropriate distribution function, such as a beta distribution, may be fitted to the data.

Provided the data are not forced to fit some inappropriate distribution, the choice between these options can depend on convenience in subsequent analysis.

The approach to handling abundant data can be illustrated in relation to an assessment of the distribution of sugarbeet yields (expressed as yield of sugar) for a particular production environment. In this case, historical data are not wholly relevant since there have been major technical changes in sugarbeet production methods, including the progressive introduction of better varieties. Therefore, instead of relying directly on past yields to assess a yield distribution, a simulation model has been used to forecast the yield under current technology, accounting for the impact of management and weather during the growing period. Because of the need to account for several aspects of weather, which cannot be treated as stochastically independent, the model was run using historical weather data. Historical weather data for 100 computer-created years were used to give a sample of possible yields, expressed in terms of sugar production per hectare. They are listed in Fig. 4.1, displayed as an array in Excel.

The first option for dealing with the sample data is to represent the information embodied in the data points in terms of moments. In this case, population estimates of the first, second and third moments, along with the corresponding standard deviation and the skewness coefficients (see Chapter 3, this volume), have been calculated, as follows:

- Mean = 10.6 t/ha;
- Variance = 3.16 t^2/ha^2;
- Standard deviation = 1.78 t/ha;
- Third moment = –0.271 t^3/ha^3; and
- Skewness = –0.0492.

These moments may be used directly in analysis of decision options, as discussed in Chapter 7, this volume. However, for many applications, it may be more desirable to define the full distribution.

As a first step to defining a distribution, the 100 observations have been placed in size order and plotted as a CDF, as illustrated in Fig. 4.2.

The most obvious feature of this graph is its irregular form. There is no reason to suppose that the actual distribution of yields would be anything other than a smooth curve. Evidently, many more than the actual number of 100 observations would be needed to smooth out the bumps. Reliance on historical weather data could make it impossible to increase the sample size. Thus, the first option mentioned above of letting the data 'tell their own story' by plotting them as a CDF would seem to require at least the further step of smoothing a CDF through the plotted points. The smoothing may be done via a hand-drawn curve, fitted by eye, or using appropriate software.

Such a smoothed curve is illustrated in Fig. 4.3. Note that the fitted curve has been assumed to have the sigmoidal shape commonly found in CDFs of phenomena of this type. In this case the curve was hand-drawn, which sometimes gives a better fit to the data points than can be achieved with commonly available curve-fitting software. The advantage of using software, however, is that the formula for the fitted function is then available, making subsequent analysis easier.

	A	B	C	D	E	F	G	H	I	J
1	6.041	11.223	10.985	8.783	8.857	11.371	10.268	11.688	12.687	12.226
2	8.127	10.894	11.326	9.692	8.597	11.901	11.863	11.218	12.809	13.322
3	9.278	6.584	11.251	10.092	7.840	11.978	12.696	8.841	12.764	13.733
4	9.763	8.049	10.910	10.159	9.041	11.930	12.934	11.502	12.532	13.671
5	9.850	8.882	8.634	10.065	9.664	9.897	12.748	12.925	6.907	13.348
6	9.809	9.222	10.074	9.986	9.837	10.730	12.306	13.483	8.206	12.974
7	8.305	9.209	10.846	8.894	9.690	11.025	10.281	13.550	8.910	13.721
8	9.965	8.983	11.095	9.321	9.351	10.952	11.443	13.500	9.158	13.394
9	10.907	8.462	10.965	9.362	7.623	10.681	11.960	12.017	9.087	13.521
10	11.276	10.080	10.601	9.160	10.055	10.384	11.989	12.417	8.835	9.120

Fig. 4.1. Sample of 100 observations on sugar yield from sugarbeet (t/ha) based on a simulation model.

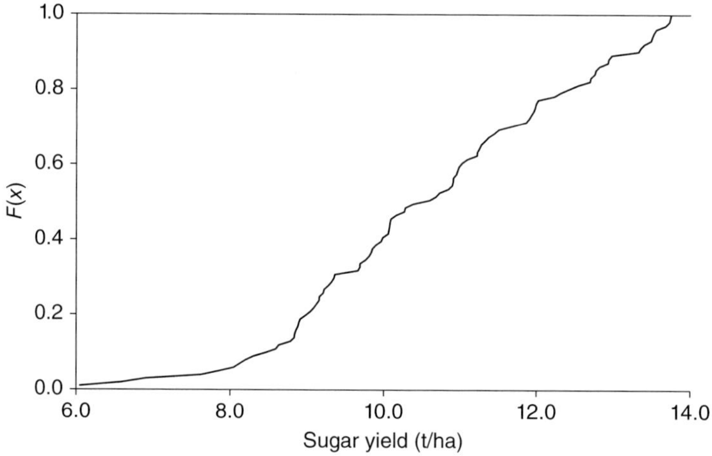

Fig. 4.2. Abundant data on simulated yield of sugarbeet plotted as a cumulative distribution function (CDF) (yield in tonnes of sugar per hectare; 100 observations).

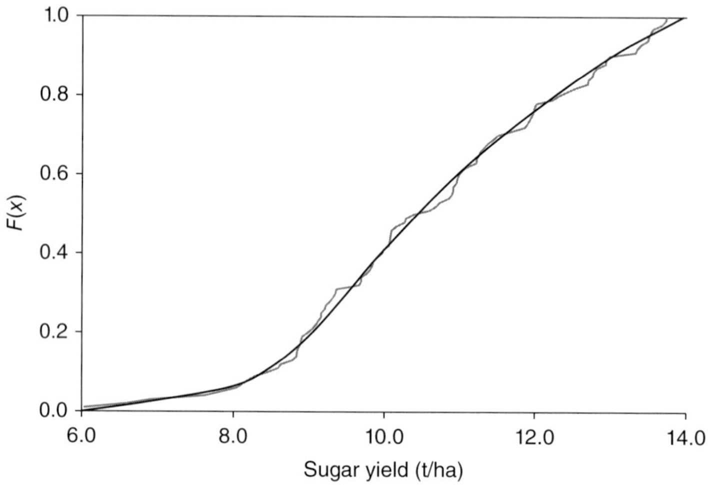

Fig. 4.3. The simulated sugar yield data approximated with a smoothed CDF.

Another option is to fit one of the wide range of standard probability distributions to the data. Sometimes the appropriate distribution to fit will be indicated by the process causing the uncertainty. For example, an exponential distribution is a waiting time distribution whose function is the direct result of assuming that there is a constant probability of an event occurring at any moment. More commonly, however, it will be a matter of picking the form of distribution that best fits the data, typically using statistics of goodness of fit combined with visual inspection of the CDF of the fitted distribution plotted against the data. In applying distribution fitting it is, of course, wise to consider whether the uncertain quantity at issue is continuous or can take only discrete values. For example, if modelling the number of live piglets weaned per sow, a discrete distribution would clearly be appropriate, while in the sugarbeet example above, a continuous distribution is to be preferred.

In the case of the sugarbeet example, we applied the distribution fitting option in Vose ModelRisk. This software offers the choice of a large number of theoretical distributions that can be fitted to data such as those in this example. The observations were treated as a sample and the software was used to try a number of distributions. While various goodness of fit statistics are provided, we chose to base the choice of function mainly on a visual inspection of the fitted CDF against the cumulative plot of the data. The result of this process is illustrated in Fig. 4.4.

A comparison of Figs 4.3 and 4.4 shows some differences in the fitted CDFs. With hand smoothing it is possible to track the empirical data more closely, although whether this is an advantage is not entirely clear.

Unlike the case of abundant data, the all-too-familiar situation is one where data are too few to provide a good basis for probability assessment. Then there is more scope for inventiveness and, of course, more scope for different interpretations, as discussed next.

Making the best use of sparse data

When data are sparse, maximum use can be made of the few observations on some continuous uncertain quantity by using the simple rule that, whatever the underlying form of probability distribution, the k-th observation from a set of n (random and independent) observations, when the observations are arranged in ascending order, is an unbiased estimate of the $k/(n + 1)$ fractile. Thus, the fractile estimates can be plotted and a CDF drawn subjectively through them.

The statement that the data points are unbiased fractile estimates means that, if we could draw a large number of samples of the same size as our sparse data set, the mean fractile values, averaged across the samples, would converge to the true values. That knowledge is of only limited help since we only have one, possibly very unrepresentative, sample to work with. It will therefore be important to use the few observations

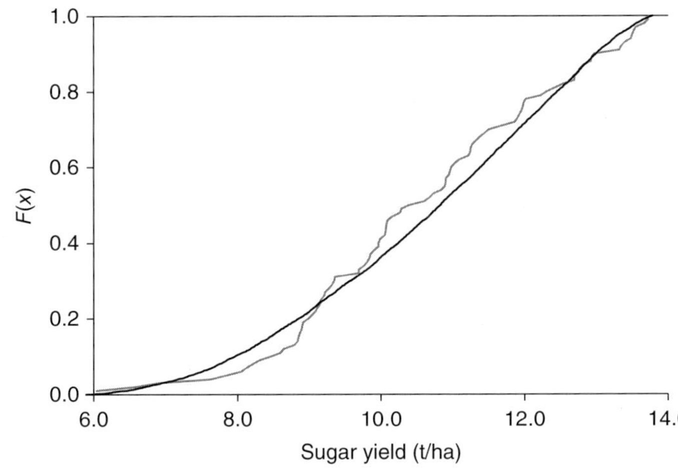

Fig. 4.4. ModelRisk graph comparing the cumulative distribution of sugarbeet yields (t/ha) with a fitted beta-binomial distribution.

in conjunction with any other relevant information and judgements. For example, many PDFs are roughly bell-shaped, implying CDFs that are smooth and S-shaped. This knowledge, combined with any further information or ideas about the distribution in addition to the few data points, will be useful in representing the location and general shape of the distribution. Estimates of the upper and lower bounds of the distribution will also be useful in deciding where the CDF is to meet the zero and one probability bounds. It will also be helpful to have an indication of the most likely or modal value, in order to locate the point of inflection of an S-shaped CDF with more confidence. Using all such additional information, it is possible to sketch a subjective CDF accounting for all the available information, including the plotted fractile values.

By way of illustration, suppose a farming partnership of husband and wife is considering going into strawberry production. They decide to base their forecast of future prices partly on market prices for the past 7 years, believing that market structure has been relatively constant over this number of years, but had been different before that, so that information for earlier years is not relevant. Their crop will be harvested late in the season owing to the geographical location of the farm and the soil type. Market records they have assembled for late-season prices for the past 7 years (adjusted for inflation) are shown in Table 4.1.

In addition to this information, the farmers express the belief, based on a study of the market outlook, that it is highly unlikely that strawberry prices will fall below \$1.75/kg or rise above \$5.00/kg.

With all these data, the next step is to arrange the observations in increasing order of magnitude and to calculate the fractile estimates. There are seven observations, so the first one, \$2.18, is an estimate of the 1/(7 + 1)-th fractile, or the 0.125-th fractile; the second value of \$2.65 is an estimate of the 2/(1 + 7) = 0.25-th fractile, and so on. To these calculated fractile estimates can be added the 0.0-th and 1.0-th fractile values corresponding to the assessed lower and upper limits on the future price. The full set of fractile estimates is shown in Table 4.2.

These fractile values are plotted as estimates of points on a CDF for the price of strawberries, as shown in Fig. 4.5. In this figure, a curve has been smoothed by eye, passing through or as close to the points as seems appropriate. The alternative, again, would be to use some curve-fitting or distribution-fitting approach, as illustrated for the sugarbeet data. However, in the case of sparse data, such computer-based approaches are generally less likely to provide as good a representation of what is known about the distribution as is a hand-drawn curve.

Table 4.1. Recent price history for strawberries.

Year	Price (\$/kg)
1	2.94
2	2.65
3	3.42
4	2.18
5	4.17
6	2.71
7	2.85

Table 4.2. Estimated fractile values for the distribution of strawberry prices.

Fractile	Estimate (\$/kg)
0.000	1.75
0.125	2.18
0.250	2.65
0.375	2.71
0.500	2.85
0.625	2.94
0.750	3.42
0.875	4.17
1.000	5.00

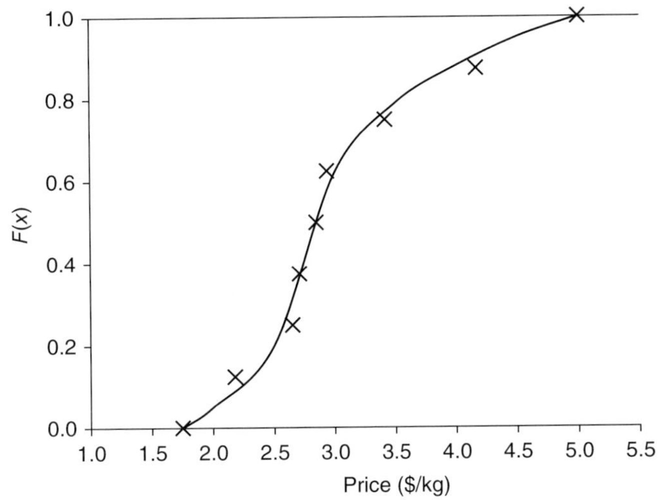

Fig. 4.5. CDF estimated from sparse data on the price of strawberries.

The CDF in Fig. 4.5 can be used for decision analysis, as appropriate. For example, the moments of the distribution can be calculated, using the method described in Chapter 3. However, it is important to keep in mind the possibility that a CDF derived from a very few data points may be found to be seriously wide of the mark, if and when more evidence is accumulated. Very sparse data situations imply a need to look carefully at the scope for improving the reliability of the estimated distributions either by collecting more information or by other means, such as use of experts, described in the following section.

Taking Expert Advice

In some cases, there are no data at all available to determine probability distributions. In other cases, circumstances may have changed significantly such that data that are available cannot be used to estimate reliable and relevant distributions. Consider as an example the probability of an outbreak of a rare contagious livestock disease. The few data available all come from regions and countries with hygiene standards and animal densities different from those in the region under consideration. Furthermore, in the study region in the past all animals were routinely vaccinated against the disease. Since routine vaccination has now been forbidden, we have to deal with unprotected animals that are susceptible to contracting the disease. Consequently, the historical data are of little use in estimating the chances of future outbreaks. Yet the consequences of a disease outbreak are so serious that some decision analysis is needed. In such a case, the decision analyst may turn to expert opinion to assess the required probability.

An expert in the case outlined above might be a veterinarian who has experience with the disease, ideally in a number of regions with varying conditions and circumstances. An animal disease epidemiologist who has studied the disease and its transmission would also be a candidate to be considered expert for this example problem. Moreover, if it makes sense to consult one expert, it presumably makes better sense to consult more than one on the grounds that several heads are better than one. The expectation is that each expert will bring to the elicitation task different knowledge and experience so that, by pooling their views in some way, a better probability distribution will result.

One option, therefore, is to consult a number of experts, individually, using one or other of the assessment methods described in Chapter 3, although preferably only after they have each undergone some training in probability assessment as described in that chapter, and have been tested to confirm that they have benefitted from that training. But consulting each expert individually has the obvious disadvantage that there is no sharing of knowledge and information among the group of experts. It would make sense, therefore, to find a way that members of the expert group can learn from each other and so refine their responses. The so-called *nominal group* approach is sometimes used for just such a purpose.

A nominal group is simply a selected group of people brought together to consider some issue. They are chosen on the basis of who they are and what they are expected to be able to contribute to the discussion. Normally, such groups are run by an experienced facilitator whose task it is to keep the discussion 'on track'. The facilitator will also try to avoid or minimize dysfunctional behaviour such as confrontational disputes between group members. There is some evidence to suggest that the best results are likely to be obtained by keeping each person's probability assessments confidential. It may work best to start with a general preliminary discussion of the issues, leading to a first round of individual assessments that are then summarized and participants told of the results without revealing individual assessments. After this feedback and more discussion, participants are asked to make reassessments of the required probabilities.

Sometimes, such groups seem to work well, but on other occasions they have been found to be disappointing, sometimes seriously bad. *Group dysfunction* can arise in a number of ways. One dominant but ill-informed member may hold too much sway, especially if high and low status people are in the group. Even without that problem, experience suggests that group dynamics can sometimes lead to extreme bias. It is not unknown for a group to come up with a recommendation that is so extreme that it would not have been supported by any one of the members individually. Such 'group think' was blamed for the incorrect advice emerging from a US security committee about weapons of mass destruction in Iraq. So, while it makes sense to consult a number of experts and then to pool relevant knowledge and ideas, this pooling may be done best without direct contact between the experts and with probability assessments made individually, not collectively.

One effective way of bringing together the views of experts and aggregating group opinion while minimizing these possible problems is to use the *Delphi method* (Linstone and Turoff, 2002). This method uses a selected panel of experts, but replaces direct debate and possible confrontation with a planned programme of sequential, individual interrogations, usually conducted by questionnaire, followed by feedback. The method may be used for a wide variety of assessment tasks, of which the assessment of uncertainty is but one. The method includes the important features of: (i) anonymity; (ii) controlled feedback; (iii) reassessment; and (iv) group response.

Anonymity is ensured by non-disclosure of the identity of the group members, or at least, the non-attribution to individuals of opinions expressed. Knowing that their identities will not be disclosed may allow participants to feel more secure in expressing their true views. Feedback is used to share information, thereby generally leading to some convergence of views. Assessments are made in a series of rounds with assessors being notified of the results after each round. The nature of the feedback is controlled by the analyst who must steer a course between too much and too little feedback. Feedback is followed by reassessment, which provides the opportunity for group members to reflect on what they have learned from others and to review their previously expressed assessments. Finally, the procedures ensure that the opinion of every panel member is obtained and can be taken into account.

There can be considerable variation within these features, especially with respect to the controlled feedback. In probability assessment, the results returned to respondents may range from reporting only the medians of each group member's assessed distributions to tabulating selected fractiles, such as the quartiles, of the different distributions. If, say, the negative tail of the distribution is especially important in affecting consequences, obviously feedback will focus on the lower fractiles.

Usually it is helpful also to include in the feedback some information about each respondent's reasons for responses, although the extent of feedback can vary. There is also variation in the number of rounds of assessment and feedback that are used. In some applications, early rounds may have to be devoted to defining the issues at stake more clearly. For example, in an uncertain situation it may not be obvious and so not agreed at the start what are the possible outcomes for which probabilities need to be assessed. But once such preliminary matters are resolved, it usually takes two or three rounds to reach a point where group members are satisfied with their answers and will not wish to revise them further.

Used for assessing probability distributions, the Delphi method is based on a twofold expectation that: (i) with repeated rounds the range of responses in terms of statistics of interest will decrease and converge towards some consensus; and (ii) the total group response will successively move towards the 'correct' or 'true' answer, meaning, in the present context, the probability distribution that best describes the uncertainty.

Despite efforts to achieve consensus, both the nominal group and the Delphi approach are likely to end with some remaining differences of opinion among the experts. If no complete group consensus is obtained, the different expert responses can be combined in several ways. For example, the assessor can put equal weights on the different distributions and determine the average or expected distribution. Or the respondents can be asked to rate themselves on the issues at hand on some suitable scale. These scores can then be 'normalized' (scaled to add to 1.0) to give weights which can be used to calculate the average distribution. Alternatively, cross-ratings or group ratings can be determined by asking the respondents to rate each other. Again, normalized weights can be calculated from these data to form a combined distribution. Note, however, that anonymity cannot be preserved using group or cross weighting. Finally, the assessor may use his or her own ratings or weights to aggregate responses. There is some research which suggests that a simple average is as good as more 'fancy' methods.

Accounting for New Information

As noted earlier, in many situations when the initial information about the uncertain events of interest is incomplete, there is an opportunity to reduce the uncertainty by collecting more data, obtaining a forecast of some sort, or conducting some kind of experiment. Such opportunities raise two questions:

1. How should prior probabilities be revised to account for new information?
2. Given that information gathering almost always has a cost, how much information should be collected?

Methods of analysis have been evolved to answer both these questions. We address the first below and leave the second until Chapter 6, this volume.

Probability updating using Bayes' Theorem

We now return to the example introduced in Chapter 2 of the dairy farmer who has to decide whether or not to insure against losses from a possible outbreak of foot-and-mouth disease (FMD). Initially the farmer has formed the following probability estimates of the chances of an outbreak, believing that there will be an outbreak in the area in the coming year with a 6% probability, therefore also believing that there will be no outbreak with a 94% probability. We can call these initial probabilities *prior probabilities*, meaning that they have been assessed prior to receiving some additional information about the future disease incidence, as explained below.

As will be shown in Chapter 6, this volume, it is possible to analyse the decision tree using such prior probabilities to find the optimal prior decision (i.e. the best choice for the farmer, based on the farmer's own prior probabilities). In this case, probabilities would also have to be obtained for the chances that the control policy implemented in the event of an outbreak will involve bans or slaughter, but we leave this extra complication aside for now.

Suppose that the dairy farmer now discovers that there is a commercial agency that specializes in animal disease risk assessments and provides predictions about the chance of disease outbreaks, including FMD, applicable to particular situations. The agency, which has had considerable experience in this type of work, bases its predictions on the advice of experts who draw on historical data and other information such as current and prospective developments in preventative measures and forecasts of animal densities. For a fee, it will provide the farmer with a report about the likely disease status in the region for the coming year. The farmer thinks that it might be worthwhile to buy this forecast.

The disease status forecast, if obtained, will boil down to one of three predictions: 'unlikely' (z_1), 'possible' (z_2) or 'probable' (z_3). The accuracy of the predictions of the agency can be judged by evaluating its past forecasting performance over a range of animal disease issues, as shown in Table 4.3.

As shown, in years when it has turned out that there has been no outbreak of the specified disease, the agency has given a prediction of 'unlikely' (z_1) in 60% of the cases, 'possible' (z_2) in 30% of the cases and 'probable' (z_3) in 10% of the cases. On the other hand, when there has been an outbreak of the specified disease, the corresponding frequencies have been 10% for z_1, 40% for z_2 and 50% for z_3. Evidently, though far from perfect, the agency's predictions do contain some information.

If the farmer believes that the past reliability of the agency in predicting the future incidence of animal diseases in the region is relevant to the decision making, these relative frequencies can be interpreted as probabilities that measure the chance of the various predictions being obtained for different eventual disease outcomes. A *likelihood probability*, often simply called a *likelihood* and written $P(z_k|S_i)$, is the probability that a prediction z_k will be obtained given that S_i turns out to be the true state. A likelihood is

Table 4.3. Accuracy of predictions about FMD outbeaks.

Type of prediction	Frequency of type of prediction given no outbreak	Frequency of type of prediction given outbreak
Unlikely (z_1)	0.6	0.1
Possible (z_2)	0.3	0.4
Probable (z_3)	0.1	0.5

the probability of z_k *conditional upon* S_i, with the conditionality indicated by the vertical line. For instance, in Table 4.3, the probability of receiving prediction z_1 given that a 'No outbreak' situation (S_1) eventuates is 0.6, that is the likelihood probability $P(z_1|S_1) = 0.6$.

At issue is how the farmer's personal prior probabilities should be updated if this new information were to be bought in the form of a prediction. *Bayes' Theorem*, a standard formula from probability theory, provides a means to learn from forecasts or experiments. The Theorem offers the farmer a logical way to adjust subjective prior probabilities, accounting for new information weighted according to the accuracy of the predictions, as measured by the likelihoods. The two types of information available to work with are summarized in the probability tree presented in Fig. 4.6.

The probability tree consists of two parts. The left part represents the prior probabilities $P(\text{No outbreak}) = P(S_1) = 0.94$, and $P(\text{Outbreak}) = P(S_2) = 0.06$. The right part shows the likelihood probabilities. The probability tree enables us to calculate the *joint probability* of observing a certain state of nature (for example S_1) and a certain prediction (for example z_1). The probability of reaching any end point in the tree is found by multiplying the probabilities attached to the two branches leading to that end point. For example, for S_1 and z_1 $P(S_1 \text{ and } z_1) = P(S_1)\ P(z_1|S_1) = 0.94 \times 0.6 = 0.564$. Likewise, for S_2 and z_3, $P(S_2 \text{ and } z_3) = P(S_2)\ P(z_3|S_2) = 0.06 \times 0.5 = 0.030$. In Fig. 4.7 all the joint probabilities have been added to the probability tree.

A check will show that the joint probabilities shown at the right-hand side of the tree add up to 1.0, as they should.

There is one further set of probabilities that is obtainable from the tree shown in Fig. 4.7. The *marginal probabilities of the forecast*, $P(z_k)$, may be obtained by adding together the respective joint probabilities. Thus:

$$P(z_1) = P(S_1 \text{ and } z_1) + P(S_2 \text{ and } z_1) = 0.564 + 0.006 = 0.570 \quad (4.1)$$

$$P(z_2) = P(S_1 \text{ and } z_2) + P(S_2 \text{ and } z_2) = 0.282 + 0.024 = 0.306 \quad (4.2)$$

$$P(z_3) = P(S_1 \text{ and } z_3) + P(S_2 \text{ and } z_3) = 0.094 + 0.030 = 0.124 \quad (4.3)$$

These probabilities show the chances of getting each type of forecast. Once again, they meet the condition of adding to 1.0.

Unfortunately, none of these probabilities provides the answer to the farmer's problem of how to update prior probabilities in the light of a disease incidence prediction. The *posterior probabilities* are needed for each disease state, conditional upon each type of prediction, that is the probability of a

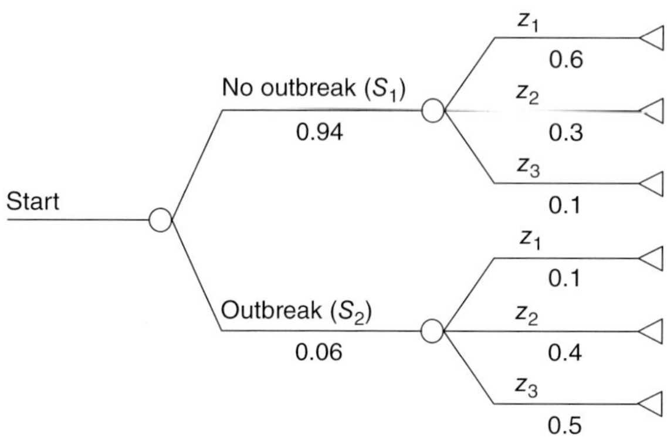

Fig. 4.6. Initial probability tree for revision of the dairy farmer's probabilities.

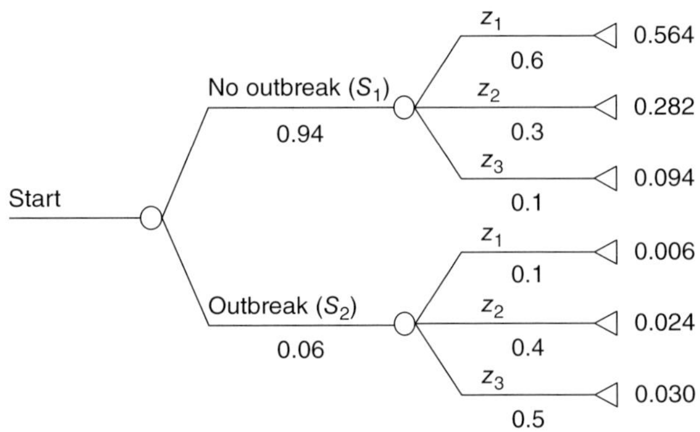

Fig. 4.7. The dairy farmer's probability tree showing the joint probabilities.

particular state of nature, S_i, eventuating given a particular prediction, z_k, denoted by $P(S_i|z_k)$. Finding these posterior probabilities is a matter of 'flipping' the probability tree so that the chance node for the prediction is on the left and the node for the disease state is on the right, as in Fig. 4.8.

The flipped tree in Fig. 4.8 has had added to it all the probabilities so far established. The marginal probabilities of the forecast are taken from the previous calculation. The joint probabilities come from Fig. 4.7 above (though in a different order) since, for instance, $P(S_1 \text{ and } z_2) = P(z_2 \text{ and } S_1) = 0.282$. We need to determine the posterior probabilities of the disease states in the flipped tree. The probabilities in the new tree must be consistent with the laws of probability. That requirement means that the joint probabilities at the right-hand side of the tree must be the product of the probabilities on the two branches leading to each terminal node, as for Fig. 4.7. For example $P(z_2 \text{ and } S_1) = P(z_2)\,P(S_1|z_2)$. Since we know $P(z_2 \text{ and } S_1) = 0.282$ and $P(z_2) = 0.306$, we can now calculate $P(S_1|z_2) = P(z_2 \text{ and } S_1)/P(z_2) = 0.282/0.306$, which equals 0.922 (approximately).

Generally, the posterior probabilities are calculated as:

$$P(S_i|z_k) = P(z_k \text{ and } S_i)/P(z_k) \qquad (4.4)$$

The results of these calculations are shown in Fig. 4.9. This flipped tree contains the posterior probabilities that the dairy farmer wants to know. For example, a forecast 'unlikely', z_1, leads to a high posterior probability

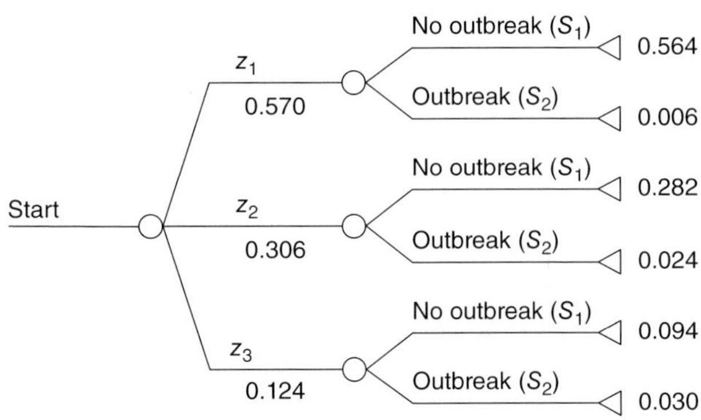

Fig. 4.8. The dairy farmer's 'flipped' probability tree.

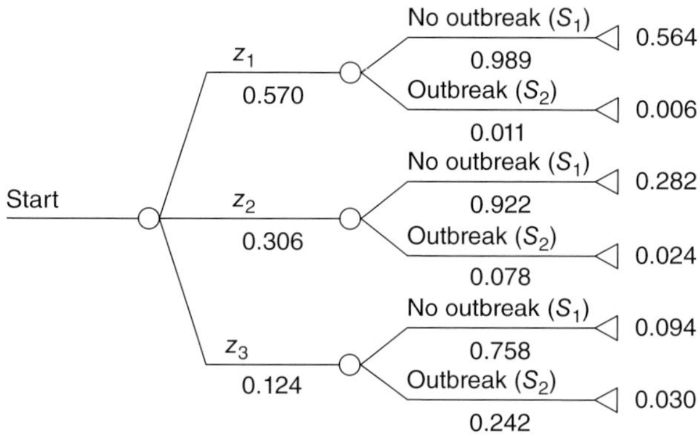

Fig. 4.9. The dairy farmer's 'flipped' probability tree showing the posterior probabilities.

of 0.989 that the disease incidence will be 'No outbreak', while a forecast of 'probable', z_3, leads to a revised probability of 0.242 that there will be an outbreak.

The calculations followed above in revision of the prior probabilities into posterior probabilities were accomplished using Bayes' Theorem, which may be written, in discrete form, as:

$$P(S_i|z_k) = P(S_i)\ P(z_k|S_i)/\{\Sigma_i\ P(S_i)\ P(z_k|S_i)\} = P(S_i)\ P(z_k|S_i)/P(z_k) \quad (4.5)$$

where $P(S_i|z_k)$ is the posterior probability of state S_i given that the k-th prediction has been observed, $P(S_i)$ is the prior probability of S_i, $P(z_k|S_i)$ is the likelihood of the k-th prediction given that S_i is the true state, and $P(z_k)$ is the marginal probability of the k-th prediction.

Bayes' Theorem applied in spreadsheet format

The probability tree format used above to apply Bayes' Theorem may add to clarity but is too cumbersome for routine use. Exactly the same equations can be implemented more efficiently in a spreadsheet. The Excel spreadsheet for the dairy farmer's probability revision via Bayes' Theorem is shown in Fig. 4.10.

The relationship between Fig. 4.10 and Eqn 4.5 is as follows. The prior probabilities $P(S_i)$ are given in cell range B5:B6. The likelihoods $P(z_k|S_i)$ for each state of nature are in cells C5:E5 for S_1 and in C6:E6 for S_2. Then the joint probabilities $P(S_i$ and $z_k)$ are calculated in cells F5:H5 for S_1 and F6:H6 for S_2 by multiplying the prior probabilities by the corresponding likelihoods. For instance, $P(S_1$ and $z_1) = P(S_1) \times P(z_1|S_1) = 0.94 \times 0.6 = 0.564$ (cell F5). The marginal probabilities $P(z_k)$ are found by adding the joint probabilities for each z_k. For example, $P(z_1) = P(S_1$ and $z_1) + P(S_2$ and $z_1) = 0.564 + 0.006 = 0.570$ (cell F7). Finally, the posterior probabilities $P(S_i|z_k)$ are in cells F8:F9 (given forecast z_1), G8:G9 (given forecast z_2) and H8:H9 (given forecast z_3). They are calculated as the joint probabilities divided by the marginals: $P(S_i$ and $z_k)/P(z_k)$. For instance, $P(S_1|z_1)$ in F8 is calculated as $P(S_1$ and $z_1)/P(z_1) = 0.564/0.570 = 0.989$.

	A	B	C	D	E	F	G	H
1	**Application of Bayes' Theorem**							
2								
3	State	Priors	Likelihoods $P(z_k \mid S_i)$			Joint probabilities $P(S_i$ and $z_k)$		
4	S_i	$P(S_i)$	z_1	z_2	z_3	z_1	z_2	z_3
5	S_1	0.94	0.6	0.3	0.1	0.564	0.282	0.094
6	S_2	0.06	0.1	0.4	0.5	0.006	0.024	0.03
7	Sum	1.00	Marginals		$P(z_k)$	0.570	0.306	0.124
8			Posteriors		$P(S_1 \mid z_k)$	0.989	0.922	0.758
9					$P(S_2 \mid z_k)$	0.011	0.078	0.242
10					Sum	1.000	1.000	1.000

Fig. 4.10. Application of Bayes' Theorem to the foot-and-mouth disease problem.

Concluding comment on probability updating using Bayes' Theorem

A study of the operation of Bayes' Theorem shows that, when a large amount of new and relevant information is available, it will swamp the prior probabilities. So posterior probabilities in abundant new data situations are defined almost wholly by the new data. So, while we argued in Chapter 3 that all probabilities for decision analysis are essentially subjective, subjective priors will be swamped by data when the data are abundant (and relevant and reliable).

The value of the Theorem arises when the likelihoods are known or can be reliably estimated, as in the case illustrated. In other cases, the likelihoods may be known from the nature of the sampling process. In cases where they are not, it may be as easy or easier for the DM to assess the posterior probabilities directly. Even then, however, it may be useful to use the Theorem to check consistency.

Accounting for Stochastic Dependency

Introduction

Most decision problems involve more than one uncertain quantity. The one-variable methods of probability elicitation outlined above can be applied validly to several variables only if the variables are *stochastically independent*. Two variables are stochastically independent if the probability distribution of one does not depend on the value taken by the other. In practice, complete stochastic independence may be the exception rather than the rule. If two uncertain quantities relevant to the analysis of some decision problem are not stochastically independent, analysis assuming away (or simply ignoring) that dependency may produce results that are significantly in error and perhaps seriously misleading. On the other hand, because accounting for stochastic dependence is difficult, it may often be judged to be 'near enough' to assume dependency away when the degree of association between variables is thought to be low.

The extent of dependency between variables can often be judged from the circumstances. For example, economic logic suggests that the prices of different cereal grains will usually vary more or less together, while there may be a less close association between, say, the price of wheat and the price of sugarbeet.

If it is judged that the assumption of independence for two or more uncertain quantities is unrealistic, there is usually no alternative but to confront the inherently difficult task of joint specification, which usually involves the elicitation of joint distributions. There are fundamental problems in specifying continuous joint distributions because there are only a few mathematical forms that are tractable. Many of the functional forms used for a single uncertain variable cannot be extended to joint distributions of several variables. It is not surprising, therefore, that the forms of continuous joint distributions used in risk analyses have often been limited to the few relatively tractable cases, such as the multivariate normal – the joint distribution of several underlying normally distributed variables. However, later in this chapter we explain and discuss the concept of *copulas*, which provide a flexible framework to account for stochastic dependency, largely independent of the forms of the marginal distributions. Modelling of dependence structure and copulas, combined with Monte Carlo simulation, has increased substantially in recent years, although applications in agricultural economics are still rather rare.

Before we go into modelling multivariate distributions and copulas, two other methods of dealing with stochastic dependency will be explained. One is the case of abundant data when we can 'let the data speak', by which we mean using the data themselves directly in the decision analysis, with the stochastic dependency implicit in the data. The other method is to model the underlying structure of the dependency, also called the 'hierarchy of variables' approach.

Using abundant data: let the data speak

When there are abundant data available, these may be the basis for probabilities in the decision analysis. However, in this case, it is important first to consider the reliability and the temporal and spatial relevance of the data to the assessment of the uncertainty at hand. How many observations are there? How were they obtained and by whom? What, if anything, was done to verify and validate the data? If the data were a sample, how representative are they of the population from which they were drawn? How large were errors in collecting and reporting the data likely to be? Is the stationarity assumption justified? That will be so only if the processes creating uncertainty about the future are just the same as the processes leading to variation represented in the historical data to hand. Were the data collected on the farm or for the environment for which the decision analysis is to be performed? Or did they come from some location perhaps quite far away? Do the data represent farm-level outcomes or are they averaged or totalled to some district or regional level? If they are area-level data, they are likely to provide under-estimates of the variation to be expected at farm level.

If the data are likely to be biased for any of these or other reasons, ways to try to correct for such bias should be considered and, if possible, applied. Given access to abundant data and if these type of questions can be satisfactorily answered, the historical data set can be used directly in the decision analysis.

A historical states of nature matrix can be included in the model as a *multivariate empirical distribution*, also called a *statistical bootstrap approach*. With this approach, stochastic dependency between variables is simulated by sampling the same state of nature for all variables at each iteration. The VoseDiscrete

	A	B	C	D
1	State	Probability	Wheat	Dairy
2	1	0.45	100	300
3	2	0.55	150	310
4	Iteration	State	Wheat	Dairy
5		2	150	310
6	Formulae used:			
7	B5:	=VoseDiscrete(A2:A3,B2:B3)		
8	C5:	=HLOOKUP(C1,C1:D3,B5+1,FALSE)		
9	D5:	=HLOOKUP(D1,C1:D3,B5+1,FALSE)		

Fig. 4.11. Illustration of use of lookup table function.

function in ModelRisk coupled with the lookup function in Excel can be used for this purpose. The VoseDiscrete function generates a discrete outcome at each iteration corresponding to a row index in the states of nature matrix. Sampling is proportional to the specified probability weights of the states. These will be equal for all states if it is judged that each state is equally representative of what the future might bring; otherwise differential probabilities may be assigned to states reflecting available information or subjective views about the likely recurrence of each state in the planning period. Then a horizontal lookup table function, with the sampled row index as input, returns values for each column in the state of nature matrix for the indicated state.

As an illustration, suppose we have the minimal state of nature matrix in Excel for the gross margins (GMs) of two activities, as shown in worksheet format in Fig. 4.11. The VoseDiscrete function in cell B5 of this worksheet generates a value for the state of nature using a discrete distribution that reflects the relative chances of occurrence of the various states. At each iteration this generated value is used for all activities in the lookup functions. The horizontal lookup functions in cells C5:D5 then picks the wheat and dairy GMs from matrix C2:D3 that correspond to the state sampled in that iteration. For example, for the iteration shown in the figure, the sampled state of nature was 2 (cell B5), with associated GMs per unit for wheat and dairy of $150 and $310, respectively (i.e. the vector C3:D3 from the matrix C2:D3).

If this procedure is simulated a sufficient number of times (e.g. 1000), the empirical distribution will be replicated. Then state 1 will occur in about 45% of the cases and state 2 will occur in about 55% of the cases.

Modelling the underlying structure of dependency

This approach to assessing joint distributions is based on the proposition that there must be reasons for any stochastic dependencies between variables. By identifying those reasons, and modelling the relationships they imply, albeit perhaps rather crudely, the main features of the joint distributions of interest may be captured. This approach can yield insight into the underlying stochastic variables and their impacts on other variables important for decision analysis. For example, suppose a decision analysis requires specification of a joint distribution of crop yields. Suppose too few relevant data are available, perhaps because records have not been kept, or because technological changes make the historical information out of date. In such situations the *hierarchy of variables* approach may be used with the aims of identifying and modelling the main determinants of crop yields. These determinants might be judged to be a number of weather variables such as growing-season rainfall, growing-season solar radiation, and the incidence and timing of frosts and short-term droughts. Then, either using regression analysis applied to whatever data set is judged most relevant, or using a 'synthetic engineering' approach, based on knowledge of the biology of the crops, a set of equations is derived of the form:

$$y_i = f(w_i) + e_i \tag{4.6}$$

where y_i is the yield of crop i, w_i is the set of weather variables thought to affect the yield of this crop, and e_i is a term measuring the unexplained variance in y_i. (Usually e_i will be assumed to be independently and normally distributed, though other assumptions can also be made, tested and accommodated – see also the use of stochastic production functions and state-contingent production functions in Chapter 8, this volume.)

Using the regression approach, these equations may be estimated, for example, based on data collected at a different but broadly similar location to the one of interest for the analysis, provided that climatic information for both sites is available. Typically, there will be more complete weather data than exist for the crop yields. Consequently, the probability distribution of yields obtained by evaluating the equations for a long series of (real or simulated) weather situations may be more reliable than one based on a short and unreliable set of records of crop yields for the one farm.

Note that, because some of the same causal factors will appear in the equations for different crops, albeit with different coefficients attached to them, the method will provide an indication of the stochastic dependencies. For example, grazing yields may be found to increase with rainfall, while potato yields may decline, owing to more disease problems in wet years. Thus, the yields of these two crops might be found to be negatively correlated, using this method.

The basic idea illustrated above can be extended to more than one level of dependency. For instance, it might be judged that prices for the crops grown will be influenced by yields, with lower prices prevailing in years of high production, and vice versa. A further set of relationships could therefore be established, again showing explained and unexplained components of uncertainty, to model the joint distribution of both yields and prices.

Sometimes in constructing such hierarchies of stochastic variables it is impossible to get data on the causal factors at work. In that case, the use of surrogate variables might be considered. For example, a whole-farm risk modelling study might require the joint distribution of future prices of farm outputs and inputs. These are likely to be positively associated since most will be affected by changes in the general level of prices and costs in the economy. Yet identifying the specific causal components for individual inputs may be too difficult to be practical. In such a case a rough and ready approach is to relate all prices to some suitable index, say the index of wholesale prices, again with error terms for unexplained variability in individual input prices. Then the forecasting task becomes one of deriving a probability distribution for the wholesale price index at the future date of interest, from which the system of equations, including the error terms, can be used to infer a joint distribution of all input prices.

The structure of such a model, albeit simplified to fit on a page, is illustrated in the influence diagram in Fig. 4.12. This model was part of a more extensive one relating to the optimal financing for an Australian wheat and sheep farm (Milham, 1992). Deliberately omitted are the decision nodes that would allow the model to be manipulated to explore alternative management scenarios, as well as some additional sources of uncertainty that would have unduly cluttered the figure. The main origins of uncertainty represented are general economic conditions and local seasonal conditions. These two then influence other measures, with additional general uncertainty captured in the error terms on the estimated regression equations for the relationships between components represented in the figure.

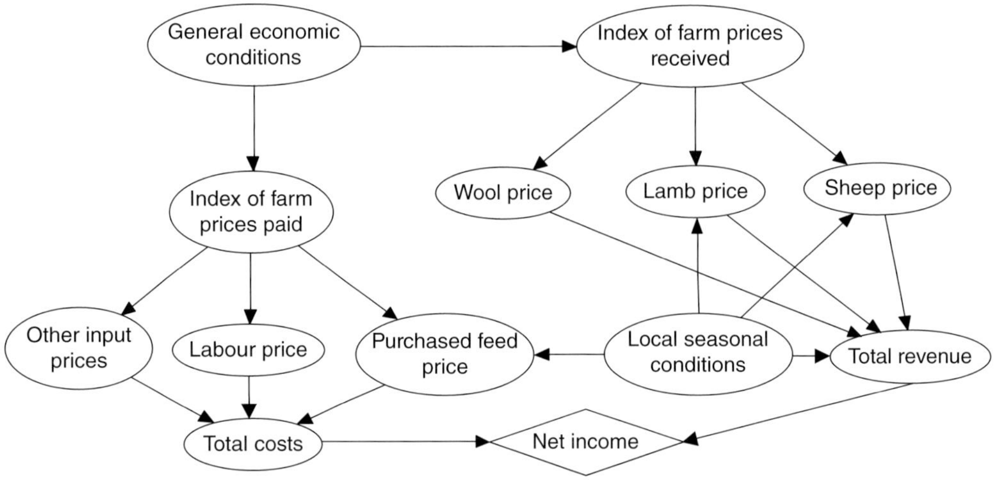

Fig. 4.12. A hierarchy of variables model of sources of uncertainty affecting net income of an Australian sheep farm.

An alternative to regression approaches such as were mainly used in the farm model illustrated in Fig. 4.12 is what was described above as a 'synthetic engineering' approach. This option usually means constructing a simulation model (or models) to represent the processes that determine the various uncertain quantities of interest, including a representation of the stochasticity in those processes. Most often, at least in agriculture, such stochastic simulation models are built to represent agrobiological processes in order to generate estimates of levels of production of crops or animals. For example, models have been developed to predict crop yields from information about relevant management, soil and weather variables, while models of animal production are typically based, inter alia, on data on nutritional intake. Given a good representation of the production processes in the simulation models, and good characterization of the distributions of the stochastic factors impinging on those processes, such as weather, then Monte Carlo sampling can be used to generate distributions of production measures of interest. Moreover, in so far as different aspects of production, or production of different animals or crops, are likely to be affected by some of the same factors on the one farm, such simulation may be used to generate joint distributions of variables of interest.

Some common limitations of many, but by no means all, such agrobiological simulation models arise because they tend to omit some of the limiting factors on production, particularly effects of variation in uncontrolled factors. As a result, they often lead to estimates of productivity that are both too optimistic for farm conditions and that do not show as much dispersion in outcomes as is experienced on farms. Thus, before any such model is used as an aid in estimating probability distributions for decision analysis, it is important that it is properly validated to be sure that it is close enough to the perceived reality.

Multivariate distributions and correlation

In a situation of abundant relevant data, it may be more convenient to model the dependency so as to be able to generate a large number of samples for stochastic analysis. Given two or more uncertain quantities

$X, Y, \ldots$, the joint probability distribution or multivariate distribution for $X, Y, \ldots$ is a probability distribution that gives the probability that each of $X, Y, \ldots$ falls in any particular range of values specified for the variables.

Correlation coefficients provide a measure of the strength and direction of dependence between two or several uncertain quantities. The most widely used correlation is the *Pearson product-moment correlation*, which is a measure of the strength and direction of the linear relationship between variables. The definition of the Pearson correlation is:

$$Corr(X,Y) = \frac{Cov(X,Y)}{\sqrt{V(X)V(Y)}} \tag{4.7}$$

where $Cov(X,Y)$ is the covariance between X and Y, $V(X)$ is the variance of X, and $V(Y)$ is the variance of Y.

When calculated from data, the formula is:

$$Corr(X,Y) = \frac{\frac{(x_i - \mu_x)(y_i - \mu_y)}{n}}{\sqrt{\frac{(x_i - \mu_x)^2}{n} \frac{(y_i - \mu_y)^2}{n}}} \tag{4.8}$$

where i is the observation index from 1 to n, μ_x is the mean of X and μ_y is the mean of Y. Correlation coefficients range from −1 for perfect negative linear dependence, through zero for independence, to +1 for perfect positive dependence.

There are limitations to the use of Pearson correlations to measure stochastic dependency, including:

1. They measure only linear dependency. Often in practice, the degree of association varies over the distribution, perhaps being strongest in one of the tails of the distribution.
2. Feasible values for correlation depend on the marginal distribution. For example, it is mathematically impossible to generate random deviates if there are two variables X and Y that have a linear correlation of 0.7 and X is a beta distribution and Y is log-normal.
3. The only joint distributions that can be constructed using Pearson correlation are the class of *elliptical distributions* (multivariate normal and some generalizations thereof such as multivariate t distributions).

The *rank correlation* measures broader forms of dependency (i.e. non-linear dependency). Rank correlation measures the direction and strength of the relationship between the ranks of the observations of two variables rather than the relationships in the actual data. There are two main varieties of rank correlation, Spearman's and Kendall's.

Spearman's rank correlation specifies the relationship between two distributions in terms of the ranks or positions of the values of each variable in their respective distributions.

Kendall's rank correlation, known as tau (τ), is a measure of concordance. If (x_1, y_1), (x_2, y_2), …, (x_n, y_n) is a set of values of the joint uncertain quantities X and Y, respectively, such that all the values of (x_i) and (y_i) are unique, then each pair of observations (x_i, y_i) and (x_j, y_j) are said to be *concordant* if the sign for both elements agree: that is, if both $x_i > x_j$ and $y_i > y_j$ or if both $x_i < x_j$ and $y_i < y_j$. They are said to be *discordant*, if $x_i > x_j$ and $y_i < y_j$ or if $x_i < x_j$ and $y_i > y_j$. If $x_i = x_j$ or $y_i = y_j$, the pair is neither concordant nor discordant. Kendall's τ is defined as:

$$\tau = \frac{n_c - n_d}{n_o} = \frac{((x_i - x_j)(y_i - y_j) > 0) - ((x_i - x_j)(y_i - y_j) < 0)}{\left(\frac{1}{2} n\,(n-1)\right)} \tag{4.9}$$

where n_c is number of concordant pairs, n_d is number of discordant pairs, n_o is the total number of pairs, and n is the number of observations.

The rank correlation methods are distribution-free because rank correlations can be specified for any pair of variables, regardless of the type of marginal distribution defined for each of them. Ranking methods also 'smooth out' outliers in the data, which may be considered to be an advantage. Rank correlation coefficients range from −1 for perfect negative monotonic dependence (i.e. not a linear dependence), to +1 for perfect positive monotonic dependence. Values of τ around zero imply independence.

Calculations of Pearson's and Spearman's rank correlation are illustrated in a simple example below. Suppose we have the observed and adjusted GMs (in \$1000/ha) for winter wheat (X) and sugarbeet (Y) as in the states of nature matrix A2:C14 in Fig. 4.13. Applying Eqn 4.8 to the data yields a Pearson's correlation coefficient of 0.49. Using the ranks of the data as input in Eqn 4.8 yields Spearman's rank correlation of 0.75.

To calculate Kendall's tau we first sort the original data (X and Y, matrix A1:B13 in Fig. 4.14) in decreasing order of X. Then all possible $(x_i - x_j)$ in Eqn 4.9 will be 1. Next we check the signs of the corresponding $(y_i - y_j)$ values to determine the numerator in Eqn 4.9. We put the Y vector (except the last observation) on the vertical axis of a new matrix and the Y vector (except the first observation) on the

	A	B	C	D	E	F	G
1				Pearson's corr			Spearman's rho
2	State	X	Y	$(x_i - \mu_x)$ $*(y_i - \mu_y)$	X rank	Y rank	$(x_{ri} - \mu_{rx})$ $*(y_{ri} - \mu_{ry})$
3	1	1.60	2.00	0.01	5	7	−1
4	2	2.10	1.80	0.01	8	5	−2
5	3	3.00	3.00	0.82	10	10	13
6	4	1.20	1.80	0.22	3	5	4
7	5	1.50	1.50	0.35	4	3	8
8	6	4.00	2.20	0.32	11	9	12
9	7	1.80	1.60	0.15	6	4	1
10	8	2.00	3.00	−0.15	7	10	2
11	9	0.50	0.80	2.03	1	1	29
12	10	2.50	3.50	0.50	9	12	14
13	11	1.00	1.10	1.07	2	2	19
14	12	4.70	2.00	−0.06	12	7	4
15	Mean	2.16	2.03		6.50	6.25	
16	Sum			5.27			105
17	SDs	1.18	0.76		3.45	3.34	
18	Covariance			0.44			8.71
19	Pearson corr.			0.49			
20	Spearman's rho						0.75
21	Some formulae used:						
22	D3:	=(B3-\$B\$15)*(C3-\$C\$15)			E3:	=RANK(B3,\$B\$3:\$B\$14,1)	
23	D16:	=SUM(D3:D14)			G3:	=(E3-\$E\$15)*(F3-\$F\$15)	
24	D18:	=D16/COUNT(D3:D14)			G18:	=G16/COUNT(G3:G14)	
25	D19:	=D18/(B17*C17)			G20:	=G18/(E17*F17)	

Fig. 4.13. Calculating Pearson's product-moment correlation and Spearman's rank correlation.

horizontal axis of the same matrix. Then, for each pair of observations, subtract the value of the horizontal axis of the matrix from the value of the vertical axis, and find the sign of this difference. The results are in D1:O12 in Fig. 4.14. Then in columns P and Q we count line by line positive and negative signs, and sum them up in cells P13 and Q13. Then, we take the difference of these two numbers and divide by the total number of pairs and find Kendall's τ for this example to be 0.56, as shown in cell Q15.

As this simple example shows, different dependency measures can give quite different estimates. The reason for the differences in estimates in this simple case can partly be explained by the plot of the data in Fig. 4.15. First, it is evident that there is a high degree of uncertainty, since only 12 observations are included in this simple example. Second, the dependency does not appear to be linear. Non-linearity may explain the low estimate of Pearson's correlation, which measures only linear dependency, compared with the estimated coefficients of Spearman and Kendall, which better reflect non-linear dependency.

While the two rank correlation coefficients measure broader forms of dependency (i.e. non-linear), they too provide only single summary statistics quantifying the degree of dependency; they give no information about the structure of that relationship. In a stochastic simulation, for example, we may need

	A	B	C	D	E	F	G	H	I	J	K	L	M	N	O	P	Q
1	*X*	*Y*		*Y*	2.2	3.0	3.5	1.8	3.0	1.6	2.0	1.5	1.8	1.1	0.8	nc	nd
2	4.70	2.00		2.0	−1	−1	−1	1	−1	1	0	1	1	1	1	6	4
3	4.00	2.20		2.2		−1	−1	1	−1	1	1	1	1	1	1	7	3
4	3.00	3.00		3.0			−1	1	0	1	1	1	1	1	1	7	1
5	2.50	3.50		3.5				1	1	1	1	1	1	1	1	8	0
6	2.10	1.80		1.8					−1	1	−1	1	0	1	1	4	2
7	2.00	3.00		3.0						1	1	1	1	1	1	6	0
8	1.80	1.60		1.6							−1	1	−1	1	1	3	2
9	1.60	2.00		2.0								1	1	1	1	4	0
10	1.50	1.50		1.5									−1	1	1	2	1
11	1.20	1.80		1.8										1	1	2	0
12	1.00	1.10		1.1											1	1	0
13	0.50	0.80			Sum											50	13
14					Total number of pairs (n_0)												66
15	Some formulae used:				Kendall's tau												0.56
16	E2:	=SIGN($D2-E$1)				Q2:	=COUNTIF(E2:O2,"<0")					Q14:	=0.5*COUNT(Q2:Q13)*(COUNT(Q2:Q13)-1)				
17	F2:	=SIGN($D2-F$1)				P13:	=SUM(P2:P12)					Q15:	=(P13-Q13)/Q14				
18	P2:	=COUNTIF(E2:O2,">0")				Q13:	=SUM(Q2:Q12)										

Fig. 4.14. Calculating Kendall's rank correlation.

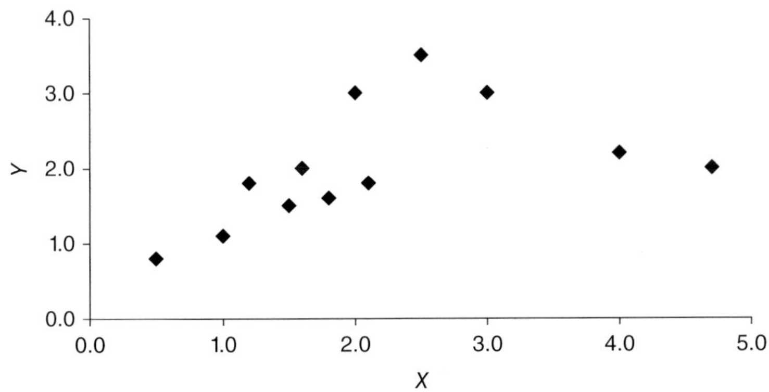

Fig. 4.15. Scatter plot from the paired sample of winter wheat (*X*) and sugarbeet (*Y*).

more information about the structure of dependency than just a correlation measure. Except in those rather rare cases where elliptical distributions can be appropriately fitted, it may be judged to be better to capture or represent the dependency structure using a copula, as described below.

Copulas

A *copula* (from the same root as 'copulate') is a function for 'joining together' two or more marginal distributions (McNeil *et al.*, 2005, Chapter 5). With copula-based multivariate models we can separate modelling the marginal distribution of individual variables from modelling the dependency that 'couples' these distributions to form a joint distribution. This separation allows for a much greater degree of flexibility, compared with other ways of representing dependency.

Copulas are frequently used in Monte Carlo simulation, as is outlined in Chapter 3. Copulas allow generation of sets of two (for the bivariate case) or more uniform [0,1] random deviates u_i that, when transformed via $F^{-1}(u_i)$, where $F(u_i)$ is the cumulative function for the specified marginal distributions $f(x_i)$, generate two or more values for x_i that have particular required stochastic dependency.

To illustrate how multivariate stochastic simulation analysis with copulas can be implemented, the simple example of winter wheat (named X) and sugarbeet (named Y) from the section above is used in conjunction with a normal copula.

Normal copula

The bivariate (for this case) normal copula can be implemented as follows:

1. Specify the univariate marginal distributions, which can be of almost any form.
2. Generate two independent random numbers $\{u_1, u_2\}$.
3. Fix $u_1^* = u_1$ and then apply the inverse conditional copula to translate u_2 into u_2^*. That is, set $u_2^* = C_{2|1}^{-1}(u_2|u_1^*)$, so that $\{u_1^*, u_2^*\}$ are simulations of the copula with uniform marginals. $C_{2|1}^{-1}$ means the inverse of the conditional copula function, C where the subscript means it is for finding u_2 given a value of u_1.
4. Feed the correlated uniform marginals from step 3 into the inverses of the marginal distributions specified in step 1 to obtain a sample from the joint distribution of the uncertain variables themselves.

In Fig. 4.16 an example with a normal copula simulated in Excel is illustrated. In step 1 we specify the univariate marginal distribution for variables X and Y. These marginal distributions were obtained as statistical fits in the Excel add-in ModelRisk. Chosen were a gamma distribution for variable X (specified in B21:B22) and a Weibull distribution for variable Y (specified in C21:C22). In step 2 independent random numbers $\{u_1, u_2\}$ are generated in cells B26 and C26. The normal copula can be written:

$$C(u_1^*, u_2^*; \rho) = \mathbf{\Phi}(\Phi^{-1}(u_1^*), \Phi^{-1}(u_2^*)) \tag{4.10}$$

where ρ is the Pearson correlation, $\mathbf{\Phi}$ is the bivariate standard normal distribution function and Φ is the univariate standard normal distribution function. In step 3 u_1^* is specified in cell E26 (which comes from B26) and generates u_2^* in cell F26. The inverse of the normal copula conditional distribution is:

$$u_2^* = C_{2|1}^{-1}(u_2|u_1^*) = \Phi\left(\rho\Phi^{-1}(u_1^*) + \sqrt{1-\rho^2}\Phi^{-1}(u_2)\right) \quad (4.11)$$

The Pearson ρ for the normal copula is calculated in cell C16 (and B17). In step 4 we put the correlated uniform variates from step 3 in as probabilities in the derived marginal gamma and Weibull distributions from step 1 and simulated joint draws for variable X in cell H26 and for variable Y in cell I26. To generate 1000 draws we then simply copied line 26 downwards 999 times, down to line 1025 (the lines from 29 to 1022 are hidden in the Excel figure). The scatter plot in Fig. 4.16 shows the simulated pairs of X and Y generated by the gamma and Weibull marginals and the normal copula. Whether this scatter represents the actual dependency well enough for the purpose of the decision analysis is a subjective judgement.

Note the difference between the multivariate normal distribution and the normal copula. A multivariate normal distribution has univariate normal marginal distributions, while a normal copula can be applied with almost any marginal distributions, as illustrated. Modelling our case example with a multivariate normal distribution would probably be less than ideal, since the marginal distributions are non-normal, and the multivariate normal would not account for the negative tail dependency evident in Fig. 4.15. Just how less than ideal would depend on the specificities of the decision problem being addressed. As in so many aspects of decision analysis, strong generalizations about pragmatic approximative methods versus more formally exact methods are not readily possible.

Clayton and other copulas

Besides the normal copula, there are many other copulas with a wide choice of functional forms and settings. Copulas can be symmetric or asymmetric and exhibit or not exhibit strong tail dependence. Some can join only two marginal distributions; others can work with multiple distributions.

Copulas built on some multivariate distributions such as the multivariate normal and multivariate *t*-distributions are restricted to symmetric dependence structures. In many applications it seems reasonable that there is a stronger dependence in one of the tails than the other. Such asymmetries can be captured with Archimedean copulas such as the Clayton and Gumbel copulas. The Clayton copula has positive lower tail dependence, while the Gumbel copula has a positive upper tail dependence. Bivariate Archimedean copulas are easy to calibrate, given relevant data, since there is a direct relationship between the bivariate copula functions and Kendall's tau, described above.

Here we illustrate just one example, the Clayton copula for the same data and marginal distributions as used above to illustrate a normal copula. The Clayton copula is calibrated using ModelRisk and simulated in Excel (see Fig. 4.17). ModelRisk provides an option to calibrate the copula directly, based on the data. Using the graphical interface provided, a range of copulas can be compared visually (and some goodness of fit statistics are also shown and can be used to assist in ranking the copulas).

The formula for the Clayton specification in ModelRisk is in cells B23:C23 (a vector) in Fig. 4.17. The dependency parameter for Clayton, α, is directly estimated in ModelRisk. The relationship between α and τ is shown in cell C20. The correlated uniform variates generated from the copula are in B23:C23. These are then put in as probabilities in the inverse cumulative functions of the derived marginal gamma and Weibull distributions and the first simulated joint draws for variable X and Y are in cells E23 and F23, respectively. To construct 1000 draws we simply copied line 23, 999 times downwards, down to line 1022 (the lines from 26 to 1019 are hidden in the Excel figure, Fig. 4.17).

	A	B	C
1	State	*X*	*Y*
2	1	1.60	2.00
3	2	2.10	1.80
4	3	3.00	3.00
5	4	1.20	1.80
6	5	1.50	1.50
7	6	4.00	2.20
8	7	1.80	1.60
9	8	2.00	3.00
10	9	0.50	0.80
11	10	2.50	3.50
12	11	1.00	1.10
13	12	4.70	2.00
14	Pearson correlation matrix		
15		*X*	*Y*
16	*X*	1	0.49
17	*Y*	0.49	1
18			
19	Marginal distributions		
20		*X* - Gamma	*Y* - Weibull
21	Alpha	3.32	2.87
22	Beta	0.65	2.28
23			
24			

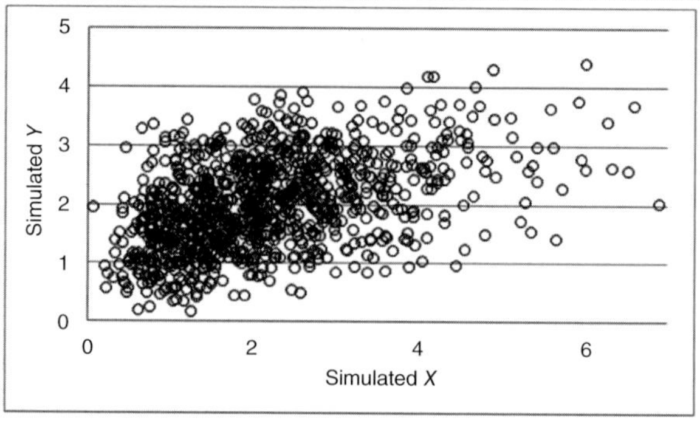

Some formulae used:

```
C16:     =CORREL(B2:B13,C2:C13)
B21:B22: ={VoseGammaFitP(B2:B13)} (a vector)
C21:C22: ={VoseWeibullFitP(C2:C13)} (a vector)
B26:     =RAND()          C26  =RAND()          E26:  =B26
F26:     =NORM.S.DIST($C$16*NORM.S.INV(E26)
         +ROOT(1-$C$16^2)*NORM.S.INV(C26);TRUE)
H26:     =VoseGamma($B$21,$B$22,E26)
I26:     =VoseWeibull($C$21;$C$22;F26)
```

	A	B	C	E	F	H	I
25	Draws	Random draws		Normal Copula		Sim. *X*	Sim. *Y*
26	1	0.75	0.03	0.75	0.16	2.82	1.25
27	2	0.82	0.38	0.82	0.57	3.13	2.15
28	3	0.17	0.63	0.17	0.44	1.07	1.87
1023	998	0.02	0.15	0.02	0.03	0.45	0.65
1024	999	0.68	0.56	0.68	0.64	2.50	2.29
1025	1000	0.50	0.86	0.50	0.83	1.95	2.78

Fig. 4.16. Calibrating and simulation of the normal copula (in Excel) for the example data.

The scatter plot in Fig. 4.17 provides a basis for a subjective judgement as to whether the dependency is modelled well enough for the purpose of the decision analysis.

Comparing the scatter plots in Figs 4.16 and 4.17 illustrates that different copulas (and copula settings) can produced different representations of the dependency. Differences in the way dependency is modelled for one or other of the tails of the joint distribution may be critical in some decision analysis where extreme outcomes have serious, even disastrous consequences. The two copulas illustrated show a marked difference in the tails, making it clear that the choice of an appropriate copula for the modelling requires careful consideration.

Some final comments about copulas

This presentation has been selective and focused only on the main descriptions and properties of some copulas that are important for decision analysis and stochastic simulation. For a more detailed but easy-to-read treatment of copulas, see McNeil *et al.* (2005, Chapter 5) and Alexander (2008, Chapter II.6). Examples of the application of copulas to agricultural economics research problems are provided by Richardson *et al.* (2000) and Woodard *et al.* (2011).

	A	B	C	D	E	F
1	State	*X*	*Y*			
2	1	1.60	2.00			
3	2	2.10	1.80			
4	3	3.00	3.00			
5	4	1.20	1.80			
6	5	1.50	1.50			
7	6	4.00	2.20			
8	7	1.80	1.60			
9	8	2.00	3.00			
10	9	0.50	0.80			
11	10	2.50	3.50			
12	11	1.00	1.10			
13	12	4.70	2.00			
14	Marginal distibutions				Some formulae used:	
15		*X* - Gamma	*Y* - Weibull		C20:	=2*C19*(1-C19)^-1
16	Alpha	3.32	2.87		B16:B17:	={VoseGammaFitP(B2:B13)} (a vector)
17	Beta	0.65	2.28		C16:C17:	={VoseWeibullFitP(C2:C13)} (a vector)
18					B23:C23:	=VoseCouplaBiClaytonFit($B$2:$C$13;FALSE;FALSE) (a vector)
19	Kendall's tau (from Figure 4.14)		0.56		E23:	=VoseGamma($B$16,$B$17,B23)
20	Alpha (Clayton)		2.552		P23:	=VoseWeibull($B$16,$B$17,B23)
21						
22	Draws	Clayton Copula			Sim. *X*	Sim. *Y*
23	1	0.45	0.44		1.82	1.89
24	2	0.05	0.03		0.65	0.69
25	3	0.86	0.53		3.43	2.07
1020	998	0.30	0.48		1.41	1.97
1021	999	0.31	0.18		1.43	1.29
1022	1000	0.56	0.57		2.10	2.15

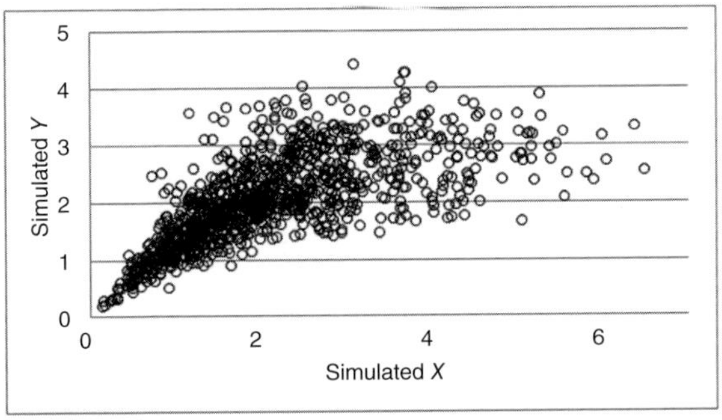

Fig. 4.17. Calibrating and simulation of the Clayton copula (in ModelRisk and Excel) for the example data.

Copulas are included in a number of commercial software packages for stochastic simulation. In ModelRisk, which was used in the examples above, it is relatively straightforward, when relevant and reliable data exist, to estimate and simulate both multivariate parametric and empirical copulas (empirical copulas have not been dealt with in our short overview). A challenge for would-be users is to acquire enough knowledge to use such software effectively.

Sparse or absent data situations

If data are sparse or absent, the best obtainable joint probability judgements about important uncertain variables should be used. Two approaches in addition to the hierarchy of variables approach are outlined below.

Visual impact method for joint distributions of two variables

In the case where there are only two variables *X* and *Y*, that are jointly distributed, it may be possible to assess the distribution by allocating probability counters across a two-way table. However, most people find this quite hard as it is not easy to consider simultaneously the entire two-way table with, say, 10 × 10 entries.

A reasonable alternative may be to do the allocation of probability counters in three steps. Suppose n_{xy} is the number of counters allocated to interval x for variable X and y for variable Y. Then define $n_x = \Sigma_y n_{xy}$ (i.e. the number of counters corresponding to the marginal probability of the interval x, independently of the level of Y).

The first step is to take one of the two variables, assumed to be X, and to use the visual impact method to set the range of X and then allocate the probability counters over chosen intervals for X (i.e. to assign n_x for all x) in the same fashion as illustrated for milk yield in the previous chapter (see Fig. 3.5 in Chapter 3). In doing this, the level for the other variable Y is unspecified and should be thought of as taking any possible value.

The second step is to re-allocate the counters assigned to each interval for X (i.e. n_x) over the possible intervals for Y within the set range of Y to obtain the n_{xy} values such that:

1. $n_x = \Sigma_y n_{xy}$ (i.e. the total number of counters in each interval for X remains as previously determined).
2. The number of counters for each possible interval for Y, n_{xy}, reflects the beliefs of the probability assessor about the joint probability.
3. The implied marginal probability distribution for Y, calculated as $n_y = \Sigma_x n_{xy}$, also reflects the assessor's beliefs about the uncertainty of Y.

Note that, to have enough counters for re-allocation, the numbers of counters allocated in step 1 may be increased by a suitable multiple such that the relative numbers of counters for all intervals for X remain the same.

The third step involves a thorough check of all n_{xy} values in the table to make sure the numbers of counters really reflect the beliefs of the probability assessor. If necessary, the probability assessor may remove or add a few counters to some entries before the final count is made. The joint probabilities p_{xy} are obtained by dividing the number of counters in each entry by the total number of counters used in the table: $p_{xy} = n_{xy}/\Sigma_x\Sigma_y n_{xy}$.

For example, Fig. 4.18 shows the elicitation of the joint distribution for the price a farmer can get for pig meat from selling fattening hogs at 110 kg live weight ($/kg) and for piglets sold at 25 kg live weight ($ per head). Suppose that the relevant range in hog prices is $1.2–2.6/kg (split in seven intervals of $0.2), and in piglet prices $40–90 per head (split in five intervals of $10).

The first step is to ask the probability assessor to allocate probability counters to each of the seven intervals for hog prices (n_x). Suppose the assessor assigns two counters to the interval $1.2–1.4, five to the interval $1.4–$1.6, and so on (see Fig. 4.18). In total 34 counters are used.

The second step starts with the numbers of counters allocated to each hog price interval (step 1) as the basis. To facilitate re-allocation, these numbers are increased by a multiple of five. These new numbers of counters are given in Fig. 4.18 in the right-most column of the upper part of the figure. In step 2, the probability assessor is asked to allocate the given numbers of counters in each row (i.e. hog price interval) across the columns for piglet price intervals. For instance, the ten (= 5 × 2) counters of the hog price interval $1.2–1.4 should be allocated across the five possible piglet price intervals. The result of step 2 is shown in the figure with the total number of 170 counters assigned to hog price and piglet price combinations.

In the third step the assessor is asked to review the allocations of counters carefully to make sure they really represent the assessor's own beliefs about the joint probability distribution. As can be seen in the lower part of Fig. 4.18, the assessor added and moved a few counters resulting in a final allocation of 180 counters.

Once the allocation has been finalized, the joint probabilities are found by dividing each entry (i.e. combination of hog price and piglet price interval) by the total number of counters used. For example, the joint probability of the hog price interval of $1.2–1.4 and the piglet price interval of $40–50 is calculated as 3/180 = 0.017. The joint probabilities are summarized in Table 4.4. Note that the allocation across

Step 1

Hog prices	Counters	Σ
1.2–1.4	••	2
1.4–1.6	•••••	5
1.6–1.8	••••• ••••	9
1.8–2.0	••••• •••	8
2.0–2.2	••••• •	6
2.2–2.4	•••	3
2.4–2.6	•	1
Total		34

Step 2

Hog prices	Piglet prices					
	40–50	50–60	60–70	70–80	81–90	Σ
1.2–1.4	••••	••••	•	•		10
1.4–1.6	•••••	••••• •••••	••••• •	•••	•	25
1.6–1.8	••••• ••••• •	••••• ••••• •••••	••••• ••••• ••••	••••	•	45
1.8–2.0	••••• ••	••••• ••••• •••	••••• ••••• ••••	••••	••	40
2.0–2.2	••	••••• ••	••••• ••••• ••	••••• •	•••	30
2.2–2.4		•	•••	••••• ••	••••	15
2.4–2.6				••	•••	5
Totals	29	50	50	27	14	170

Step 3

Hog prices	Piglet prices					
	40–50	50–60	60–70	70–80	81–90	Σ
1.2–1.4	•••	••••	•••	•		11
1.4–1.6	••••• ••	••••• •••••	••••• •••	•••	•	29
1.6–1.8	••••• ••••• ••	••••• ••••• •••••	••••• ••••• •••	••••	•	45
1.8–2.0	••••• ••	••••• ••••• ••••	••••• ••••• ••••	•••••	••	42
2.0–2.2	••	••••• ••	••••• ••••• ••	••••• •	•••	30
2.2–2.4		•	•••	••••• ••	•••••	16
2.4–2.6			•	••	••••	7
Totals	31	51	54	28	16	180

Fig. 4.18. Three-step elicitation of a joint probability distribution for hog prices ($/kg) and piglet prices ($ per head) using a visual impact method.

Table 4.4. Joint probability distribution for hog prices and piglet prices.

Hog prices ($/kg)	Piglet prices ($ per head)					
	40–50	**50–60**	**60–70**	**70–80**	**81–90**	**Totals**
1.2–1.4	0.017	0.022	0.017	0.006	0.000	0.061
1.4–1.6	0.039	0.056	0.044	0.017	0.006	0.161
1.6–18	0.067	0.083	0.072	0.022	0.006	0.250
1.8–2.0	0.039	0.078	0.078	0.028	0.011	0.233
2.0–2.2	0.011	0.039	0.067	0.033	0.017	0.167
2.2–2.4	0.000	0.006	0.017	0.039	0.028	0.089
2.4–2.6	0.000	0.000	0.006	0.011	0.022	0.039
Totals	0.172	0.283	0.300	0.156	0.089	1.000

[a]Row and column totals do not appear to sum exactly to the totals shown because of rounding.

the table implies both the joint distribution and the marginal distributions for each variable. Thus, this assessor's marginal distribution for the hog price can be read from the table as (0.061, 0.161, 0.250, 0.233, 0.167, 0.089, 0.039) for the intervals shown, and her marginal distribution for the piglet price is (0.172, 0.283, 0.300, 0.156, 0.089). The assessor must be comfortable with both the joint and these two marginal distributions. With such comfort reached, such probability data can then be combined with the acts and payoff data to complete the decision analysis.

Eyeballing copula scatter plots

In many cases with two or more variables, the elicitation of the joint probability distribution using a visual impact method such as the one just described is likely to be too tedious or complex for most assessors, and some more manageable approach is needed. One such alternative is to use software such as

ModelRisk to generate scatter plots of different copulas and copula settings for chosen univariate marginal distributions until one copula and setting is found that is regarded as a good match to the decision maker's (or analyst's) subjective view about stochastic dependency between pairs of uncertain quantities. Obviously, the person undertaking the difficult 'eyeballing' task would need to have a good understanding of the nature and ideally the causes of the dependency.

Applying this approach to the example above relating to prices for pig meat and piglets, we can start with the elicited unconditional marginal distributions for the two uncertain quantities, as reported in Fig. 4.18 and Table 4.4. We then smoothed these discrete subjective distributions using ModelRisk to fit a gamma (40.2;0.046) distribution for hog prices and a log-normal (61.7;11.91) distribution for piglet prices. By putting these marginal distributions into the bivariate copula simulation procedure in ModelRisk, we could then try different copulas and copula settings to find the scatter plot that fitted the farmer's subjective view about stochastic dependency between the two prices. If the farmer believes there is stronger positive dependence in the positive tail than in the negative tail, asymmetric copulas, such as the normal or Student-*t*, would not be suitable. Rather an asymmetric copula, such as the Clayton and Gumbel copulas, should be tried. After several trials of copulas and copula settings and consideration of the resulting scatter plots, we assume for the present purpose that the farmer opted for a Clayton copula with the settings shown in Fig. 4.19. That is, the farmer believes that the scatter plot shown adequately represents her view regarding stochastic dependency between hog and piglets prices. This copula can then be used in simulations to model alternative choice options in the decision analysis.

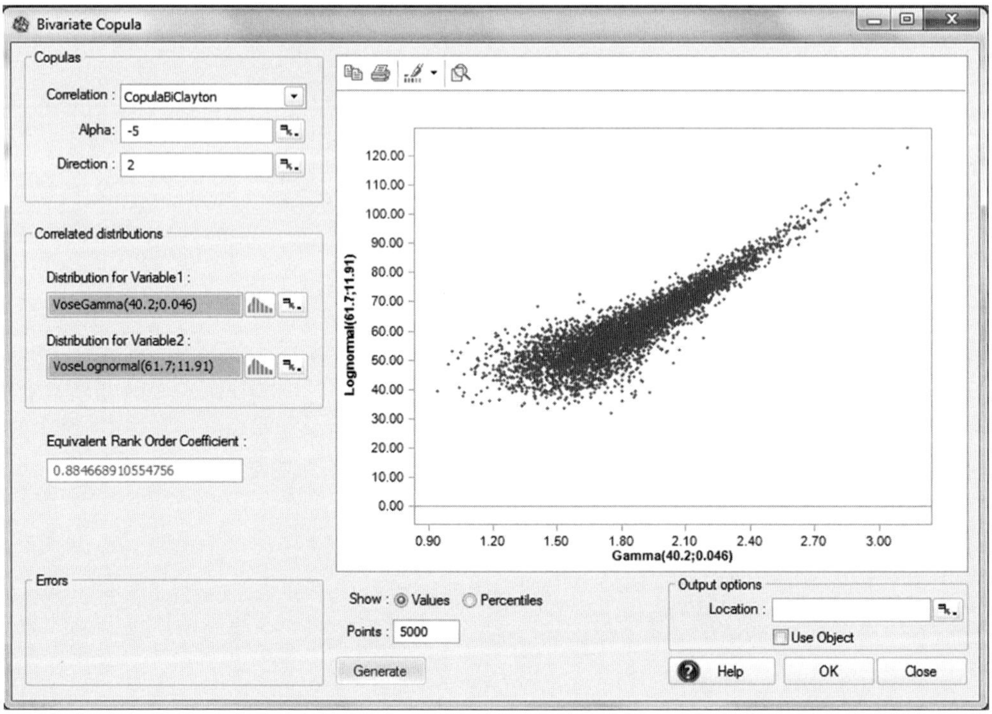

Fig. 4.19. Eyeballed copula using ModelRisk.

Selected Additional Reading

There are many texts that deal with the estimation of probability distributions from data, only a minority of which take a Bayesian approach. Very few emphasize the subjectivity of such methods. Winkler (1972) makes a good starting point for those seeking to pursue these issues further. Clemen (1996) deals with the use of data for probability assessment in Chapter 10. For one review of the topic, see Wright and Ayton (1994). A comprehensive treatment of statistical aspects of decision analysis is to be found in Pratt *et al.* (1995). Much of our treatment of probability assessment using smoothing of CDFs comes from Anderson *et al.* (1977, Chapter 2).

Rather few authors have addressed the issue of how to minimize bias in probability assessment, and almost none in an agricultural context. However, for some pioneering work see Winkler and Murphy (1968) on the evaluation of probability assessors, and Murphy and Winkler (1970) on the use of scoring rules. An excellent introduction to the psychological aspects of decision making under risk can be found in Gardener (2008). Hubbard (2010) makes a powerful case for more systemmatic and careful elicitation procedures including training assessors with calibration to minimize over-confidence and other forms of bias.

References

Alexander, C. (2008) *Market Risk Analysis*, Volume II, Practical Financial Econometrics. Wiley, Chichester, UK.

Anderson, J.R., Dillon, J.L. and Hardaker, J.B. (1977) *Agricultural Decision Analysis*. Iowa State University Press, Ames, Iowa.

Clemen, R.T. (1996) *Making Hard Decisions: an Introduction to Decision Analysis*, 2nd edn. Duxbury Press, Belmont, California.

Gardener, D. (2008) *Risk: the Science and Politics of Fear*. Scribe, Melbourne, Australia.

Hubbard, D.W. (2010) *How to Measure Anything: Finding the Value of Intangibles in Business*, 2nd edn. Wiley, Hoboken, New Jersey.

Linstone, H.A. and Turoff, M. (eds) (2002) *The Delphi Method Techniques and Applications*. Addison-Wesley, Reading, Massachusetts. (Republished as digital pdf file in 2002.)

McNeil, A.J., Frey, R. and Embrechts, P. (2005) *Quantitative Risk Management: Concepts, Techniques, and Tools*. Princeton University Press, Princeton, New Jersey.

Milham, N. (1992) The financial structure and risk management of woolgrowing farms: a dynamic stochastic budgeting approach. MEc dissertation, University of New England, Armidale, New South Wales, Australia.

Murphy, A.H. and Winkler, R.L. (1970) Scoring rules in probability assessment and evaluation. *Acta Psychologica* 34, 273–286.

Pratt, J.W., Raiffa, H. and Schlaifer, R. (1995) *Introduction to Statistical Decision Theory*. MIT Press, Cambridge, Massachusetts.

Richardson, J.W., Klose, S.L. and Gray, A.W. (2000) An applied procedure for estimating and simulating multivariate empirical (MVE) probability distributions in farm-level risk assessment and policy analysis. *Journal of Agriculture and Applied Economics* 32, 299–315.

Winkler, R.L. (1972) *An Introduction to Bayesian Inference and Decision*. Holt, Rinehart and Winston, New York.

Winkler, R.L. and Murphy, A.H. (1968) Good probability assessors. *Journal of Applied Meteorology* 7, 751–758.

Woodard, J.D., Paulson, N.D., Vedenov, D. and Power, G.J. (2011) Impact of copula choice on the modeling of crop yield basis risk. *Agricultural Economics* 42, 101–111.

Wright, G. and Ayton, P. (eds) (1994) *Subjective Probability*. Wiley, Chichester, UK.

5 Attitudes to Risky Consequences

Introduction

In Chapter 2 we have shown how a simple risky decision problem, such as that faced by the dairy farmer thinking about insuring against foot-and-mouth disease (FMD), can be solved. The key step was to transform the risky consequences of an event fork into the DM's certainty equivalents (CEs). However, the assessment of CEs can become very tedious if there are many such risky event forks. Moreover, the introspective capacity needed to decide on CEs rises with the number of branches emerging from the fork. As explained in Chapter 2, the central notion in decision analysis is to break this assessment of consequences into separate assessments of beliefs about the uncertainty to be faced, and of relative preferences for consequences. In Chapters 3 and 4 we dealt with the former of these assessments. Now it is time to look in more detail at how preferences for consequences can be assessed and how those preferences can be encoded.

In Chapter 2 we laid the theoretical foundation for this chapter on utility theory via the presentation of the axioms of the subjective expected utility hypothesis. Readers might find it useful to review the axioms and other assumptions detailed there. In particular, the assumption of *asset integration* noted in Chapter 2 is central to the treatment that follows, implying that DMs will be prepared to view financial losses and gains from some risky business decision as being equivalent to changes in net assets or wealth. Later, we give some justification for this assumption.

We shall also be arguing that most people are *risk averse* most of the time. That does not mean that they refuse to take risks. Rather it means that the CE they assign to any risky prospect will be less than the expected money value (EMV) of that prospect. So someone who is averse to risk will prefer a risky prospect to a sure thing only if the CE of the former is greater than the actual value of the latter; risk averters do take risks, provided there is an incentive to do so. (Of course, in reality, there may not be a sure thing available, so decision making becomes a matter of choosing between alternative risky prospects.) But people vary in their degree of risk aversion. In this chapter we show how the degree of risk aversion of a DM can be assessed and quantified.

Subjective expected utility

As explained in Chapter 2, it can be shown that, for a person who accepts the axioms outlined there, there exists a utility function U that assigns a utility value for that person to any risky prospect. Moreover, this utility function has the very useful property that, if one risky prospect has a higher utility than another, it will be the more preferred, and vice versa. Perhaps more importantly, the *subjective expected utility*

(SEU) of a risky prospect, calculated as the (subjective) probability-weighted average of the utilities of the possible payoffs, is equal to the utility of that prospect. In other words, there is no need to consider the distribution of utility values or measures of dispersion such as the variance of utility. The SEU of any risky prospect tells the whole story about the individual's attitude to the entailed risk, provided that the expectation was calculated using that person's subjective probabilities and risk preferences.

The application of the SEU hypothesis can be illustrated by means of a simple example. Consider the following decision problem in which there is a once-only choice to be made between a_1 and a_2, with consequences depending on which of two equally likely uncertain events S_1 or S_2 occurs. The second column of the following table shows the DM's subjective probabilities that states S_1 or S_2 will occur and the next two columns show the money consequences (in thousands of dollars) from choosing either a_1 or a_2 depending on which of the two uncertain events occurs.

S_i	$P(S_i)$	a_1	a_2
S_1	0.5	2000	1250
S_2	0.5	500	1250
EMV		1250	1250

Payoffs are terminal wealth in thousands of dollars.

Note that a_1 is risky and a_2 is a sure thing. Because the EMVs of the two prospects are equal, we can say that any risk averter will prefer a_2 to a_1. Only a person indifferent to risk, who would base choice on the EMVs, would be indifferent between the two options. However, because almost everyone dislikes risk, at least for non-trivial money consequences, we shall concentrate on risk aversion. What follows applies with only minor modifications for DMs who are not risk averse.

As we have seen, if we progressively reduce the \$1,250,000 (hereafter denoted by \$1250k) payoffs for the sure thing represented by a_2, there will come a point where the DM is indifferent between the two options. Suppose that, in the example above, the CE for some individual is \$1000k. The SEU hypothesis means that the SEU of the risky prospect a_1 is equal to the utility of the \$1000k for sure for this person, i.e.:

$$U(a_1) = 0.5\ U(2000\text{k}) + 0.5\ U(500\text{k}) = U(a_2) = U(1000\text{k}) \tag{5.1}$$

As this simple example indicates, the SEU hypothesis integrates the two components of utility (preference) and subjective probability (degree of belief). Leaving aside for the moment the issue of how we assess the specific form of the utility function in Eqn 5.1, the axioms imply that this function lets us calculate a single index of preference for any risky prospect. We can use the computed utility values to rank alternative risky prospects, enabling risky choice to be rationalized.

The scale of utility functions

It is important to note that the scale used to measure utility is arbitrary. Technically, the utility scale is defined only up to a positive linear transformation – it is always possible to redefine $U_t = a + bU$, $b > 0$. Such a positive linear transformation is used to change the scale and origin when converting temperatures between the Celsius and Fahrenheit systems. Purely for convenience, a utility function $U(w)$ is often defined such that the lowest value for the payoff w of interest has a utility of zero, and the highest value of interest has a utility of 1.0 (though any other origin and scale will serve).

Among the consequences of the arbitrary nature of utility scales are:

1. It makes no sense to talk of one prospect yielding, say, twice as much utility as another (just as it makes no sense to say that one day is, say, twice as hot as another).
2. Interpersonal comparisons of utility are not possible (although some interpersonal comparisons of risk aversion are, as explained later in this chapter).

The power of the SEU hypothesis

Decision theory, based on maximization of SEU, has both prescriptive and descriptive power. *Prescriptive analysis* means working out for a specific DM or for a group of DMs (e.g. an agribusiness manager or, a group of farmers), which actions should rationally be taken. *Descriptive analysis* starts with the proposition that people, such as farmers, tend to act as if they are SEU maximizers – it provides a behavioural theory of choice under uncertainty. Of course, if people really do act to maximize expected utility all the time, the prescriptive theory has no power! In reality, many decisions are too complex for intuitive choice to be a good guide to rationality, yet the SEU hypothesis often helps in understanding behaviour, even if it does not reliably predict how people behave in all cases.

Prescriptive decision analysis is primarily a theory of individual choice and so is most relevant when applied to help an individual DM, such as a farmer, to resolve a difficult risky choice. It has been used prescriptively to help policy makers and planners with their choices, and to suggest, for example, which technology options might best be recommended to target groups of risk-averse farmers. It has also been used as a behavioural theory to explore, for example, lags in adoption of new farming technologies or likely responses of farmers to risk-reducing measures such as crop insurance or price stabilization. The focus of the treatment in this book is mainly on the prescriptive use of decision analysis.

Utility Function Elicitation

Before we can talk about how to elicit utility functions we need to discuss the choice of utility function argument. The utility of what? In principle, it is possible to elicit a utility function for any desirable commodity and a disutility function for any undesirable one. In this chapter we confine the discussion to cases where the consequences of some decision are measured in terms of a single attribute which is expressed in money terms, designated herein as dollars. That means that any important non-monetary aspects of consequences have to be valued in money terms before utility can be assessed. In Chapter 10 we describe utility functions with several attributes, some of which may be measured in money and some not. Moreover, for reasons that will become clearer later in the chapter, we deal initially with wealth as the measure of performance of interest. Later we look at assessing the utility of income or losses and gains.

A number of different approaches have been developed to elicit the required information from DMs to be able to encode their preferences for wealth into a suitable utility function. One way of eliciting such a function is a direct approach whereby a DM is asked to rate his or her relative preferences for consequences. Thus, if the utility of the best outcome is defined as having a utility value of, say, 1.0, and the utility of the worst outcome is defined as having a utility of zero, the DM could be asked to assess on this 'utility thermometer' personal utility values for a sufficient number of intermediate points to define a utility function. The questions may be put directly, or the DM may be asked to consider the ratio of best to worst outcomes needed to establish indifference with the particular outcome being assessed. So, for example, if some outcome is rated as equivalent to a 60:40 mix of best to worst outcomes, the implied utility value is 0.6 × (utility of best outcome) + 0.4 × (utility of worst outcome) which is equal to 0.6(1) + 0.4(0) = 0.6.

While such a direct approach has the merit of simplicity, and may work well for some people and some situations, on other occasions it will be less satisfactory. Some DMs will find the point-blank nature of the required assessments too challenging. Moreover, there are concerns that utility assessments made assuming the absence of risk, as in the direct approach, may not be applicable for risk analysis. An assessment method that is more often preferred, as it overcomes this difficulty, is described next.

Elicitation via certainty equivalents

Consider the CE used before:

S_i	$P(S_i)$	a_1	a_2
S_1	0.5	2000	1000
S_2	0.5	500	1000
EMV		1250	1000

As before, assume payoffs are terminal wealth expressed in thousands of dollars and the DM is indifferent between a_1 and a_2. From Eqn 5.1 we have:

$$0.5\ U(2000\text{k}) + 0.5\ U(500\text{k}) = U(1000\text{k}) \tag{5.2}$$

We can impose a utility scale by assigning utility values to the two extreme levels of consequences. Suppose we let $U(500\text{k}) = 0$ and $U(2000\text{k}) = 1$. Then the above equation can be re-written as:

$$0.5\ (1) + 0.5\ (0) = U(1000\text{k}) \tag{5.3}$$

so that $U(1000\text{k}) = 0.5$.

This one CE tells us quite a lot about the DM's attitude to risk. First, we can tell that this person is averse to risk because the CE of \$1000k is less than the EMV of \$1250k. It seems that this DM is prepared to sacrifice some expected return from the risky prospect in order to be able to have the sure thing. Moreover, the magnitude of the difference tells us quite a lot about how risk averse this person is. If we are prepared to make some assumptions about the form of the implied utility function, to be discussed later in this chapter, this one CE may be all that we need to know. Of course, to proceed on that basis, it would be important to be confident that the established CE was correct in the sense that the DM was content to accept this one indifference relationship as representative of a personal overall risk attitude.

The strong assumptions needed about the form of the utility function might suggest that a single CE provides too little information and it might be better to elicit a few more CEs from the DM to get some more utility values. This can be done in a number of different ways. von Neumann and Morgenstern (1947) proposed holding constant the two risky payoffs in the payoff table given above for eliciting the first CE and varying the probabilities. However, there can be a problem with the von Neumann–Morgenstern method if the DM has difficulty in handling different probabilities. So there is merit in holding the probabilities constant at the so-called *ethically neutral* 50:50 level, as is done below.

The two elicitation methods based on deriving CEs with ethically neutral probabilities are what Anderson *et al.* (1977) called the ELCE (equally likely certainty equivalent) method and the ELRO (equally likely risky outcomes) method. Before explaining more about these methods, we need to introduce some shorthand notation. We can represent a risky prospect with discrete payoffs in the format $(x_1, x_2, \ldots; p_1, p_2, \ldots)$, indicating a set of possible payoffs $x_1, x_2, \ldots$ with corresponding probabilities $p_1, p_2, \ldots$, summing to 1.0. Then we introduce the symbol ~ and assign to it the meaning 'is indifferent between'. To illustrate, in the CE elicitation example above, at indifference we can write:

$$(500\text{k}, 2000\text{k}; 0.5, 0.5) \sim (1000\text{k}; 1.0) \quad (5.4)$$

The ELCE method

Using this notation, the ELCE procedure for eliciting a utility function is set out in Table 5.1. In this table and for the subsequent explanations the notation *a*, *b*, *c*, etc. indicates different values of the payoff measure *w*. For now, suppose that *a* is the lowest payoff of interest and *b* is the highest.

The procedure has the advantage of allowing incorporation of a check question for $U(c)$ versus $U(c')$, as shown. If the two results diverge widely, it will be necessary to ask the subject to review all previous answers to try to obtain a more consistent set of preferences. Should this fail, the validity of the whole elicitation process is thrown into doubt.

We can illustrate the ELCE method for the dairy farmer who has to decide about buying insurance to protect against the risk of an FMD outbreak (Chapter 2, this volume). In the decision tree presented in Chapter 2, payoffs were measured in terminal wealth, with initial wealth of $500,000. We therefore assume that the CE illustrated above actually applies to this farmer's risk preference, and so develop the application of the ELCE method from the starting point already reached. That means that, in setting a scale, we assume that the farmer's lowest terminal wealth of interest is $250,000, and

Table 5.1. Sequence of elicitation of certainty equivalents (CEs) for the ELCE (equally likely certainty equivalent) method of estimating a utility function.

Step	Elicited CE	Utility calculation
1	Setting a scale	$U(a) = 0; U(b) = 1$
2	$(c; 1.0) \sim (a, b; 0.5, 0.5)$	$U(c) = 0.5\,U(a) + 0.5\,U(b) = 0.5$
3	$(d; 1.0) \sim (a, c; 0.5, 0.5)$	$U(d) = 0.5\,U(a) + 0.5\,U(c) = 0.25$
4	$(e; 1.0) \sim (c, b; 0.5, 0.5)$	$U(e) = 0.5\,U(c) + 0.5\,U(b) = 0.75$
5[a]	$(c'; 1.0) \sim (d, e; 0.5, 0.5)$	$U(c') = 0.5\,U(d) + 0.5\,U(e) = 0.5$

[a]Optional check.

the highest \$1,000,000. Moreover, we arbitrarily set $U(250k) = 0$ and $U(1000k) = 1$. Suppose that the sequence of elicited CEs is as shown in Table 5.2. Then the corresponding utility values are as also shown in the table.

These utility values are plotted step-by-step in Fig. 5.1. After step 1, two utility values are known, i.e. $U(250) = 0$ and $U(1000) = 1$. After step 2, a third utility value is elicited ($U(500) = 0.5$), and so on. Step 5 is a check and, for perfect consistency, the elicited CE should have been \$500k whereas it was found to be \$490k, which is judged to be acceptably close.

A utility function, for now smoothed by hand through the utility points obtained, is shown in Fig. 5.2. In reality, even more irregularity in the location of the utility points is likely to be encountered than is shown in Fig. 5.2.

The basic assumption underlying the elicitation procedure is that preferences revealed in the stark, simplified choices of the ELCE method are more likely to represent real preferences than those revealed by actual behaviour in more complex real situations.

Table 5.2. Example of application of the ELCE method of estimating a utility function (wealth in thousands of dollars).

Step	Elicited CE	Utility calculation
1	Setting a scale	$U(250) = 0$; $U(1000) = 1$
2	(500; 1.0) ~ (250, 1000; 0.5, 0.5)	$U(500) = 0.5 (0) + 0.5 (1) = 0.5$
3	(365; 1.0) ~ (250, 500; 0.5, 0.5)	$U(365) = 0.5 (0) + 0.5 (0.5) = 0.25$
4	(735; 1.0) ~ (500, 1000; 0.5, 0.5)	$U(735) = 0.5 (0.5) + 0.5 (1) = 0.75$
5	(490; 1.0) ~ (365, 735; 0.5, 0.5)	$U(490) = 0.5 (0.25) + 0.5 (0.75) = 0.5$

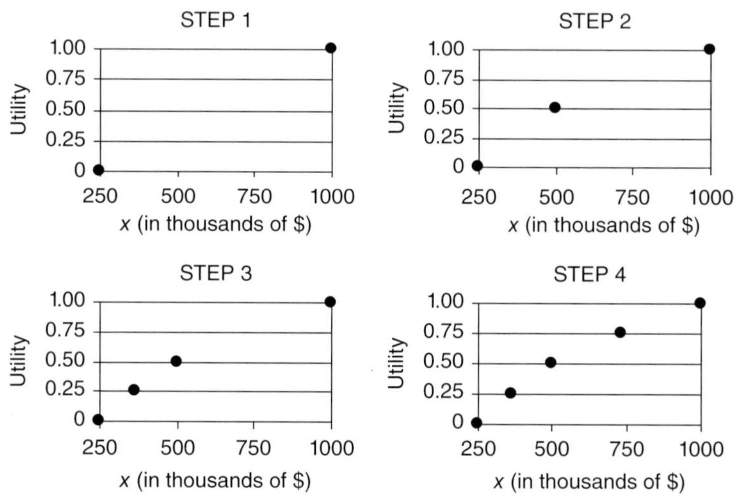

Fig. 5.1. Certainty equivalents (CEs) and utilities after each step from the ELCE (equally likely certainty equivalent) elicitation example.

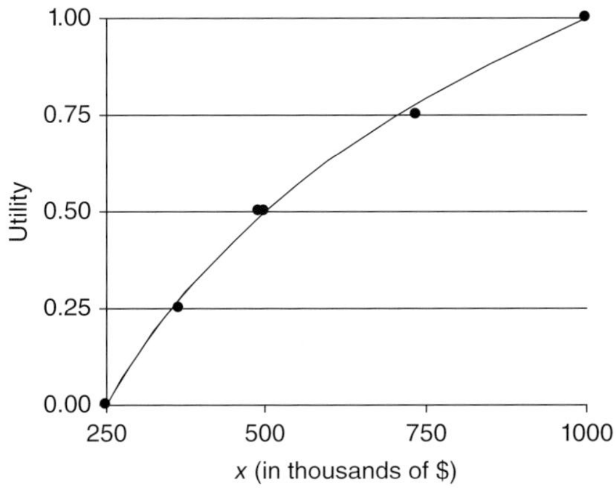

Fig. 5.2. Utility function from the example ELCE elicitation (data points shown include the check step shown in Table 5.2).

An advantage of the ELCE method is that it is based on the ethically neutral probabilities of 0.5. This means that no bias will be introduced if the DM likes or dislikes particular probabilities different from 0.5. Moreover, most people find 50:50 risky prospects much easier to conceptualize than prospects with other probability ratios. On the other hand, this reliance on equally likely outcomes has two disadvantages, which we now discuss in turn.

First, it may be noted that the ELCE method requires comparison of two prospects, one of which is risky and the other not. If the DM has an aversion to, or a preference for gambling, bias will be introduced. Indeed, in the extreme case of a person who believes that gambling is morally reprehensible, the ELCE method may not work at all. If gambling preference or aversion is suspected, the ELRO method may be used. Second, the ELCE method (and also the ELRO method) can only be applied to consequences that can be varied in continuous fashion, such as wealth or income. In cases where continuity does not apply, such as where consequences are classified into a number of discrete categories, the von Neumann–Morgenstern method, outlined earlier, is usually the best option.

The ELRO method

The ELRO method is based on comparing equally likely but risky outcomes in the format:

$$(a, y; 0.5, 0.5) \sim (b, x; 0.5, 0.5) \tag{5.5}$$

with $a < b$ and $x < y$ a reference interval. The outcome b is varied to find indifference. Then, setting an origin such as $U(a) = 0$ and defining a scale by assuming $U(y) - U(x) = 1$ leads to $U(b) = 1$. Substituting b for a then finding a new indifference relationship allows a new point on the utility function to be established, and so on. Some manipulation of the format will usually be needed to cover the whole range of values of wealth of interest.

The ELRO is not widely used, perhaps because it is a bit trickier to implement than the ELCE method. Also, it demands more introspective effort from the DM who has to weigh up four levels of outcome across two prospects for each indifference relationship, compared with three with the ELCE method. For these reasons, the method is not described further or illustrated here. Interested readers are referred instead to Anderson *et al.* (1977, pp. 75–76).

Risk Aversion

Risk attitudes implied by the shape of the utility function

The shape of the utility function reflects the preferences of the DM. For example, if the utility function has a positive slope over the whole range of payoffs, the implication is that more payoff is always preferred to less. Preferences of this kind are normal for money, but may not apply for other things. For example, many people enjoy walking for pleasure, but the utility does not necessarily always increase with distance – a very long walk may be too tiring.

In the language of mathematics, the characteristic that more money is preferred to less may be written:

$$U^{(1)}(w) > 0$$

where $U^{(i)}(w)$ is the i-th derivative of the utility function, $U(w)$, for wealth, w. (Other 'goods' such as income can be substituted for wealth here.) So, if the first derivative of the utility function for wealth is positive for all w, it represents the situation that more is always preferred to less.

Similarly, risk aversion is indicated by a utility function that shows decreasing marginal utility as the level of the payoff is increased. Three possible attitudes to risk can be characterized in terms of the second derivative:

1. $U^{(2)}(w) < 0$ implies risk aversion.
2. $U^{(2)}(w) = 0$ implies risk indifference.
3. $U^{(2)}(w) > 0$ implies risk preference.

These three cases are illustrated in Fig. 5.3. Of these three shapes, the one illustrating risk aversion is the one likely to be most commonly encountered.

The SEU hypothesis implies that the utility of any risky prospect is its expected utility. This utility value can be converted through the inverse utility function into a CE, as explained below. Ranking prospects by CE is the same as ranking them by expected utility (i.e. in the order preferred by the DM). Moreover, the difference between the CE and the expected value of a risky prospect,

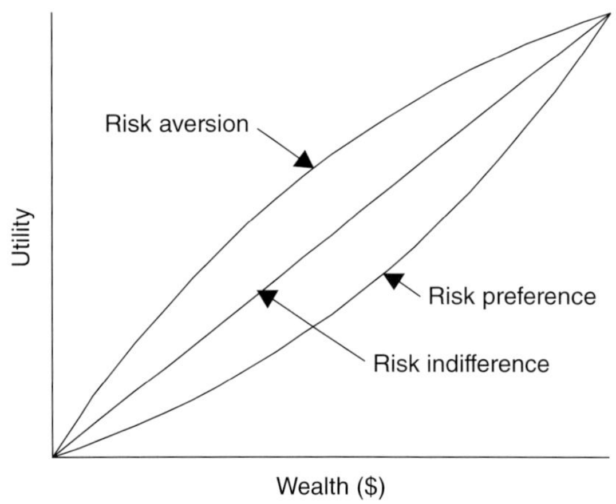

Fig. 5.3. Risk attitudes and the shape of the utility function.

known as the risk premium (RP), is a measure of the cost of the combined effects of risk and risk aversion:

$$RP = EMV - CE \tag{5.6}$$

The three cases listed above can therefore also be described in terms of risk premium:

1. $RP > 0$ ($CE < EMV$) implies risk aversion.
2. $RP = 0$ ($CE = EMV$) implies risk indifference.
3. $RP < 0$ ($CE > EMV$) implies risk preference.

Measures of risk aversion

For a number of purposes it is often useful to be able to quantify the degree of risk aversion. However, measuring risk aversion is not simple. As we have seen, risk aversion is reflected by the curvature of a person's utility function. Measuring that curvature is problematic because a utility function is defined only up to a positive linear transformation. We need a measure of curvature that is constant for such a transformation.

The simplest measure of risk aversion that is constant for a positive linear transformation of the utility function is the absolute risk aversion function:

$$r_a(w) = -U^{(2)}(w)/U^{(1)}(w) \tag{5.7}$$

where $U^{(2)}(w)$ and $U^{(1)}(w)$ represent the second and first derivatives of the utility function, respectively (Pratt, 1964; Arrow, 1965, p. 33). It is generally accepted that the *absolute risk-aversion coefficient* $r_a(w)$ will decrease with increases in w since people can better afford to take risks as they get richer.

Absolute risk aversion is a much used and abused concept. First, note that it is a function, not a constant as often implied. Moreover, although robust enough to be unaffected by a positive linear transformation of the utility function, absolute risk aversion as measured by $r_a(w)$ depends on the monetary units of w. Thus, risk-aversion measures derived in different currency units are not comparable. It is not valid to transport a coefficient estimated for US farmers in US dollars to an analysis of a non-US farm management problem where outcomes are expressed in local currency units or worse, perhaps thousands of them.

The currency units problem is overcome using the *relative risk-aversion function*, defined as:

$$r_r(w) = w r_a(w) \tag{5.8}$$

The *relative risk-aversion coefficient* $r_r(w)$ is independent of the units of w and so can be used in international comparisons of risk aversion, only remembering that, like $r_a(w)$, $r_r(w)$ is a function, not a constant and may change with w.

The absolute risk-aversion function may be categorized according to how it changes with respect to increasing wealth as increasing absolute risk aversion (IARA), constant absolute risk aversion (CARA) or decreasing absolute risk aversion (DARA). Correspondingly, the relative risk aversion may be categorized as increasing, constant or decreasing with wealth (IRRA, CRRA and DRRA, respectively).

Constant absolute risk aversion (CARA) means that preferences are unchanged if a constant amount is added to or subtracted from all payoffs. Constant relative risk aversion (CRRA) means that preferences among risky prospects are unchanged if all payoffs are multiplied by a positive constant.

While there is general agreement that $r_a(w)$ declines as wealth increases (i.e. DARA), there is less agreement on how $r_r(w)$ is likely to be affected by increases in wealth. Arrow (1965, p. 36) argued, on theoretical and empirical grounds, that it would generally be an increasing function of w. However, he also noted that, while some fluctuations are possible, the actual value is likely to hover around 1, being, if anything, somewhat less for low levels of wealth and somewhat higher for high levels. Similarly, Eeckhoudt and Gollier (1996, p. 46) hypothesized that, if wealth increases, relative risk aversion does not decrease. On the other hand, Hamal and Anderson (1982) found that, in extremely resource-poor farming situations, relative risk aversion could reach values as high as four or more – quite contrary to what Arrow had hypothesized. While such disagreement implies the need for more study of magnitudes of relative risk aversion among various DMs, it nevertheless seems reasonable to assume that $r_r(w)$ is less likely to change appreciably as w changes, than is $r_a(w)$. We return to this issue later in the chapter.

Expectations about risk aversion and choice of functional form

Different algebraic specifications of a utility function will often give similar measures of goodness of fit to elicited utility points, yet may imply appreciable differences in risk-taking behaviour in terms of the risk-aversion functions discussed above. Moreover, some simple and popular functions do not have the properties regarded as sensible – the familiar quadratic, for example, is not everywhere increasing and also exhibits increasing, not decreasing, absolute risk aversion as wealth increases.

Functional forms that are commonly used are indicated below, classified according to what they imply for risk attitudes.

Constant absolute risk aversion (CARA)

Negative exponential:

$$U = 1 - \exp(-cw),\ c > 0 \tag{5.9}$$

for which $r_a(w) = c$ (constant) and $r_r(w) = cw$. Although CARA is not usually regarded as a desirable property, the convenience of this functional form means that it has found extensive use in decision analysis. For example, the function can be estimated from a single CE, and it is particularly useful in analyses where the distribution of returns may be assumed to be normal, as explained in subsequent chapters.

There can be numerical problems in evaluating this function for large values of w in association with a relatively large value of c. Utility values very close to 1.0 may be rounded off to this value in computation. Since CARA means that risk aversion is unchanged for the addition or subtraction of any constant from all payoffs, it may help to reduce all payoffs by the same amount. Use of double-precision arithmetic, if available, may also help.

Constant relative risk aversion (CRRA)

CRRA functions are necessarily also DARA. Two main functions are commonly used:

Logarithmic:

$$U = \ln(w),\ w > 0 \qquad (5.10)$$

for which $r_a(w) = w^{-1}$ and $r_r(w) = 1.0$.

Power:

$$U = \{1/(1 - r)\}\ w^{(1 - r)},\ w > 0 \qquad (5.11)$$

for which $r_r(w) = r$ and $r_a = r/w$. This special form of the power function is preferred over the simpler $U = w^c$, as it directly incorporates r as the constant coefficient of relative risk aversion for wealth. When the argument of the function is income or marginal gains and losses, not wealth, r is called the *partial risk-aversion coefficient*. For values of r close to 1.0, Eqn 5.11 must be replaced by the logarithmic function of Eqn 5.10. Note that this function is unsuitable for cases where some payoffs are negative.

As with the CARA function, there is sometimes a problem in applying this function for some values of w and r. The problem may be reduced by scaling the range of w by dividing all payoffs by the maximum payoff – legitimate under CRRA – and by using double-precision arithmetic, if available. For values of r of 4 or more the function implies very high marginal utility for low values of w with a sharp fall to give essentially zero marginal utility for higher values. Such properties of preference seem unlikely, suggesting that such extreme risk aversion is seldom plausible.

Other functional forms

There are many other function forms that can be and have been used to represent attitudes to risk, some of which exhibit greater flexibility in their implications for risk aversion, should that be perceived as a virtue. Functions with the flexibility to represent increasing or decreasing risk aversion have been suggested by Farquhar and Nakamura (1987) and Bell (1988). The so-called *polynomial-exponential* family of functions includes the *sumex* and the *expo-power* utility function. We have chosen not to discuss these functions here on the grounds that their use is relatively rare and experience shows that the simpler forms given above usually serve equally as well in most cases. Interested readers are referred to the cited literature for full details.

Utility Function Fitting

Once a functional form has been chosen, it is usually a relatively simple task to fit that function to the data. In most cases, simple curve-fitting software such as ordinary least-squares regression available in Excel will do the job, although for some functional forms such as exponential functions iterative non-linear methods may be required. (Bear in mind that any positive linear transformation of the above functional forms can be used to facilitate curve fitting.) However, a word of warning is appropriate – the purpose of the curve fitting is to find the equation of a curve that is already partly defined by the elicited utility points, not to fit a curve to a scatter of points representing random deviations from some underlying but unknown relationship. Consequently, statistical measures of goodness of fit, such as R^2, do not have their

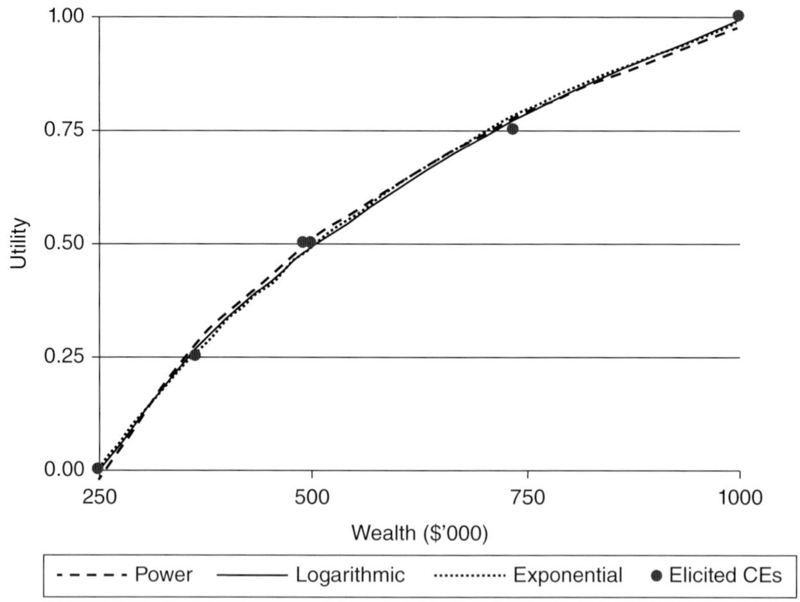

Fig. 5.4. Results from plotting three different forms of utility function (power function, logarithmic function and negative exponential function) to the elicited CEs (data points shown include the check step shown in Table 5.2).

usual interpretation. Rather the adequacy or otherwise of the fitted curve is often best judged by eye, comparing the shape of the curve with the plotted points obtained from the preference elicitation procedure.

By way of illustration, we have used regression to fit three different functions to the data points elicited using the ELCE method, as plotted in Fig. 5.2. The three functions were the power function, the logarithmic function and the negative exponential function. The three curves are plotted in Fig. 5.4.

Figure 5.4 shows that there is little to choose between these three functional forms in this case. This is a quite common but by no means universal finding, suggesting that the functional form used will often be much less important in applied decision analysis than the extensive discussion of this issue in the theoretical literature implies.

From Expected Utility to Certainty Equivalent

There are some advantages in using the CEs of the alternative risky prospects instead of expected utility values for decision analysis. They are more easily understood and interpreted than expected utility values. Also, the magnitudes of differences between alternatives can be assessed; the arbitrary nature of utility scales means that it is not possible to weigh one utility value against another (except to rank them within a specific analysis), whereas CEs can be quantitatively compared. When comparing independent alternatives, one CE can be expressed as a proportion of another, or two CEs can be subtracted, yielding results that can be interpreted simply and directly. It follows that, even when a utility function is known, or has been approximated by methods such as those indicated above, it will often still be preferable to convert expected utilities to CEs. There are two main ways of doing this:

1. If the utility function exists as a plotted curve, it may be straightforward to read from a given expected utility value on the vertical axis to a value on the horizontal axis representing the CE.
2. If the utility function has been estimated via a particular algebraic equation, it will usually be possible to solve the equation for w (or whatever other argument was used for the function) given a value of U. (If algebraic manipulation to invert the function is difficult, a trial-and-error numerical approach via a spreadsheet will give an answer without too much trouble.)

Problems in Utility Function Elicitation

Responding to the questions described above for utility function elicitation is not easy – not everyone has the required introspective capacity. In particular, people who are unused to dealing with abstract concepts may be unable to respond to the hypothetical questions required. Not surprisingly, inconsistencies have often been observed.

In some situations people have been shown to exhibit extreme risk aversion for risky prospects with quite modest losses and gains (Rabin, 2000; Rabin and Thaler, 2001). Indeed, it is often found that the smaller the amounts at stake, the more risk averse are the responses. Rabin convincingly showed that the degree of risk aversion for small payoffs (of the order of plus or minus $100) commonly exhibited (usually for hypothetical choices) is inconsistent with the proposition, widely accepted by economists, that risk aversion is a reflection of the diminishing marginal utility of wealth. In other words, in situations where there are roughly equal chances of winning or losing around $100, many people exhibit *loss aversion*, meaning that they are apparently extremely risk averse about such decisions – much more so than can be explained by any feasible utility function for wealth. On the face of it, loss aversion is inconsistent with the SEU hypothesis to the extent that Rabin and Thaler claimed the theory to be as dead! However, there is another explanation, which we prefer.

Curiously, many of the same people who exhibit loss aversion also exhibit risk preference for small-stake risky prospects with low probabilities of very large gains, such as buying a lottery ticket. (Risk preference means people will accept prospects with less than fair odds and hence negative expected values.) The important point is that most of the people who exhibit either loss aversion or risk preference reconsider their choice and act more rationally if they have the chance to make the same or similar choices repeatedly. Thus, for example, a 50:50 chance of winning $110 versus losing $100 will often be rejected, but the same people often say they would accept the prospect if able to play several times. Similarly, except for addicted gamblers, most people who gamble will limit the amount they stake; they may buy a $10 lottery ticket, but will not buy 100 or 1000 such tickets.

For a prescriptive theory of choice it is clear that both loss aversion and risk preference are not sensible general guiding rules. Over many small risky decisions, risk preference will almost certainly lead to net losses, and loss aversion will almost certainly mean forgone opportunities to gain, with little or no risk of loss. Moreover, it is clear that in practice most business DMs do not persistently act in these irrational ways; if they did they would go broke! In what follows, therefore, we assume that rational DMs will generally view losses and gains as changes in wealth (which is what they are) and will choose to evaluate them in those terms. This is our assumption of asset integration, outlined in Chapter 2.

In light of the above, it is not surprising that plausible utility functions are most often obtained from elicitation of the utility of wealth, and that utility functions for income are often more convincing than those for losses and gains. The truth is, however, that few people are able to articulate their risk preferences consistently. Some alternative approaches are therefore examined next.

Alternatives to Direct Elicitation of Utility Functions

Estimating risk aversion from observed behaviour

Although we emphasize the normative use of decision analysis throughout this book, it is possible to look at things the other way round. Suppose farmers (and other DMs) generally act more or less consistently with the SEU hypothesis. In that case we could observe behaviour and draw from the decisions taken, inferences about the attitude to risk of an individual or community of DMs. The inferences drawn depend on assumptions about the nature of the production and decision environment, including the structure of preferences and perceptions about the associated uncertainty.

There are broadly two approaches to inferring something about risk aversion from observed behaviour. Most studies in agricultural economics have used econometric methods applied to cross-section or panel data. Of course, such studies normally lead to an estimate of the average risk attitude of the farmers in the sample used. Typically, a model of some risky production process is developed and estimated to describe the optimal level of input use or output produced. Models of this kind are outlined in Chapter 8, this volume. Such models will typically include some stochastic elements to represent the risk faced by the DMs and also include coefficients to be estimated in order to reflect DMs' risk responses. For example, it might be assumed that farmers make production and resource allocation choices to maximize an indirect utility function in terms of the mean and variance of returns, with the regression coefficient on the variance term presumed to reflect risk aversion. There is a significant literature reporting variants of and improvements to this basic approach (see Moschini and Hennessy (2001) and Chavas (2004) for an overview). Perhaps we can expect to see other such studies in future using the state-contingent analyses proposed by Chambers and Quiggin (2001) and described in Chapter 8.

The methods in the second category are typically based on analysis of farm enterprise mix decisions within a portfolio selection framework, usually implemented by mathematical programming (Chapter 9, this volume). Again, typically but not always in a mean–variance context, the extent to which individual farmers choose to trade away expected returns for a reduction in variability of returns can be estimated, leading to conclusions about the degree of risk aversion, this time usually on an individual basis (Robison *et al.*, 1984; Lien, 2002).

While both approaches have their adherents, both tend to suffer from two basic weaknesses. First, they depend on a strong assumption that the analyst and the DMs share the same view of the uncertainty to be faced. In particular, it is implicitly assumed in almost every study that the probabilities used in the modelling, typically based on some historical series of observations of key uncertain phenomena, are the same probabilities that the DMs used in allocating resources or choosing enterprise mixes. The discussion in Chapters 3 and 4 may have convinced readers that such congruence of beliefs would seldom be plausible given the generally different access to information of farmers and research workers.

Second, both types of modelling are subject to specification errors. Reality is far more complex than can be represented in an econometric or mathematical programming model. Moreover, the effects of these specification errors tend to be rolled into the estimates of risk aversion, throwing still more doubt on the reliability of the results.

For these reasons, and also because of the strongly normative orientation of our presentation, we choose not take the discussion of these methods any further.

Assumptions about the degree of risk aversion

In the past, it has been common to assume, often implicitly, that all DMs are indifferent to risk. For example, such an assumption is necessary to justify the many farm budgets that are done with no accounting at all for risk. Yet assuming no risk aversion seems to be a second-best option when we know that risk aversion is widespread. A more sensible course, if there is no other information at all, might be to assume a relative risk-aversion coefficient of 1.0, which is the value suggested by Arrow (1965) as most likely. The constant relative risk-averse function for $r_r(w) = 1$ is $U = \ln(w)$, the so-called 'everyone's utility function' postulated by Daniel Bernoulli as long ago as 1738.

If this seems to be too strong an assumption, Anderson and Dillon (1992) have proposed a rough-and-ready classification of degree of risk aversion, based on the magnitude of the relative risk-aversion coefficient. Their classification is:

- $r_r(w) = 0.5$, hardly risk averse at all;
- $r_r(w) = 1.0$, somewhat risk averse (normal);
- $r_r(w) = 2.0$, rather risk averse;
- $r_r(w) = 3.0$, very risk averse; and
- $r_r(w) = 4.0$, extremely risk averse.

To locate yourself on this scale, consider how much of your present wealth (or equivalently, long-run future income) you are prepared to stake for a 0.5 chance of a 20% increase in wealth. The implications of different maximum stakes are set out in Table 5.3.

While it is a matter for individual judgement, the numbers in this table suggest to us that values of $r_r(w)$ somewhat higher than 1.0 may be more common than has been implied in the literature.

Table 5.3. Implications for relative risk aversion of the maximum amount staked for a 50% chance to increase wealth by 20%.

Maximum stake as a percentage of current wealth	Implied relative risk-aversion coefficient
20%	0
18%	0.5
17%	1.0
14%	2.0
12%	3.0
11%	4.0

Suppose a value for $r_r(w)$ can be approximated as described above that can be assumed to be more or less constant for local variation in wealth. Then if $r_a(w)$ is required, it may be derived using the formula $r_a(w) = r_r(w)/w$. Experience suggests that choice of exact form of a utility function is seldom important for decision analysis provided the degree of risk aversion (absolute, relative or partial, as appropriate) is consistently represented. So risk-aversion coefficients derived as described can be used with reasonable confidence in a function of any convenient form. In the work of research stations and extension agencies estimates of the range in risk aversion, derived consistently with the above methods and relationships for some target group of farmers, can be used in risk analysis for such specific contexts. Appropriate methods for such analyses are described in Chapter 7, this volume.

Rationalizing Risk Aversion

In Chapters 3 and 4 we argued that a rational person would want to achieve consistency in subjective probability judgements. We suggest that it is also rational to strive for consistency in attitudes to risk – a view supported by Meyer (2001). It makes no sense, for example, to be much more risk averse to a risky prospect expressed in terms of losses and gain compared with the assessment of the same prospect expressed in terms of wealth. Moreover, we also argued earlier in this chapter that extreme aversion to small losses, found by Rabin (2000) to be common behaviour, is irrational. We therefore suggest that it will usually be best to elicit a utility function (or to estimate a coefficient of risk aversion) for wealth, then to use this as a basis for assessing the appropriate risk attitude for risky prospects expressed in terms of income or losses and gains.

To apply this approach we need to understand the logical relationships between risk-aversion measures when payoffs are expressed in different but related ways.

Consistency of risk aversion across payoff measures

We start by considering the case of a decision problem with payoffs expressed as losses and gains. Under the assumption of asset integration, a loss or gain should be viewed as a change in wealth of the person experiencing that loss or gain. Hence we can write:

$$w_t = w_0 + x \tag{5.12}$$

where w_t is terminal wealth after the event, w_0 is initial wealth and x is the loss or gain. If we assume (for the moment) that either w_0 is known for sure or that x and w_0 are stochastically independent (not correlated), then we should expect a rational person to make the same choice whether the risky outcomes are expressed in terms of wealth or gains/losses.

Recall that CARA means that preferences are unchanged if a constant is added to or subtracted from all payoffs – the exact situation we have here. Therefore, if we do not want preference to change whether we express outcomes in terms of w or x, and assuming x is small relative to w so that $w_0 \approx w_t \approx w$, we can specify that:

$$r_a(w) = r_a(x) \tag{5.13}$$

In other words, we should apply a utility function with the same absolute risk-aversion coefficient to losses and gains as applies to wealth at the current level of wealth.

Taking the argument a little further, as before, $r_r(w) = wr_a(w)$ so $r_a(w) = r_r(w)/w$. Moreover, $r_r(x) = xr_a(x)$ by definition, but now we know that $r_a(x) = r_a(w)$ so:

$$r_r(x) = xr_a(w) = (x/w)r_r(w) \tag{5.14}$$

In other words, in assessing risky choices expressed in terms of losses and gain, it is not correct to apply the same relative risk-aversion coefficient as for wealth. Moreover, the smaller is x relative to w, the smaller is the applicable relative risk-aversion coefficient. The relative risk-aversion function $r_r(x)$ in Eqn 5.14 is also sometimes called the *partial risk-aversion function* since it refers to only part of the payoff as shown in Eqn 5.12.

Now let's consider risky choice where payoffs are expressed in terms of income. At least two types of risky choices affecting farm income can be imagined. One is where the income next year (or in some single year in the future) is uncertain. This is the typical situation when doing, say, annual farm planning. The uncertainty in the outcomes stems largely from the year-to-year unpredictability of yields, prices and costs that affect farmers' incomes. This type of uncertainty contrasts with longer-term uncertainty as when a farmer may be contemplating a major investment, perhaps associated with a dramatic change to the farming system. Here the uncertainty is about the long-run level of income. The distinction between the two is similar to the distinction Friedman (1957) drew between *transitory income* and *permanent income* in his work on the consumption function.

Drawing further on Friedman's ideas, it seems clear that transitory income can be treated in decision analysis very much like losses and gains. We can write:

$$w_t = w_0 + y - c_p \tag{5.15}$$

where y is transitory income and c_p is Friedman's permanent consumption, assumed constant. Defining $x = y - c_p$ converts Eqn 5.15 into Eqn 5.12, so the matter will not be pursued further since identical conclusions apply as for risk aversion with payoffs as losses and gains.

Now consider what happens when it is long-run or permanent income that is risky and the focus of attention. It seems reasonable to assume that a rational person will view their wealth as equal to the capitalized value of future (permanent) income flows with the capitalization factor calculated over the relevant future time horizon. In that case we can write:

$$w = ky \tag{5.16}$$

where w is current wealth, y is the annual permanent income and k is the appropriate capitalization factor, $k > 1$. Then, since wealth is viewed (ignoring for simplicity issues of terminal asset valuation) as a fixed multiple of permanent income (and vice versa), a rational individual will assign the same proportional risk premium to a given risky prospect whether the payoffs are expressed in wealth or in terms of the equivalent permanent income. This implies that:

$$r_r(w) = r_r(ky) = r_r(y) \tag{5.17}$$

where $r_r(.)$ is the relative risk-aversion function. In other words, the same relative risk-aversion coefficient is applicable to the current levels of wealth and permanent income.

In terms of absolute risk aversion, since $r_a(w) = r_r(w)/w$ and $r_a(y) = r_r(y)/y$, then $r_a(y) = r_r(w)/y$ or:

$$r_a(y) = (w/y)r_a(w) \tag{5.18}$$

Finally, since $k = w/y$ from Eqn 5.16, then:

$$r_a(y) = kr_a(w) \tag{5.19}$$

In other words, $r_a(y)$ is k times as large as $r_a(w)$ where k is the relevant capitalization factor. Evidently, as Meyer (2001) also notes, we cannot use the same absolute risk-aversion coefficient for long-run income as for wealth.

A suggested approach to rationalizing risk aversion

What do all the above calculations mean in practice? We suggest that the following would be a sensible way to approach the consistent assessment of risky prospects.

We start by assuming that an estimate of relative risk aversion for wealth can be obtained with reasonable confidence. It may be obtained by applying Eqns 5.7 and 5.8 to a utility function for wealth, evaluated at the current wealth level, or it may be an assumed value obtained as indicated, for example, in Table 5.3. We also assume that this coefficient will be approximately constant for all but large changes in wealth. We can therefore derive the coefficient of absolute risk aversion at the current level of wealth from Eqn 5.8, dividing r_r by w. To assess risky prospects with payoffs that are small relative to w, this value of r_a can then be used as the coefficient of a CARA utility function of the form of Eqn 5.9. On the other hand, if the risky prospects to be assessed have payoffs that are reasonably large relative to w, it will be best to evaluate them in terms of terminal wealth. That can be done using the elicited utility function for wealth or the directly determined coefficient of relative risk aversion for wealth implemented in a CRRA utility function of the form of Eqn 5.11 (or Eqn 5.10 if $r_r(w) \approx 1.0$).

By way of a postscript to the above discussion, it should have become apparent that there are likely to be very substantial difficulties in inferring anything about the appropriate degree of risk aversion if payoffs are expressed in ways other than those canvassed above. It becomes difficult indeed to see how an appropriate degree of risk aversion can be consistently derived for such measures as gross margin per hectare of crop, per kilogram of milk produced or per dollar invested. Still worse are attempts to derive an appropriate degree of risk aversion for comparing distributions of, say, crop yield per hectare or biodiversity. Yet it is not unusual to come across examples of results expressed and analysed in just such partial terms!

It is also worth noting that the measure of income used needs to be clear. Is it income from farming or total household income? Does income as measured include or exclude imputed values for own production consumed (if this is important) and for family labour or other family-owned resources used on the farm? It may be important to specify whether the income is measured before or after tax, at least for countries where there is a progressive marginal rate of tax on income. Unless there are other provisions in place to adjust the tax payable, a progressive income tax system penalizes taxpayers whose incomes fluctuate from year to year relative to taxpayers with more stable incomes. A progressive marginal income tax system therefore has some similarity with risk aversion in its impact because even risk-neutral DMs have an incentive to sacrifice some expected pre-tax income to reduce year-to-year fluctuations.

The Importance of Risk Aversion

Because of actual or perceived problems in using utility analysis, some critics have suggested that decision analysis, as expounded in this and similar texts, is too seriously flawed to be useful. That begs the question of what to put in its place, since at least as many problems can be identified for most of the proposed alternatives. However, notwithstanding the efforts set out above to rationalize risk aversion via a consistent theory of utility, we believe that the importance of risk aversion has often been exaggerated, particularly by critics of the approach. We show why below.

In Eqn 5.6 we defined the risk premium as:

$$RP = E - CE \tag{5.20}$$

where *E* is now used for EMV for brevity. As noted earlier, for a risk-averse DM, *RP* will be positive and its magnitude will depend on both the distribution of outcomes and the DM's attitude to risk. For what follows it is convenient to compute the proportional risk premium, *PRP*, defined as:

$$PRP = RP/E \tag{5.21}$$

i.e. the proportion of the expected value of the risky project absorbed by the risk premium in computing *CE*. The more risk averse is the DM or the more uncertain the risky prospect, the higher will be the *PRP*.

An indication of the implications for risky choice of different degrees of risk aversion can be obtained from the approximation (Freund, 1956):

$$CE \approx E - 0.5r_a V \tag{5.22}$$

where r_a is the appropriate absolute risk-aversion coefficient (assumed constant) and V is variance of payoff. Then the approximate risk premium is given by:

$$RP \approx E - CE = 0.5r_a V \tag{5.23}$$

Multiplying through by E/E^2 gives the proportional risk premium representing the proportion of the expected payoff of a risky prospect that a DM would be willing to pay to trade away all the risk for a sure thing:

$$PRP \approx RP/E = 0.5r_a E(V/E^2)\ 0.5r_r C^2 \tag{5.24}$$

where C is the coefficient of variation of the risky prospect, equal to the standard deviation divided by the mean. For example, if $r_r = 2$ and $C = 0.2$, $PRP = 0.04$. Similarly, if $r_r = 4$ and $C = 0.3$, $PRP = 0.18$. Note, however, that, for reasons explored above, magnitudes of r_r such as 2 or 4 are only likely to apply for risky prospects expressed in terms of permanent income or total wealth.

The impact of risk aversion will be different from the above for the more usual case of DMs assessing a *marginal* additional risky prospect (Anderson, 1989). We now assume that initial wealth w_0, equivalent to the capitalized value of future earnings without the additional activity, is uncertain. If a marginal risky prospect is evaluated in terms of gains and losses, x, relative to w_0, Anderson and Hardaker (2003) have shown that:

$$PRP_x \approx r_r(w_0)C[x]\{0.5ZC[x] + \rho C[w_0]\} \tag{5.25}$$

where PRP_x is the risk premium as a proportion of $E[x]$, $C[.]$ is the coefficient of variation, Z is the relative size of the marginal risky prospect, approximated by $E[x]/E[w_0]$ and ρ is the correlation between w_0 and x.

Table 5.4, based on the above approximation, shows the risk deduction from expected income for a situation of a somewhat typical Dutch farmer. For the basic case the farmer is assumed to have $r_r(w) = 1.0$. It is assumed that the coefficient of variation of the additional activity, $C[x]$, is 30% (relatively risky) and that it has a relative size, Z, of 0.1, based on an expected payoff equal to 10% of the expected initial level of net assets w_0. The correlation between x and w_0, ρ, is set at 0.5, while the coefficient of variation of existing wealth, $C[w_0]$, is 20%. As the table shows, the proportional risk premium, PRP_x, for this basic case is about 3.5%. In other words, if the expected value of income from the additional activity is 100, the CE of this income is 96.6. Whether this difference would lead to a different choice obviously depends on the circumstances, but such a difference might often be viewed as unimportant.

The right-hand part of the table includes some sensitivity analysis based on changing each of the above assumptions, one at a time. Perhaps most interesting is the effect of halving $C[x]$, the coefficient of variation of the new prospect. This result might be thought of as a comparison of two alternative additions to the farm, one much more risky than the other. The difference between them in terms of the index of CEs is still quite small, and in practice might well be outweighed by any difference in expected value, assumed to be zero for this calculation.

On the basis of such calculations it can be argued that risk aversion will often not be nearly as important as getting the expected value calculation right, at least in commercial farming in more-developed countries. On the other hand, it is of course possible to construct cases where risk aversion is important, for example by increasing the relative risk-aversion coefficient and the proportional size of the additional activity. With r_r at 2.0 and Z at 1.0, PRP_x is 15%.

Similarly, this analysis indicates that risk aversion should seldom be assigned much importance in public decision making. In applying Eqn 5.25 in this context, the proportional size of risk, Z, will almost always be very small since few public choices will have payoffs that will significantly affect w_0, which can now be viewed as current national wealth. Moreover, the diversified nature of most economies means that

Table 5.4. Approximate risk deductions and CEs of a marginal addition to a typical Dutch farm.

		Sensitivity[a]		
Coefficient	**Basic case**	**Base plus**	**PRP_x[b]**	**CE index**
r_r	1.000	2.00	0.069	93.1
$C[x]$	0.300	0.15	0.016	98.4
Z	0.100	0.20	0.039	96.1
ρ	0.500	0.00	0.005	99.6
$C[w_0]$	0.200	0.10	0.020	98.1
PRP_x	0.035			
Index of $E[x]$	100.0			
Index of $CE[x]$	96.6			

[a]Evaluated using the basic case values except for one at a time changes in each coefficient to the values shown.
[b]PRP_x, Proportional risk premium.

the coefficient ρ, measuring the correlation between a marginal public project and the rest of national income, will usually be small. We address some exceptions to these general remarks in Chapter 13.

The cost of ignoring risk aversion

As shown by Eqn 5.25 and illustrated above, the *PRP* may be a small proportion of the expected value for some transiently risky prospect that constitutes only a part of the risk faced by, say, a farm household. While it is an empirical matter, many such marginal risks in diversified agricultural systems may have near-zero values of *PRP*, so that choices can be based on expected values alone. Even when *PRP* values are somewhat larger, the ranking of alternatives based on expected payoffs may be the same as that based on expected utility.

To assess the costs of ignoring risk aversion we need to compare the indicated optimal choice in some risky decision when accounting for risk aversion with the choice indicated when assuming risk indifference. In other words, we need to compare the choice under the SEU maximization rule with that under the maximization of EMV. These two decisions can conveniently be compared in terms of their CEs. We can then define the cost of ignoring risk aversion, *CIRA*, as:

$$CIRA = CE_{U^*} - CE_{E^*} \tag{5.26}$$

where CE_{U^*} and CE_{E^*} are the CEs under SEU maximization and EMV maximization, respectively.

Then, somewhat analogously with our treatment of the risk premium, and assuming that both CEs are positive (which will normally be the case if the payoffs are measured in terms of terminal wealth), we can define the proportional cost of ignoring risk aversion, *PCIRA*, as:

$$PCIRA = CIRA/CE_{U^*} \tag{5.27}$$

It is apparent that *CIRA* and *PCIRA* will depend on the particular decision context, including the choice options and the degree of risk aversion of the DM. However, we can illustrate the relationships involved in an E,V context as shown in Fig. 5.5. We assume a portfolio selection problem such as the choice of a cropping and stocking plan for a farm. Thus, the set of feasible plans is effectively infinite, with near perfect divisibility of activity levels. We also assume an indirect utility function of the form:

$$U = CE = E - 0.5r_aV \tag{5.28}$$

which implies linear E,V indifference curves as shown.

From this construction it is evident that *PCIRA* will be smaller the 'flatter' are the gradients of the E,V indifference lines U_1 and U_2. The gradient of these lines is $0.5r_a$, which declines as risk aversion diminishes. *PCIRA* also depends on the shape of the E,V-efficient frontier, particularly on its gradient in the region of the two optima as well as on the location of the frontier as it affects CE_{U^*}. These are all empirical matters, making generalization impossible. However, experience shows that, at least for some Australian farm-planning applications, *PCIRA* is typically very small (Pannell *et al.*, 2000).

Given the added degree of difficulty and complexity in assessing risk aversion and incorporating it into the analysis, the problem for analysts, of course, is to distinguish *ex ante* those cases where risk aversion really matters from the bulk of everyday decisions for which it can safely be ignored. While it may be easy to distinguish cases that lie firmly in one category or the other by informed guesswork, more marginal

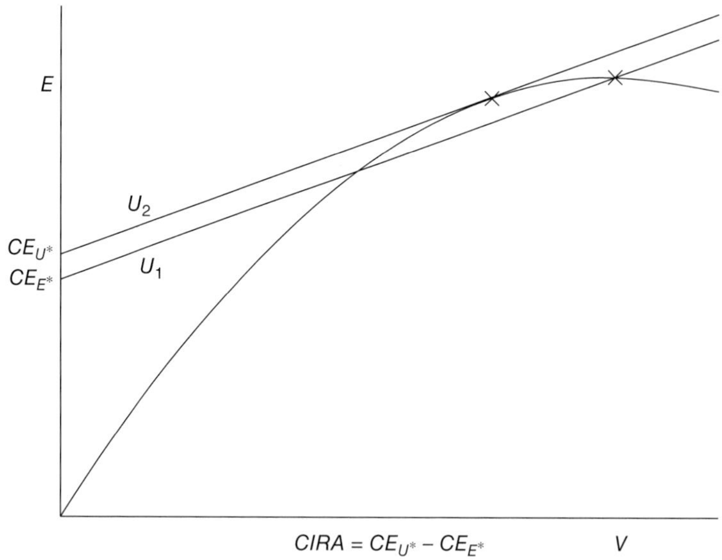

Fig. 5.5. Graphical representation of the cost of ignoring risk aversion, *CIRA*.

cases are difficult to deal with at present. As more experience is gained in evaluating *PCIRA* for a range of situations, it may be possible to make better informed guesses in future.

If empirical testing shows *PCIRA* to be small for a range of operational farm management situations, it is good news for analysts who need not worry too much about measuring risk aversion and for scientists who can focus more intently on developing technologies that improve expected returns without worrying too much about stability.

More generally, because *PCIRA* may be relatively small in many farm decisions, Anderson and Hardaker (2003) have argued that analysts have paid too much attention to risk aversion, at least relative to efforts to get good specifications and cogent updates of the probability distributions of outcomes (see also Hardaker, 2000; Hardaker and Lien, 2010). If these distributions are mis-specified or incompletely estimated, the estimate of *E* will be biased, which may matter much more than an error in calculating *RP* through using the wrong risk-aversion coefficient. Moreover, the focus on risk aversion and its importance may have been a source of confusion in that attention has been directed to reducing variability of returns rather than on finding the most risk-efficient option (erroneously minimizing *RP* rather than maximizing *CE*).

Group Decision Making

One of the important consequences of the arbitrary scale used to measure utility is that interpersonal comparisons of utility are not possible. This fact has some important implications for the analysis of any risky decision that is to be taken by a group of people such as a family, a committee or board, or a government. Related issues arise when the decision, no matter who takes it, will have significant implications for numbers of other people, and when, in reaching a decision, the DM wants to account for the impact of that choice on the preferences or welfare of those affected. It would be ideal in such situations to be able to obtain a group utility function. Unfortunately, Arrow (1963), in his famous *Impossibility Theorem*, showed that no such function can exist without violating some seemingly innocent conditions. Moreover, matters get worse when the role of subjective probabilities in decision making is recognized. As Raiffa (1968, p. 230) showed, no matter what procedure is used to combine the beliefs and preferences of the members of a group, provided only that it is not a dictatorship, it is always possible to construct an example in which members of the group all agree on what they individually see is best, yet the result of applying the rule to combine beliefs and preferences leads to a different recommendation. If the above difficulties seem insurmountable, the consolation is that many groups seem able to go on making risky choices with no apparent concern for the lack of relevant theory!

What, then, is to be done to rationalize group decisions, which, after all, are probably the rule rather than the exception in agriculture? Consideration of social choices in which policy makers reach decisions that affect large numbers of people is deferred to Chapter 13, this volume. Here we deal with decisions in which a group choice has to be made. Anderson *et al.* (1977, pp. 139–140) list some possible approaches:

1. There have been some attempts to standardize utility functions, in a rather similar fashion to the way utility functions for different attributes are standardized, as explained in Chapter 10, this volume. Though clearly in violation of Arrow's Impossibility Theorem, such methods may nevertheless be acceptable to some groups in some circumstances. An example might be when a panel is charged with ranking a set of alternatives, such as research project proposals or applicants for a job, when a simple scoring rule with equal weights may be agreed upon. Such a rule may be interpreted as a multi-person utility function.
2. If the group is well integrated, as many families are, individual members will be concerned about the welfare of the others. A single DM acting for the group may therefore be able to express such concern in terms of a higher level utility function that incorporates a personal understanding of the preferences of the others, and not just the DM's own individual preferences. It seems unlikely that such altruism will often reach the point where all inter-personal differences in preferences are eliminated, but at least such differences will be partly accommodated.
3. Groups function to make decisions all the time, often seemingly with little conflict. This observation suggests the possibility of simply seeking to elicit a utility function (and probabilities to go with it) from the group as a whole, without worrying, or even inquiring, about how the required judgements are made. Anderson *et al.* (1977) call this the 'under the boardroom door' method.
4. Finally, and related to the first option, the group may agree to use some well-defined decision rule such as majority rule, or election of a leader whose higher level preferences are to be used to make a choice.

In principle at least, the choice of one of these alternatives, or of any other that may be devised, is, of course, a matter for the group itself. A decision analyst charged with assisting the group might be able to help rationalize the procedures used by the group to articulate its preferences. The task for the decision analyst, however, is to learn how the group functions so as to be able to analyse important risky decisions faced by the group as effectively as possible.

Conclusion

Accounting for risk in the analysis of agricultural decisions is much harder than pretending it doesn't exist. Risk analysis in agriculture has stumbled in the past because of difficulties in assessing and categorizing managers' attitudes to risk. While no complete solutions to these problems are offered in this chapter, it is argued that risk aversion may not be as important for many choices as is commonly believed. Moreover, as described above, there are some 'rough-and-ready' ways to estimate the relevant range of risk aversion for some designated target group. Methods of stochastic efficiency analysis (Chapter 7, this volume) then allow at least something to be said about better and worse solutions.

Some risk analyses that have been based on brave assumptions about the degree of risk aversion have overlooked some of the complexity in making the move from utility of wealth to utility of gains and losses or the utility of income. Very few such analyses have recognized that risk aversion for permanent income is likely to be much more important than is risk aversion for transitory income.

Risk analysis is, and will remain, the art of the possible. But successful artistry needs to be founded on a good knowledge of principles and technique. Clearly there is much more work to be done, but in this chapter we have sought to show some plausible and practical ways in which the tricky issue of risk aversion can be made more operational.

Selected Additional Reading

A relatively large number of references directly relevant to our treatment are cited above, providing extensive opportunities for further reading. However, the topic of utility theory is a large one and the literature is extensive. Much of this literature deals with behavioural aspects of the theory, to which we have given little attention in this chapter. For a comparison of the different approaches, see Bell *et al.* (1988). A series of papers in Edwards (1992) generally confirms that SEU maximization is the best normative tool but is a poor descriptive tool, a view supported by Meyer (2001). Among the alternatives to the SEU model presented above are *prospect theory* (Kahneman and Tversky, 1979) and *generalized expected utility* (Quiggin, 1993). Readers seeking to model behaviour rather than to prescribe good risky choices may find these alternative theories helpful.

References

Anderson, J.R. (1989) Reconsiderations on risk deductions in public project appraisal. *Australian Journal of Agricultural Economics* 28, 1–14.

Anderson, J.R. and Dillon, J.L. (1992) *Risk Analysis in Dryland Farming Systems*. Farming Systems Management Series No. 2, Food and Agriculture Organization of the United Nations (FAO). FAO, Rome.

Anderson, J.R. and Hardaker, J.B. (2003) Risk aversion in economic decision making: pragmatic guides for consistent choice by natural resource managers. In: Wesseler, J., Weikard, H.-P. and Weaver, R. (eds) *Risk and Uncertainty in Environmental and Natural Resource Economics*. Edward Elgar, Cheltenham, pp. 171–188.

Anderson, J.R., Dillon, J.L. and Hardaker, J.B. (1977) *Agricultural Decision Analysis*. Iowa State University Press, Ames, Iowa.

Arrow, K.J. (1963) *Social Choice and Individual Values*, 2nd edn. Wiley, New York.

Arrow, K.J. (1965) *Aspects of the Theory of Risk-Bearing*. Yrjö Jahnssonin Säätiö, Academic Bookstore, Helsinki.

Bell, D.E. (1988) One switch utility functions and a measure of risk. *Management Science* 34, 1416–1434.

Bell, D.E., Raiffa, H. and Tversky, A. (1988) *Decision Making: Descriptive, Normative and Prescriptive Interactions*. Cambridge University Press, Cambridge.

Chambers, R.G. and Quiggin, J. (2001) *Uncertainty, Production, Choice, and Agency: the State-Contingent Approach*. Cambridge University Press, Cambridge.

Chavas, J.-P. (2004) *Risk Analysis in Theory and Practice*. Elsevier Academic Press, San Diego, California.

Edwards, W. (ed.) (1992) *Utility Theories: Measurements and Applications*. Kluwer, Boston, Massachusetts.

Eeckhoudt, L. and Gollier, C. (1996) *Risk: Evaluation, Management and Sharing*. Harvester Wheatsheaf, Hemel Hempstead, UK.

Farquhar, P.H. and Nakamura, Y. (1987) Constant exchange risk properties. *Operations Research* 35, 206–214.

Freund, R.J. (1956) The introduction of risk into a programming model. *Econometrica* 24, 253–261.

Friedman, M. (1957) *A Theory of the Consumption Function*. Princeton University Press, Princeton, New Jersey.

Hamal, K.B. and Anderson, J.R. (1982) A note on decreasing absolute risk aversion among farmers in Nepal. *Australian Journal of Agricultural Economics* 26, 220–225.

Hardaker, J.B. (2000) Some issues in dealing with risk in agriculture. Working Paper Series in Agricultural and Resource Economics No. 2000–3, Graduate School of Agricultural and Resource Economics (GSARE), University of New England, Armidale, New South Wales, Australia. Available at www.researchgate.net (accessed 1 June 2014.)

Hardaker, J.B. and Lien, G. (2010) Probabilities for decision analysis in agriculture and rural resource economics: the need for a paradigm change. *Agricultural Systems* 103, 345–350.

Kahneman, D. and Tversky, A. (1979) Prospect theory: an analysis of decisions under risk. *Econometrica* 47, 263–291.

Lien, G. (2002) Nonparametric estimation of decision makers' risk aversion. *Agricultural Economics* 27, 75–83.

Meyer, J. (2001) Expected utility as a paradigm for decision making in agriculture. In: Just, R.E. and Pope, R.P. (eds) (2001a) *A Comprehensive Assessment of the Role of Risk in U.S. Agriculture*. Kluwer, Boston, Massachusetts, pp. 3–19.

Moschini, G. and Hennessy, D.A. (2001) Uncertainty, risk aversion, and risk management for agricultural producers. In: Gardner, B. and Rausser, G. (eds) *Handbook in Agricultural Economics*, Volume 1A. Elsevier Science, Amsterdam, pp. 87–153.

Pannell, D.J., Malcolm, B. and Kingwell, R.J. (2000) Are we risking too much? Perspectives on risk in farm modelling. *Agricultural Economics* 23, 69–78.

Pratt, J.W. (1964) Risk aversion in the small and in the large. *Econometrica* 32, 122–136.

Quiggin, J. (1993) *Generalized Expected Utility Theory: the Rank Dependent Model*. Kluwer, Boston, Massachusetts.

Rabin, M. (2000) Risk aversion and expected utility theory: a calibration theorem. *Econometrica* 68, 1281–1292.

Rabin, M. and Thaler, R.H. (2001) Anomalies: risk aversion. *Journal of Economic Perspectives* 15, 219–232.

Raiffa, H. (1968 reissued in 1997) *Decision Analysis: Introductory Readings on Choices under Uncertainty*. McGraw Hill, New York.

Robison, L.J., Barry, P.J., Kliebenstein, J.B. and Patrick, G.F. (1984) Risk attitudes: concepts and measurement approaches. In: Barry, P.J. (ed.) (1984) *Risk Management in Agriculture*. Iowa State University Press, Ames, Iowa, pp. 11–30.

von Neumann, J. and Morgenstern, O. (1947) *Theory of Games and Economic Behavior*. Princeton University Press, Princeton, New Jersey.

6 Integrating Beliefs and Preferences for Decision Analysis

Decision Trees Revisited

In Chapter 2 we introduced the notion of a decision tree to represent a risky decision. Recall that decision problems are shown with two different kinds of forks, one kind representing decisions and the other representing sources of uncertainty. We represented decision forks, where a choice must be made, by a small square at the node, and we represented event forks, the branches of which represent alternative events or states, by a small circle at the node. We showed how a decision tree can be resolved working from right to left, replacing event forks by their certainty equivalents (CEs) and selecting the optimal branch at each decision fork.

We now return to the simple example relating to insurance against losses from foot-and-mouth disease (FMD) to show how probabilities and utilities are integrated into the analysis. For convenience, the original decision tree developed in Chapter 2 (Fig. 2.2) is repeated here as Fig. 6.1. Note that the uncertainty about the future incidence of the disease is represented in the tree by the event fork with branches for 'No outbreak' and 'Outbreak'. To measure the uncertainty here we need to ask the farmer for subjective probabilities for these two events. Suppose that, as explained in Chapter 3, the farmer assigns a probability of 0.94 to there being no outbreak and a complementary probability of 0.06 to an outbreak occurring. Similarly, the farmer is uncertain about what policy for control of the disease might be implemented if an outbreak occurs, as shown by the event forks further to the right in Fig. 6.1. Again, the farmer is able to assign some subjective values to these conditional probabilities of 0.5 and 0.5 (i.e. the farmer thinks that, should an outbreak occur, it is equally likely that the policy adopted will be 'Bans only' or 'Slaughter'). For convenience we can write these probabilities on the respective branches of the tree, as we shall illustrate in a moment.

Before that, we also need to reflect the farmer's attitude to risk. In Chapter 5, we showed how to elicit from the farmer sufficient information to estimate a utility function for terminal wealth. For the purpose of illustration, we assume that this farmer's attitude to risk is adequately reflected for this decision problem using the negative exponential utility function $U = 1 - \exp(-r_a W)$ with $r_a = 0.003658$. Using this function, the utility values of the payoffs in the decision tree above are as follows:

$U(300\text{k}) = 0.6663$

$U(490\text{k}) = 0.8334$

$U(492.8\text{k}) = 0.8351$

$U(500\text{k}) = 0.8394$

The tree is now redrawn in Fig. 6.2, showing the probabilities and utilities, and also with the 'Insure' branch pruned back since the same utility is attained from it regardless of what happens thereafter.

As before, resolution of the decision tree first requires the replacement of the terminal event fork with a single value. By the SEU hypothesis, the utility of this event fork is equal to its expected utility (EU), calculated as in Eqn 6.1:

$$0.5 \times 0.8334 + 0.5 \times 0.6663 = 0.7499 \quad (6.1)$$

We can now prune this rightmost event fork, replacing it with its utility value. The process is repeated for the remaining event fork, replacing it with its utility value, calculated from the probabilities and utilities shown as 0.8340. We are then able to prune away this event fork, ending with the simple tree in Fig. 6.3.

Now the best choice is clear. Since the utility from buying insurance is greater than that from not insuring, the farmer should purchase insurance cover.

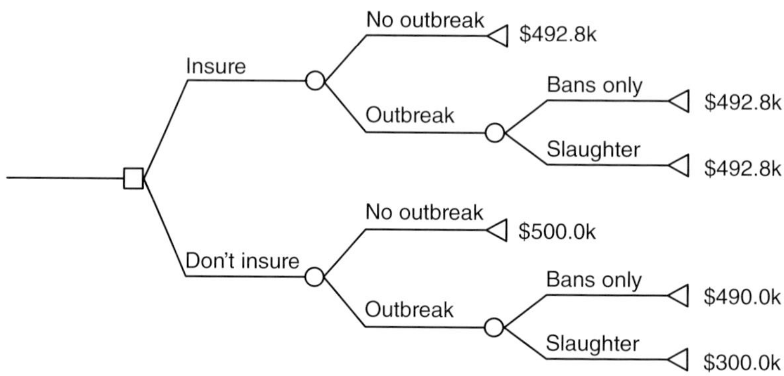

Fig. 6.1. A decision tree for the disease insurance example.

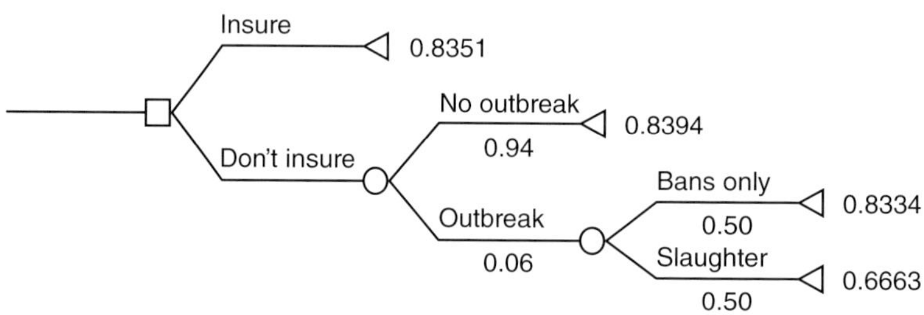

Fig. 6.2. Decision tree for insurance problem showing probabilities and utilities.

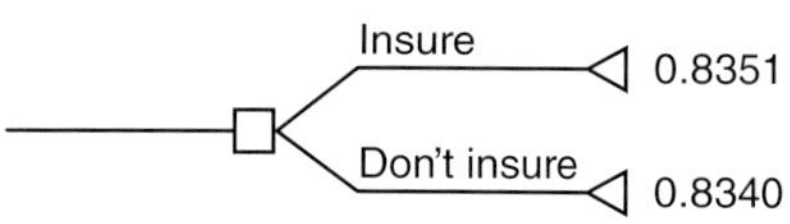

Fig. 6.3. Decision tree for insurance problem showing the insurance decision with payoffs in expected utility values.

To show the equivalence between this approach and that presented in Chapter 2 based on CEs, we can calculate the CE of the decision not to insure, with its associated uncertainty. As explained in Chapter 5, there is a one-to-one correspondence between EU and CE. For the decision not to insure, we can write:

$$1 - \exp(-r_a CE) = 0.8340 \qquad (6.2)$$

and hence

$$CE = -\ln(1 - 0.8340)/r_a \qquad (6.3)$$

which, for the value of r_a given above, is equal to $491,000, as in Chapter 2.

Resolution of decision trees using probabilities and a utility function

As illustrated with the above simple example, the method of resolving a decision tree using probabilities and utilities is as follows:

1. Assign probabilities to each event branch, making sure that these are consistent with beliefs about the uncertainties and coherent with probability principles.
2. Calculate money payoffs for each terminal node by summing costs and returns along the relevant act–event sequence.
3. Convert these money payoffs to utility values using the DM's utility function. Alternatively, utility values can be expressed as CEs, obtained from the inverse utility function or by direct assessment by the DM.
4. Working back from the terminal branches, replace each chance node by the corresponding EU or CE value. Then resolve decision forks by selecting the branch with the highest EU or CE.
5. Mark off dominated acts at each stage so that the optimal path through the tree is apparent.

The result of solving the above simple tree using the DATA program by TreeAge is shown in Fig. 6.4. Note that the overlay of the results of resolving the decision tree at each node is part of the output of the software.

The optimal choice at the first decision fork is shown to be to insure, with an EU of 0.8351. The corresponding CE is, of course, $492.8k, since insuring is assumed to give a sure payoff of this amount. The alternative branch has been marked as sub-optimal by the lines drawn across it. The method of analysis illustrated here is sometimes called *averaging out and folding back*. Event forks are averaged out to

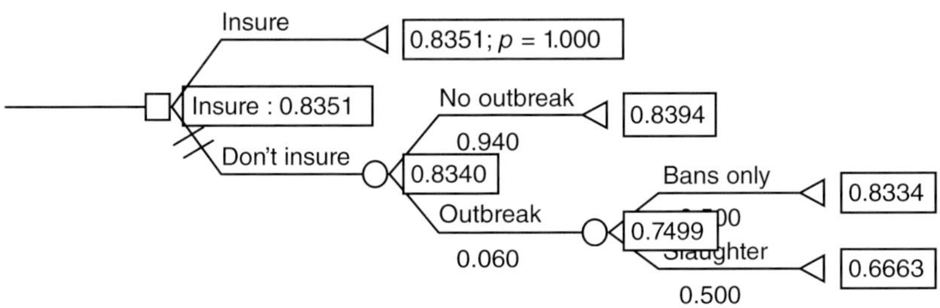

Fig. 6.4. Rolled-back decision tree for insurance decision problem produced using TreeAge DATA software.

be folded back by replacing them with EU or CE values. Branches representing sub-optimal acts are pruned away, and then decision forks that have been cut back to only one branch may be folded back using the EU or CE value of that branch. The logic used is identical to Bellman's principle of optimality (Chapter 11, this volume). It is safely assumed that the optimal choice at decision forks to the right of the tree will be unaffected by the choices made earlier in the tree. In fact, many dynamic programming problems can be expressed as decision trees, and vice versa (Chapter 11).

A more realistic example

Persisting with the issue of the feared FMD, we now assume that an outbreak has occurred in a certain region of a country. FMD is feared because it spreads rapidly among cattle, sheep and pigs and causes high losses. These losses result from the need to destroy affected animals to reduce further spread of the disease, and from other disease-control measures such as slaughter of contact herds, bans on livestock movements and preventative vaccination. To contain the overall losses and maintain animal health, it is critical to eradicate FMD as soon as possible. The national government agency that is responsible for controlling highly contagious animal diseases must choose an effective policy. As the size of the losses is small relative to the overall wealth of the society, risks can be widely spread, at least potentially, and policy DMs might therefore decide not to treat risk aversion as significant. The problem can thus be analysed using expected money values (EMVs) as the choice criterion. Policies can be compared on the basis of discounted losses, the policy with the lowest expected national economic loss being optimal.

At the beginning of an outbreak, the government agency has to make a choice between two options. The first is to attack the disease with an 'area policy', which includes tracing and killing all animals from diseased herds and also contact herds. The second option is the more risky but less expensive one called 'farm policy' whereby only herds that are diagnosed with FMD are slaughtered. Some 6 weeks after the start of an outbreak, the agency will review the success of its policy. With the area policy, it is judged that there is a probability of 0.75 that the FMD outbreak will have been eradicated. In that case, the total losses are assumed to be \$40.2 million. However, there remains a 0.25 chance that the disease will still not be under control, so that new herds are being infected. In that case, the agency has another set of options: either to continue the area policy or to start a 'vaccination policy'. The latter includes preventative vaccination of all animals in that region so that the animals are protected against contracting the disease. Vaccination is relatively costly: it involves not only vaccines but also labour to vaccinate the animals. Moreover, vaccinated animals can only be sold for slaughter, not to other farmers, as they cannot be proved to be free of the disease. If the agency continues the area policy, then uncertainty about when the disease will be eradicated is assumed to be represented by two possibilities: either the outbreak will be cleared up in about 10 weeks (incurring assumed total losses of \$59.7 million), or in about 14 weeks (with assumed losses totalling \$78.1 million). The two possibilities might be judged to be equally likely ($p = 0.5$ for both). On the other hand, a vaccination policy is assumed to result in eradication within 8 weeks with a probability of 0.8, or within 10 weeks with $p = 0.2$. The associated total discounted losses involved are taken to be \$53.3 and \$82.4 million, respectively.

Initially the agency could also opt for a farm policy. This policy is presumed to eradicate the outbreak with $p = 0.35$, and the outbreak then only costs \$15.4 million. But there is a probability of 0.65 that the disease will not be under control after 6 weeks. Then there is only one alternative left: vaccination.

This is presumed to eradicate the disease in 9 weeks (p = 0.9) or in 12 weeks (p = 0.1), with $89.7 and $129.8 million as total discounted losses, respectively.

The decision tree for this problem is shown in Fig. 6.5. It was solved using the TreeAge DATA software.

The optimal solution is shown in Fig. 6.6. As this figure shows, at the initial decision node, the expected loss with the area policy equals $44.9 million, while the expected loss of the farm policy is $66.3 million. This means that the optimal decision for the agency is to choose the area policy. Then the agency has to wait to see whether the outbreak is eradicated after 6 weeks. If it is, the job is done with a total loss of $40.2 million. If not, then the optimal decision is to apply the vaccination policy, which has a lower expected loss to eradicate the outbreak (i.e. $59.1 million) than the continuation of the area policy ($68.9 million).

Associated with this solution is a probability distribution of possible outcomes. Choice of the area policy leads to the possibility of eradication in 6 weeks with p = 0.75 and a loss of $40.2 million. The probability of not eradicating the disease in 6 weeks is therefore 0.25. This outcome requires a vaccination policy to be implemented, which has p = 0.8 of eradicating the disease in 8 weeks, with a total loss of $53.3 million.

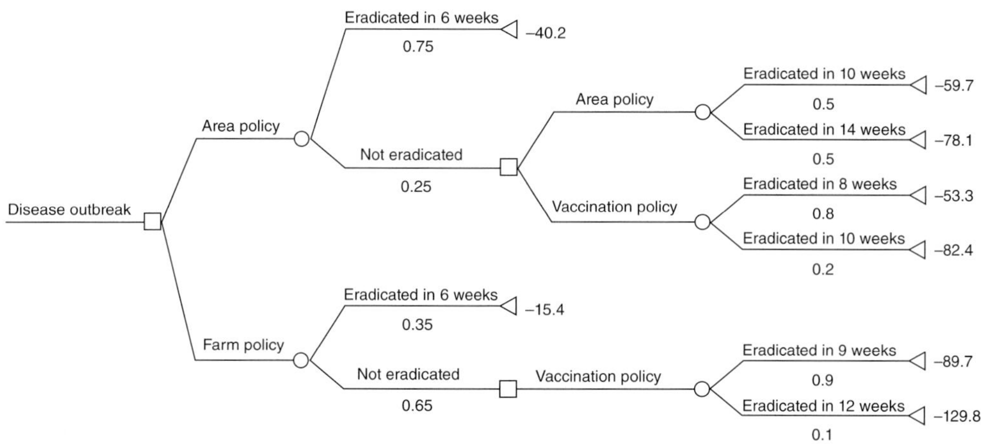

Fig. 6.5. Decision tree for foot-and-mouth disease (FMD) control problem.

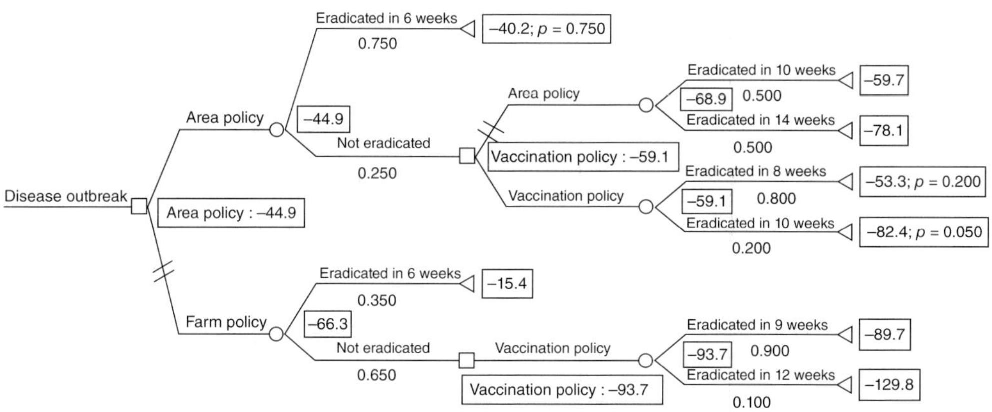

Fig. 6.6. Solution of the decision tree for FMD control problem.

The probability of reaching this terminal node in the tree is 0.25 × 0.8 = 0.2. Finally, vaccination will be effective only after 10 weeks with $p = 0.2$ and with a loss of \$82.4 million. The probability of reaching this terminal node in the tree is 0.25 × 0.2 = 0.05. These probabilities and associated outcomes can be arranged as a cumulative probability graph as in Fig. 6.7. A graph of this kind for a discrete distribution is called a cumulative mass function (CMF) and is equivalent to the more familiar cumulative distribution function (CDF) for a continuous distribution.

Note that this solution to the decision tree provides not only a resolution of the initial choice problem of whether to adopt an area policy or a farm policy, but also provides information on the best choice later in the tree, according to the outcome of the risky event forks. Such a set of 'if–then' rules comprises a *decision strategy*. This ability to generate such a strategy is one of the strengths of this form of analysis. Indeed, it is necessary to have the full strategy in order to decide what to do initially. On the other hand, once the initial decision has been made, and supposing that an area policy is chosen, as was found to be optimal, it would be possible to reconstruct the relevant sub-tree in more detail to better model the subsequent decisions and possible events. For example, one or more intermediate decision stages before the 6-weeks point could be introduced, or the eventual uncertainty of the policy options could be reflected more precisely with more branches drawn for the terminal chance nodes.

What to do about 'bushy messes'

In practice, many decisions are more complicated than the two examples given above might imply. Both decision and event forks may have many more than two branches, and some trees may extend over many more decision stages than those illustrated. For example, in the disease insurance problem, the farmer may have the option to purchase a range of levels of insurance cover, with or without a deductible. Some decision variables and some uncertain quantities bearing on the decision outcomes may, in fact, be continuous, implying an infinite number of possibilities. Clearly, it is impossible to draw a decision tree with an infinite number of branches. In all such cases, simplifications, often of major proportions, have to be made.

Attempts to take account of all the complexity that surrounds a typical real decision problem soon results in a decision tree with so many branches that it becomes what Raiffa (1968) aptly called 'a bushy mess'. Indeed, such trees may be so bushy that they are practically impossible to draw. So the puzzle is, what to do about bushy messes?

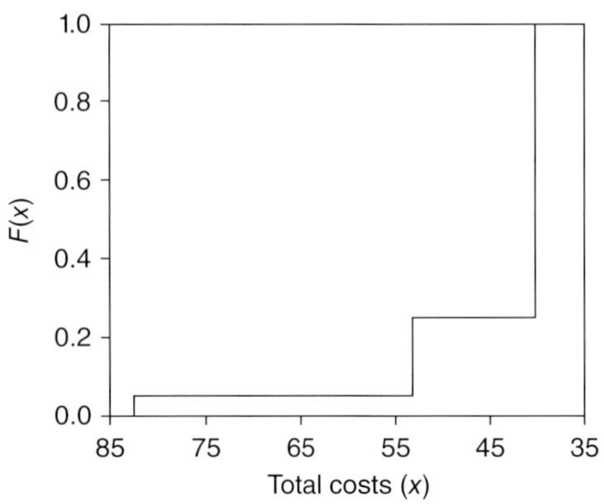

Fig. 6.7. Cumulative probability distribution associated with the optimal strategy for FMD control (costs in million dollars).

As in so much of decision analysis, the solution requires the analyst to exercise a degree of judgement and common sense. The number of branches at each fork may be reduced to just two or three, and some future decision options or less important sources of uncertainty may be left out completely. There are, however, a couple of guiding principles to keep in mind when making these, essentially subjective, simplifications.

The purpose of analysing a decision tree is to resolve the immediate decision, not to work out a decision strategy for the present and all future circumstances and choices. After the first decision has been taken, the tree can be reformulated, if appropriate, to decide what to do next, as discussed above. This means that the representation of the decision problem needs to be just good enough to allow the immediate choice to be the right one. Later choice options and later events need to be represented in the tree only with as much precision as is needed for that purpose. In particular, the further into the future are decision and event forks, the less likely is it that they need to be represented with great precision to get the immediate choice right.

There are other dimensions of 'just-good-enough' thinking that can assist in reducing the cost of decision analysis, which generically might be termed 'sensitivity analyses'. In the spirit of what is suggested above, acts and events, especially those later in the tree, might in the first instance be deliberately specified crudely, spanning the realms of the possible but with highly simplified representations. These initial assumptions can then be varied to explore their impact on the initial decision. Such sensitivity analysis can readily be undertaken in a systematic way, especially with the aid of modern software, so that changes to each of the key parameters in turn can be made to find how the results are affected. Or sensitivity analysis can be used to find *break-even values*, also called *change points*, at which different decisions are found to be optimal. On the other hand, a more subjective or ad hoc approach to sensitivity analysis may be valuable in indicating the directions for refinements that will improve the accuracy and cogency of an analysis, and also in indicating where further refinement is unnecessary. When the cost of being somewhat wrong is low, there is clearly little to be gained from further efforts to refine assumptions or the structure and detail of the decision tree.

A special situation prevails when there is some systematic structure to repeated stages of a decision tree. For example, in the FMD control example it would be possible to review the degree of success of a particular strategy on a week-by-week or day-by-day basis, not merely at the time intervals represented in Fig. 6.6. In some situations, such repetitive decisions may be modelled as a Markov process, as described in Chapter 11.

Outline decision trees

Real decision problems are often complex in structure and considerable effort may be involved in understanding and describing this structure. Moreover, as just discussed, the result may be a tree that quickly becomes a bushy mess. At the initial stage it is often useful to represent the decision by an *outline decision tree* in which some or all acts and events are represented conventionally by *decision fans* or *event fans*, respectively, each showing only one of the many possible branches. For example, a farmer in a mixed farming area has to decide at the start of the farming year what areas of various crops to plant and what numbers of different types of livestock to keep in the face of uncertain seasonal conditions, yields and product prices. After the initial farm plan has been implemented, the seasonal conditions become known, and the farmer has some scope to vary the farm plan, for example: (i) by increasing or reducing livestock numbers; (ii) by using more or less input on crops and pasture; and (iii) by conserving more or less fodder. After these adjustments are made, the eventual levels of output and sale prices become known. At each

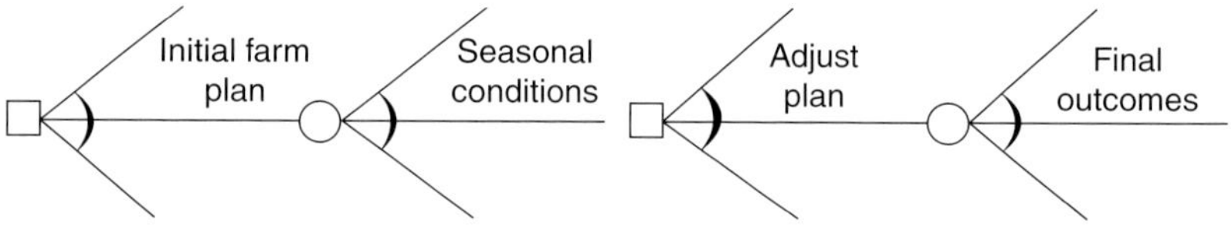

Fig. 6.8. Example of an outline decision tree.

decision and chance node there are too many branches to draw explicitly. However, the dynamic structure of the decision problem is made clear by the outline decision tree in Fig. 6.8.

Outline decision trees enable the structure of the decision problem to be clarified, especially in terms of the sequence of acts and events.

Later, if appropriate, an outline decision tree can be developed into a full tree that includes decision and event forks with a sufficient number of discrete options to allow the choice problem to be analysed adequately. Alternatively, the process of clarifying the structure of a complex decision problem using an outline decision tree may be an important preliminary step to some other form of decision analysis such as stochastic simulation or mathematical programming methods.

Payoff Table Approach

General approach

Many simple decision trees can equivalently be represented in tabular format. A payoff table can be constructed showing the alternative choices (acts) as column heads, the possible events or states as row heads, with their associated probabilities, and the payoffs for each act–event combination in the body of the table. The construction of such tables as computer worksheets simplifies calculation.

For example, the payoff table for the FMD insurance example is illustrated as an Excel worksheet in Fig. 6.9.

Naturally, the indicated optimal decision is as found in the decision tree analysis above. The choice between the specification of the problem as a decision tree or as a payoff table is a matter of convenience and personal preference. However, the payoff table approach may be less adaptable to more complex situations. When a problem has several stages, such as the FMD control problem illustrated in Fig. 6.5, it is usually best represented as a decision tree to show clearly the sequence of decisions and events.

Incorporating new information

In Chapter 3 we discussed the importance of making subjective prior probabilities as 'objective' as possible by collecting additional information. Yet information-gathering usually has a cost – either a direct cost, or a cost of delaying the main decision. Moreover, in advance of collecting the new information, it is impossible to know for sure what may be discovered. A decision to buy new information, in the form of a forecast, experiment, survey, etc., is just another decision under uncertainty and can be analysed in

	A	B	C	D	E
1	**Expected money value analysis**				
2	In terminal wealth in \$ thousands with $W = (W_0 + X)/10^3$				
3	Events		Probability	Don't insure	Insure
4	No FMD		0.94	500.0	492.8
5	FMD + bans only		0.03	490.0	492.8
6	FMD + slaughter		0.03	300.0	492.8
7			EMV	493.7	492.8
8					
9	**Expected utility analysis**				
10	Utility function $U = 1 - \exp(-r_a W)$,				
11			r_a = 3.658E-03		
12	Events		Probability	Don't insure	Insure
13	No FMD		0.94	0.8394	0.8351
14	FMD + bans only		0.03	0.8334	0.8351
15	FMD + slaughter		0.03	0.6663	0.8351
16			EU	0.8341	0.8351
17			CE (\$$10^3$)	491.0	492.8

Fig. 6.9. Worksheet of disease insurance example. CE, Certainty equivalent; EMV, expected money value; EU, expected utility.

the same way as other risky choices. We can illustrate what is involved by returning yet again to the problem faced by the dairy farmer of whether or not to insure against FMD.

Decision tree representation

As discussed in Chapter 4, the dairy farmer has the opportunity to buy information about the chances of an FMD outbreak in the coming year. The farmer is wondering whether she should buy the information, in the form of a prediction from a professional consulting organization, at a cost of \$3000. The farmer will be willing to buy the prediction only if the EU with the extra information (after accounting for the cost of obtaining it) is greater than without the information (i.e. 0.8351, see Fig. 6.9). How can the farmer solve this problem? Again, a decision tree turns out to be a valuable tool, as can be seen in Fig. 6.10.

Recall also from Chapter 4 that, if the farmer buys a forecast, one of the following three predictions is obtained: 'unlikely' (z_1), 'possible' (z_2) or 'probable' (z_3). In Chapter 4 we saw how the farmer could update personal prior probabilities in the light of these predictions. The decision tree includes both the marginal probabilities of the forecasts, i.e. $P(z_1) = 0.570$, $P(z_2) = 0.306$ and $P(z_3) = 0.124$, as well as the calculated posterior probabilities. The latter are summarized for convenience in Table 6.1.

The utility values at the far right of the decision tree are calculated on the basis of the respective terminal wealth values accounting for the cost of obtaining the prediction. For example, after receiving a prediction z_1, the decision to insure would result in a terminal wealth position of \$492.8k, less the costs of obtaining the prediction of \$3000, or a net position of \$489.8k. This is independent of whether or not an outbreak occurs and what measures might be put in place by the government agency to control the outbreak. Using the same utility function as before, the utility value of \$489.8k is calculated as:

$$1 - \exp(-0.003658 \times 489.8) = 0.8333 \qquad (6.4)$$

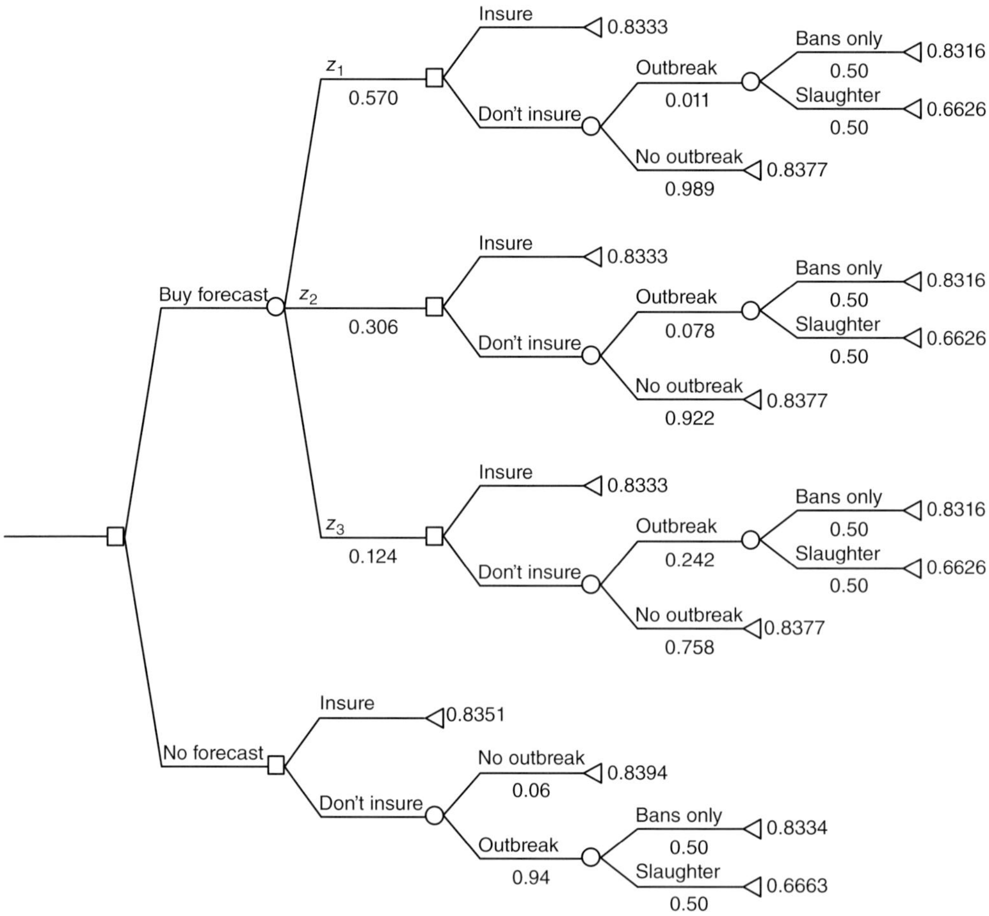

Fig. 6.10. Decision tree for the disease insurance example including purchase of a forecast.

Table 6.1. Prior and posterior probabilities for the dairy farmer.

State of nature	Prior $P(S_i)$	Posterior given prediction $z_1 P(S_i\|z_1)$	Posterior given prediction $z_2 P(S_i\|z_2)$	Posterior given prediction $z_3 P(S_i\|z_3)$
No outbreak (S_1)	0.94	0.989	0.922	0.758
Outbreak (S_2)	0.06	0.011	0.078	0.242

Likewise, the decision not to insure after buying a forecast, followed by an outbreak with bans imposed, would give a terminal wealth position of \$490k less the \$3k for the forecast, or \$487k, which has a utility value of 0.8316.

The next step is to determine the optimal solution, that is, should the farmer buy the prediction at a cost of \$3000, and if yes, under what circumstances (after which predictions) should insurance then be purchased? To answer these questions we have to calculate the EU for each decision strategy. If the decision

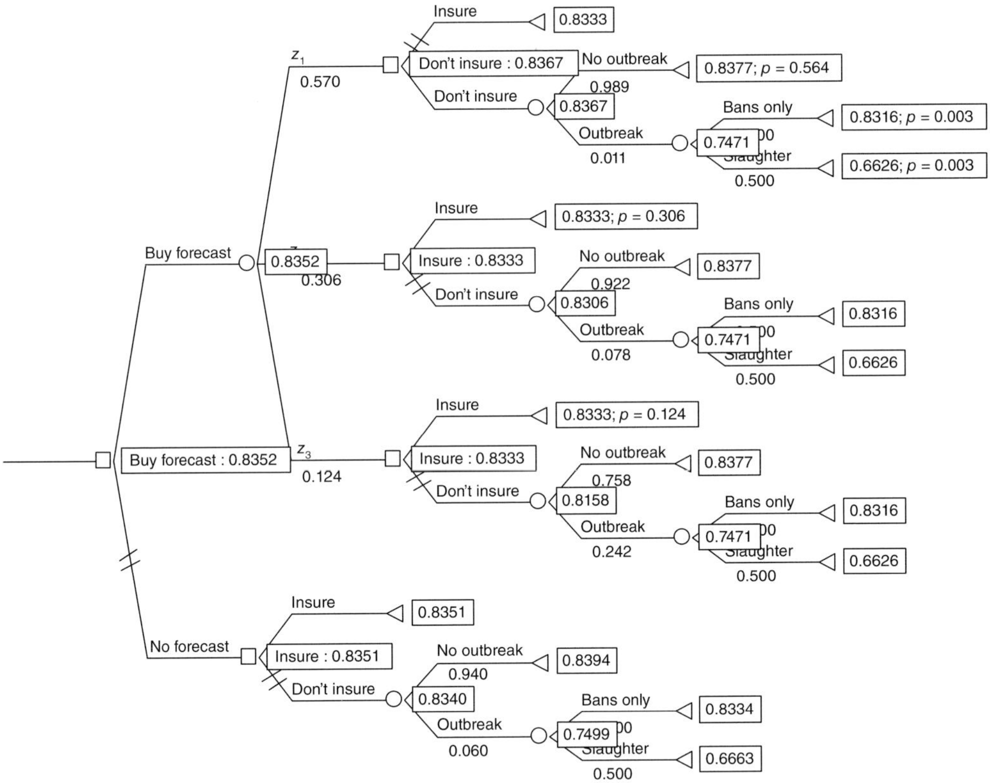

Fig. 6.11. Solution of decision tree for the disease insurance example including purchase of a forecast as produced by TreeAge DATA software.

tree is solved (averaging out and folding back with a click of the mouse using TreeAge DATA) we get the solution shown in Fig. 6.11.

Buying the forecasts results in an EU of 0.8352, just higher than the utility of not buying the information (i.e. 0.8351). So, the farmer should buy the forecast and then wait to see what prediction (z_1, z_2 or z_3) is received. If the prediction is 'unlikely' (z_1), the optimal decision is not to insure, which gives an EU of 0.8367 compared with 0.8333 for insuring. If the farmer gets the prediction 'possible' (z_2), or 'probable' (z_3), the best action is to insure, with the utility consequences shown in Fig. 6.11. In summary, the optimal (EU maximizing) actions are to buy the additional information and then not to insure if z_1 is observed, but to insure if z_2 or z_3 is observed.

Stochastic Simulation

While decision tress and payoff tables are useful for the analysis of some risky choice problems, others, particularly but not solely those where the uncertain variables are best modelled with continuous probability distributions, may be best handled with *stochastic simulation*.

Another name, *stochastic budgeting*, is often used with much the same meaning as stochastic simulation. A stochastic budget will typically have a deterministic equivalent in the form of a conventional budget under assumed certainty, while a stochastic simulation model may or may not have a deterministic equivalent. In other words, stochastic budgeting can be regarded as a sub-category of stochastic simulation.

In the following sections, stochastic simulation is introduced and illustrated with a simple example. Finally, the way stochastic simulation can be combined with optimization is briefly outlined.

Stochastic Simulation as a Tool in Decision Analysis

Scope and principles

Simulation may be defined as the use of an analogue in order to study the properties of the real system. For the kind of simulation of concern here, the analogue is in the form of a mathematical model whereby the real system is represented in the form of a set of equations and parameters. Such simulation models are commonly used to analyse so-called 'what-if' questions about a real system. Such a model typically represents the relationships between the inputs and outputs of the real system and allows for the effects of changing control or decision variables to be explored. The method is sufficiently flexible to allow the incorporation of complex relationships between variables and hence to mimic aspects of the performance of complex real systems such as exist in agriculture.

In stochastic simulation, selected variables or relationships incorporate random or stochastic components (by specifying probability distributions) to reflect important parts of the uncertainty in the real system.

As described in Chapter 4, coping with variables that are stochastically dependent means that the joint distribution has to be elicited or approximated. The assessed joint distribution can then provide the stochastic component of the simulation model that can then be used to generate the probability distributions of the selected output variables or consequences of interest.

Repetitive *Monte Carlo sampling* is used to generate values from specified input distributions. Each evaluation of the model with a set of random drawings from the specified distributions is called an *iteration*. At each iteration a set of samples is obtained representing a possible combination of values of the specified stochastic elements that could occur. The resulting values of the output variables are computed and stored. With enough iterations, the distributions of each of the output variables will converge to a stable distribution. Experiments can be performed to repeat the evaluation for different settings of the decision variables.

Since it is often difficult, if not impossible, for a DM to get a complete intuitive overview of all possible consequences of all but the simplest real-life risky decisions, a stochastic simulation model can help to make a systematic assessment of what might happen. Typically there are many inputs, interactions and non-linearities, all combined with uncertainty and variation in a complex way. For example, one might want to analyse a grazing system on a cattle farm, accounting for: (i) uncertain weather; (ii) seasonal pasture growth; (iii) interaction of pasture and grazing stock; (iv) impacts of nutrition on animal production; (v) uncertain output and prices; and (vi) the impact of all this on management decisions. In cases such as this, stochastic simulation is often the only way to model the complexity.

The purpose of stochastic simulation in risk analysis is to determine probability distributions of consequences for alternative decisions to enable the DM to make a good, well-informed choice. A common approach is to simulate the consequences of a range of alternative decisions in order to be able to compare the outcome distributions. In many business and economic decision situations, the outcome of each decision alternative can be reflected by the distribution of a single performance measure, such as terminal wealth or income. When an appropriate utility function is available, the stochastic consequences can be distilled down to a single measure of the utility or CE for each choice alternative analysed. If no utility function is available, the simulated outcome distributions might be compared, for instance, by applying the stochastic efficiency methods explained in Chapter 7. When there is more than one output measure of interest, as might apply, for example, when environmental considerations are important, the multi-attribute methods discussed in Chapter 10 may be appropriate.

Stochastic simulation software

The practice of stochastic simulation has been made much easier following the availability of specialist stochastic simulation software either as stand-alone products or as add-ins for spreadsheet software such as Microsoft Excel. Typically such software provides options for stochastic sampling from a wide range of probability distributions used to describe uncertain values. Packages vary somewhat in how well they cope with stochastic dependency, but most accommodate at least some aspect of dependency. Typically, relevant statistics of the input and output distributions are generated and reported, often accompanied by a graphical interface.

It can be an advantage to select simulation software that works as an add-in to a familiar spreadsheet program such as Excel. Many analysts are likely to be already familiar with the latter package, making the transition to stochastic simulation easier than having to learn a whole new application. On the other hand, specialist stand-alone simulation software has some advantages, notably the capability to allow the development of the simulation model on the computer screen using system dynamics graphical notation. In either case, an option that might be desirable is to be able link the developed model to some form of optimizing algorithm, as described below.

A simple example

The example presented here has been implemented using @Risk from Palisade Corporation. To illustrate stochastic simulation using @Risk with Monte Carlo sampling, we start with a simple budgeting example. Recall the dairy farmer whose subjective distribution of milk yield per cow was elicited in Chapter 3. This farmer is now interested to know the distribution of gross margin per cow. Gross margin (GM) is defined as (milk yield × price) + net income from stock sales – variable costs. The farmer believes that the variable costs are not too uncertain, so is prepared to treat these as deterministic for the analysis, held at $1500 per cow. But future milk price and average milk yield per cow are perceived to be uncertain.

A stochastic budget for this simple problem is shown in Fig. 6.12. This budget was drawn up using @Risk for Excel. In addition to the uncertainty about the milk price, the budget takes account of

	A	B	C
1	**Example stochastic budget**		
2	Dairy cow gross margin		
3			*Formulae used:*
4	Milk production per cow (l)	7030	*=1000*RiskCum(6.7,7.4,{6.87,…,7.23}{0.1,…,0.9})*
5	Milk price (cents/l)	45.33	*=RiskTriang(0.4,0.46,0.5)*100*
6	Gross income from milk ($ per cow)	3187	*=B4*5/100*
7	Animal sales less depr ($ per cow)	600	
8	Gross income ($ per cow)	3787	*=B6+B7*
9			
10	Variable costs ($ per cow)	1500	
11			
12	Gross margin ($ per cow)	2287	*=RiskOutput() + B8-B10*

Fig. 6.12. Worksheet of dairy cow gross margin budget for @Risk analysis (variables are set at their mean values in column B and formulae used in column B are given in column C).

uncertainty about the average milk yield per cow. The farmer is assumed to be willing to describe the uncertainty in milk price in terms of a triangular distribution with a mode of 46 cents/l and lowest and highest possible values of 40 cents/l and 50 cents/l, respectively. For the present exposition, stochastic dependency between any of these uncertain quantities is ignored, as is uncertainty about the calving interval.

In Fig. 6.12 the @Risk functions used to represent the uncertain quantities are indicated in column C. For example, the function for milk yield per cow in cell B4 is defined in terms of the cumulative probabilities reported in Table 3.1. The milk price in cell B5 is represented by the triangular distribution mentioned above. In both cases, @Risk puts the means of these distributions into the corresponding cells until the simulation is started.

The results of running this stochastic budget are shown in Fig. 6.13 (1000 iterations with Monte Carlo sampling). It may be seen that the mean gross margin per cow is $2295 with a standard deviation of $165. These statistics will vary somewhat for other subsequent simulations with the same input distributions, owing to sampling differences. They could be estimated with more precision by increasing the number of iterations. The @Risk package provides some guidance on the number of iterations needed to attain a required degree of reliability in the results based on the rate of change of, say, the mean gross margin as the number of iterations increases.

Other statistics are shown in the figure. In particular, the figure contains a range of fractile values, permitting the CDF for gross margin per cow to be drawn. In fact, such a graphing option is available within @Risk. The graph produced is shown in Fig. 6.14.

Now consider an extension of the above simple model to an application in decision analysis. We suppose that the dairy farmer has just sold a previously owned dairy farm for urban development for a handsome price. There is now a sum of about $1 million available to invest after all outstanding debts have been paid, and the farmer is wondering how to invest it. Because of previous experience in dairy farming, the option of most interest is to buy another farm of the same type. However, the DM is unsure how big a farm to buy and wants to compare farm purchase with other non-farm investments, such as fixed-interest bonds that would return about 5% a year.

A study of the market prices for dairy farms suggests that, for a million dollars, the DM could afford to buy a farm that would run about 65 cows. However, by borrowing, a larger farm of up to around 95 cows could be afforded. The DM is a little anxious about borrowing because of uncertainty about possible changes in interest rates and about the profitability of dairying over the establishment phase of the new

Output from @Risk after 1000 iterations			
Variable type			
Name	Gross margin per cow	Milk production per cow	Milk price
Description	Output	Cumul(6.7,7.4,{6.87,...,7.23})	Triang(0.4,0.46,0.5)
Cell	Sheet1!B12	Sheet1!B4	Sheet1!B5
Minimum	1854.657	6.701	0.402
Maximum	2730.102	7.400	0.499
Mean	2294.920	7.034	0.454
Std deviation	164.805	0.143	0.021
Variance	27160.69	2.04E–02	4.53E–04
Skewness	–0.1086	0.3840	–0.1958
Kurtosis	2.5576	3.0536	2.3911
Errors	0	0	0
Mode	2319.519	7.012	0.462
5% Perc	2002.258	6.806	0.416
10% Perc	2061.338	6.878	0.424
15% Perc	2112.811	6.899	0.430
20% Perc	2153.697	6.924	0.435
25% Perc	2186.324	6.940	0.439
30% Perc	2213.718	6.955	0.443
35% Perc	2234.995	6.977	0.447
40% Perc	2254.676	6.992	0.449
45% Perc	2278.102	7.003	0.452
50% Perc	2301.485	7.016	0.455
55% Perc	2318.812	7.030	0.458
60% Perc	2339.572	7.047	0.461
65% Perc	2361.210	7.068	0.463
70% Perc	2386.384	7.084	0.466
75% Perc	2411.000	7.111	0.470
80% Perc	2435.496	7.145	0.474
85% Perc	2470.270	7.186	0.478
90% Perc	2507.796	7.236	0.483
95% Perc	2565.588	7.315	0.488

Fig. 6.13. @Risk simulation results for dairy cow gross margin budget (1000 iterations with Monte Carlo sampling). Perc, percentile values.

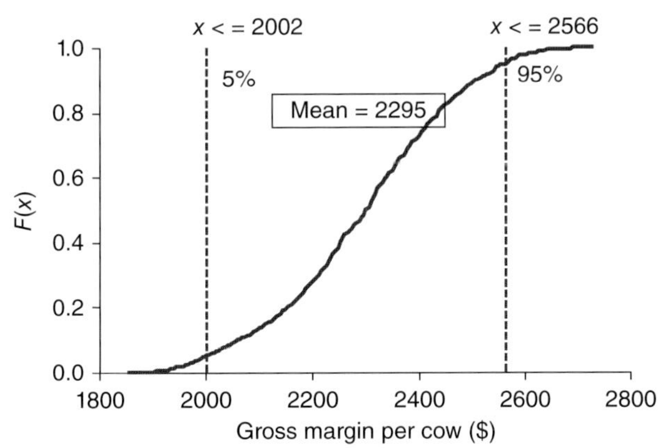

Fig. 6.14. Cumulative distribution function (CDF) of gross margin per cow from @Risk output (1000 iterations).

farm business. While it would be possible to get a loan at a fixed interest rate to avoid interest rate risk, the cost is naturally higher and the farmer decides not to consider this option initially. After taking some advice, the DM assesses that uncertainty about future interest rates on any loan is well represented by a log-normal distribution with a mean of 6% and a standard deviation of 1%.

A stochastic budget formulated for @Risk with Excel for the decision about the size of farm to buy is shown in Fig. 6.15. In this figure, all values are set at their respective means. This budget is linked to the stochastic budget for gross margin per cow, described above, that accounts for the farmer's uncertainty about milk yield and milk price. The gross margin budget is in Sheet 1 of the same Excel workbook. As shown in Fig. 6.15, the additional source of risk arising from uncertainty about interest rates is included in this part of the budget.

A Monte Carlo simulation was run to find the effects on the two measures of performance, return on total capital and return on equity. The farmer's main concern is naturally with the latter. To get a relatively reliable estimate of the full distributions of these output variables, the simulation was run for 5000 iterations for each of the alternative farm sizes of interest. In Fig. 6.15 the entry in cell B3 of the @Risk function 'RiskSimtable' followed by a list instructs @Risk about which farm sizes to use in each of three simulation comparisons. When comparing the three alternative farm sizes the random generator used in the simulation process is automatically seeded to ensure that the same sequence of random samples is drawn for each run. This makes comparison between the simulated results more reliable.

Results are summarized in Table 6.2 and the CDFs of rates of return on equity are presented in Fig. 6.16. The results show that the farmer can expect to average about 6.4% return on equity invested in dairy farming if the farm size is kept to about 65 cows, borrowing almost nothing. Relatively small increases in this mean rate can be achieved by buying a larger farm, but only at the cost of increased risk, as illustrated by the more widely spread CDFs in Fig. 6.16. The reason for the lack of much advantage from moving to a larger size of farm is clear from Table 6.2. The mean rate of return on total capital is barely higher than the expected rate of interest on a loan, so that there is only a small margin to provide an extra reward to the equity holder for the increased risks of borrowing. Given these results, the farmer

	A	B	C
1	**Debt servicing capacity according to herd size**		
2			Formulae used:
3	Herd size	65	*=RiskSimtable({65,80,95})*
4	Investment ($'000)	1010	*=100+14*B3*
5	Equity ($'000)	1000	
6	Debt ($'000)	10	*=IF((B4-B5)>0,B4-B5,0)*
7	Interest rate on loan	0.06	*=0.01*RiskLognormal(6,1)*
8	Interest on debt ($'000)	0.6	*=B6*B7*
9			
10	Gross margin per cow ($'000)	2.287	*=Sheet1!B12/1000*
11	Total gross margin ($'000)	148.66	*=B3*B10*
12	Less overheads ($'000)	84.50	*=B3*1.3*
13	Farm net income ($'000)	64.16	*=B11-B12*
14	Return on capital (%)	6.35%	*=B13/B4*
15	Margin after interest ($'000)	63.56	*=B13-B8*
16	Return on equity (%)	6.36%	*=B15/B5*

Fig. 6.15. Stochastic budget to analyse the effect of size of dairy farm on returns on capital and equity (variables are set at their mean values in column B, and formulae used in column B are given in column C).

Table 6.2. Summary of results from @Risk analysis of the dairy farm purchase problem (5000 iterations with Monte Carlo sampling).

	Herd size		
	65	**80**	**95**
Return on capital			
Mean	6.38%	6.50%	6.59%
Standard deviation	1.03%	1.05%	1.06%
Lowest	3.28%	3.34%	3.38%
Highest	9.26%	9.43%	9.55%
Return on equity			
Mean	6.38%	6.61%	6.83%
Standard deviation	1.04%	1.29%	1.57%
Lowest	3.25%	2.63%	1.92%
Highest	9.30%	10.44%	11.69%

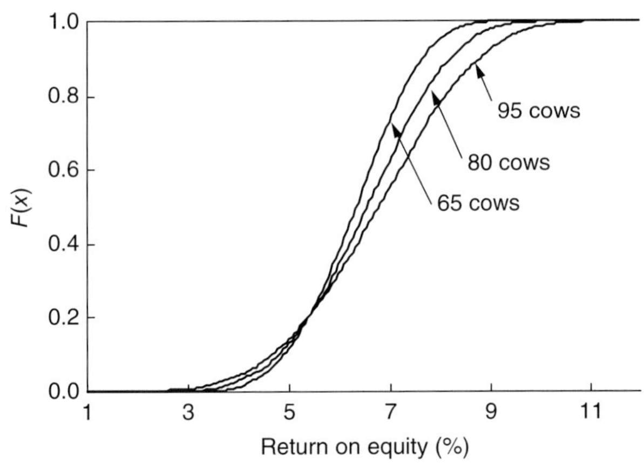

Fig. 6.16. CDFs for rate of return on equity from investing in dairy farms of different size.

might well be advised to buy a farm of about 65 cows with a relatively secure return on the investment of some $64,000 on average, but the final choice would be up to the farmer.

Combining stochastic simulation and optimization

One disadvantage of stochastic simulation is that direct optimization is generally not possible. It is necessary to use some search procedure in conjunction with the stochastic simulation model to find better settings of the decision variables. Many search methods exist for this purpose, such as: (i) gradient-type methods (quasi-Newton, etc.); (ii) direct-search methods; (iii) evolutionary algorithms (genetic algorithms, etc.); (iv) simulated annealing; and (v) tabu search (e.g. Spall, 2003). The gradient-type and direct-search methods have problems with local optima in cases with discontinuous surfaces and with surfaces containing multiple optima. Popular among the search methods are various evolutionary algorithms (Mayer *et al.*, 2005) and especially the genetic search algorithms (Mayer *et al.*, 1998; Cacho and Simmons, 1999). Simulated annealing and tabu search are 'metaheuristics', or higher level procedures that use lower level search methods to find good approximations to the unknown global optimum. Simulated annealing is especially useful when the search space is discrete. Tabu search starts with a potential solution to a problem and examines immediate neighbouring solutions in the hope of finding an improved solution.

The basic idea of genetic search algorithms is to mimic to some extent the process of natural selection via survival of the fittest. Solutions are generated by a process that involves mutation, crossovers, selection and inheritance. A set of solutions is generated by some partly random process, then these are culled, keeping only the better ones, 'offspring' solutions are then generated, typically as mixes of these culled solutions plus some more random variation, and the process is repeated. The expectation is that, after a sufficient number of generations, the best solutions will be close to the true optimum.

There exist a number of commercial software packages that implement genetic search algorithms, such as the software package RiskOptimizer provided by Palisade which works in conjunction with @Risk as a Microsoft Excel add-in. ModelRisk includes the OptQuest optimization package that combines several search algorithms into a single, composite search algorithm. The available products tend to evolve and change over time and we have chosen not to review or illustrate any here.

Concluding comments

Stochastic simulation (including stochastic budgeting) is a very flexible technique to support risky decision making. The availability of powerful PC-based software, such as @Risk and RiskOptimizer and ModelRisk and the ModelRisk OptQuest, has greatly added to the scope to apply stochastic simulation to deal with decision problems under uncertainty.

The form of model developed is restricted in principle only by the imagination of the analyst. In practice, however, as always, the costs and benefits of model development and use need to be considered. Too much complexity makes a model difficult to build, debug and use, and may give results very little better than could have been obtained from a simpler representation. It is therefore best to keep the model as simple as is judged reasonable. It is important to be critical in choice of stochastic variables in the model – too many make it complicated to account for stochastic dependencies between variables. The intention with simulation models is not to give exact answers, but to highlight relative consequences of different alternatives. Hence it is often appropriate to focus on only the main sources of uncertainty affecting outcomes.

Stochastic simulation optimization is an alternative to mathematical programming (Chapter 9) and dynamic programming (Chapter 11). Its strength is its great flexibility in use. Its main drawback is that searching for an optimum with a large model can be time consuming unless a powerful mainframe computer can be used, and there is no guarantee that the true optimal solution will be found.

Selected Additional Reading

Probably the best exposition of decision trees for analysis of risky decisions is that by Raiffa (1968). See also Clemen and Reilly (2014). There are also some useful hints to be found in the manuals for relevant software packages such as DATA from TreeAge Software Inc. or Precision Tree from Palisade Corporation. These manuals also contain information on how to extend the basic forms of analysis described in this chapter, which only hints at the full power of decision tree analysis. Vose (2008) is a good book for readers who want to learn more about stochastic simulation methods, with particular reference to the ModelRisk software. A somewhat more technical book is Law and Kelton (2000). Winston and Goldberg (2004) provide a broad treatment of operations research, including simulation approaches. A more applied book covering the same topics is Winston *et al.* (2000). Anderson (1974) reviewed earlier applications of stochastic simulation in agricultural economics, and Oriade and Dillon (1997) reviewed stochastic simulation applications in this field of study. For a more recent general treatment, see Smith (2010). For readers wanting to learn more about how to combine stochastic simulation and optimization, the book by Spall (2003) is a good starting point.

References

Anderson, J.R. (1974) Simulation: methodology and application in agricultural economics. *Review of Marketing and Agricultural Economics* 42, 3–55.

Cacho, O. and Simmons, P. (1999) A genetic algorithm approach to farm investment. *Australian Journal of Agricultural and Resource Economics* 43, 305–322.

Clemen, R.T. and Reilly, T. (2014) *Making Hard Decisions with Decision Tools*, 3rd edn. South-Western, Mason, Ohio.

Law, A.M. and Kelton, W.D. (2000) *Simulation Modeling and Analysis*, 3rd edn. McGraw-Hill, New York.

Mayer, D.G., Belward, J.A. and Burrage, K. (1998) Optimizing simulation models of agricultural systems. *Annals of Operation Research* 82, 219–231.

Mayer, D.G., Kinghorn, B.P. and Archer, A.A. (2005) Differential evolution – an easy and efficient evolutionary algorithm for model optimisation. *Agricultural Systems* 83, 315–328.

Oriade, C.A. and Dillon, C.R. (1997) Developments in biophysical and bioeconomic simulation of agricultural systems: a review. *Agricultural Economics* 17, 45–58.

Raiffa, H. (1968 reissued in 1997) *Decision Analysis: Introductory Readings on Choices under Uncertainty.* McGraw Hill, New York.

Smith, J.Q. (2010) *Bayesian Decision Analysis: Principles and Practice.* Cambridge University Press, New York.

Spall, J.C. (2003) *Introduction to Stochastic Search and Optimization – Estimation, Simulation and Control.* Wiley, New York.

Vose, D. (2008) *Risk Analysis: a Quantitative Guide*, 3rd edn. Wiley, Chichester, UK.

Winston, W.L. and Goldberg, J.B. (2004) *Operations Research: Applications and Algorithms*, 4th edn. Thomson/Brooks/Cole, Belmont, California.

Winston, W.L., Albright, S.C. and Albright, C. (2000) *Practical Management Science*, 2nd edn. Duxbury, Belmont, California.

7 Decision Analysis with Preferences Unknown

Efficiency Criteria

A difficulty often encountered in applying the SEU model lies in the elicitation of the DM's utility function. The problem may be lack of access to the relevant person, inadequate introspective capacity on that person's part, or the fact that more than one person may be involved. Similarly, in agriculture, it may be necessary to develop recommendations for a particular target group of farmers numbering perhaps some hundreds or even thousands. Efficiency criteria have been devised to allow some ranking of risky alternatives when the specific utility function (or functions) is not available.

Efficiency analysis depends on making some assumptions about preferences or, equivalently, about the nature of the utility function. Often bounds are placed on the level of risk aversion. Then, for all DMs to whom the assumptions apply, the various actions can be divided into an efficient set and an inefficient set. The inefficient set contains those actions that are dominated by (preferred less than) actions in the efficient set. The efficient set contains those actions that are not dominated. The optimal action for any individual will lie among the alternatives in the efficient set, provided that:

1. The individual's preferences are consistent with the assumptions made about the nature of the utility function in deriving the risk-efficient set.
2. The individual's subjective probability distributions for outcomes are identical to those assumed for the analysis.

The latter point is too often overlooked in efficiency analyses. When subjective probability distributions cannot be obtained from (or checked with) the DM(s), efficiency analysis is best confined to situations with abundant relevant data so that the distributions derived from those data will be reasonably uncontroversial and widely acceptable.

There is an important trade-off to be made in conducting an efficiency analysis. The fewer restrictions that are placed on the utility function, the more generally applicable the efficiency results will be, but the less powerful will be the criterion in selecting between alternatives. Weak assumptions may leave so many alternatives in the efficient set as to make the analysis of little value, while setting too restrictive assumptions runs the risk of excluding from the efficient set the alternative that is most preferred by an individual DM, or of eliminating alternatives that would be preferred by a significant number of the target group of DMs.

An example

The following simplified example concerns a farm adviser whose advice is sought by a farmer in a semi-arid region in Australia. Because of poor returns, the farmer is considering quitting beef production and going into arable farming. There are several crops that could be included in the farm plan, but they must fit into sound crop rotations to ensure sustainable production. Consequently, the adviser has prepared a short list for further consideration of six alternative rotations, all of which are believed to be sustainable. The unpredictable seasonal weather and variable product prices make the net returns from the alternative rotations uncertain. The adviser is looking to cull the alternatives to a smaller number to present to the farmer for final choice.

Making use of records from the recent past, the farmer's subjective judgements about the future and a simple stochastic agroeconomic simulation model, the adviser is able to estimate the distributions of net returns from the alternative rotations. These distributions, denoted by F–K, are summarized in Table 7.1. The distributions, expressed on a whole-farm basis, are described in terms of their respective means and variances (along with standard deviations and coefficients of variation) and selected fractile values. The means and variances could, of course, have been calculated directly from the output of the stochastic simulation model. However, to allow readers to track the calculations illustrated for themselves, the means and variances shown in Table 7.1 were obtained assuming linearly segmented cumulative distribution functions (CDFs) and hence using the formulae given in Eqns 3.9 and 3.10 in Chapter 3, respectively. The resulting approximation errors are minor and could be further reduced, if required, by taking more fractile values.

Table 7.1. Selected statistics of the alternative crop rotations (in $\$10^3$, variance in $\$^2 10^6$).

	Crop rotation					
Statistics	F	G	H	I	J	K
$E[x]$	296	339	398	444	424	418
$V[x]$	11,730	16,628	25,901	23,839	13,450	6,519
$S[x]$	108	129	161	154	116	81
$C[x]$	0.37	0.38	0.40	0.35	0.27	0.19
$f_{0.0}$	45	80	140	130	165	220
$f_{0.1}$	158	186	212	252	275	305
$f_{0.2}$	205	230	253	311	325	350
$f_{0.3}$	241	264	296	356	361	378
$f_{0.4}$	272	294	332	394	392	400
$f_{0.5}$	299	326	372	432	421	427
$f_{0.6}$	324	361	416	474	455	449
$f_{0.7}$	353	401	465	522	490	470
$f_{0.8}$	387	449	524	578	525	493
$f_{0.9}$	432	511	612	657	581	522
$f_{1.0}$	540	660	850	800	665	548

Analysis Using Moments of the Distribution

Taylor series expansion

The basis of the moment method is usually, but not always, a Taylor series expansion of the utility function. The value of a function $U(x)$ can be approximated in the region of a given value of x, taken here to be the mean, $E[x] = E$, by the expansion:

$$U(x) = U(E) + U^{(1)}(E)\ (x–E) + U^{(2)}(E)\ (x–E)^2/2! + U^{(3)}(E)\ (x–E)^3/3! + \ldots \quad (7.1)$$

where $U^{(k)}(.)$ is the k-th derivative of the function $U(.)$ and $n!$ is factorial n, i.e. $n(n - 1)\ (n - 2)\ \ldots\ (1)$.

Taking expectations and simplifying, gives:

$$U(x) = U(E) + U^{(2)}(E)\ M_2[x]/2! + U^{(3)}(E)\ M_3[x]/3! + \ldots \quad (7.2)$$

where $M_k[x]$ is the k-th moment about the mean of the distribution of x. Thus, the utility of a risky prospect is equal to the utility evaluated at the mean plus a series of products comprised of moments, corresponding derivatives of the function $U(.)$ other than the first, and inverse factorials.

Application

Provided $U^{(k)}(E)/k!$ becomes small more quickly than $M_k[x]$ gets bigger, a series with only the first two or three terms is usually an adequate approximation. Note, of course, that the derivatives of polynomial utility functions eventually disappear – only the first two terms are needed for a quadratic and the first three for a cubic. However, the derivatives do not vanish for some favoured functions such as the logarithmic and negative exponential, and for these the expansion terminated after only two or three terms could be a somewhat inexact approximation for some probability distributions of x – although generally that is not so.

In the case where the distribution of x is normal, $M_3[x] = 0$ (as for other symmetric distributions). Moreover, because the normal distribution is completely specified by the mean and variance, decision analysis using only these two moments can be exact. In the case where $U = 1 - \exp(-cx)$, then $E[U] = CE = E - 0.5cV$.

Decision analysis using moments may be convenient in cases where there are many payoffs to evaluate, or when payoffs are continuous and represented by distributions with known moments.

Mean–variance efficiency

The mean–variance or *E,V efficiency* rule is based on the proposition that, if the expected value of alternative A is greater than or equal to the expected value of alternative B, and the variance of A is less than or equal to the variance of B, with at least one strict inequality, then A is preferred to B by all DMs whose preferences meet certain conditions. Only those options that are not dominated in an *E*, *V* sense are regarded as members of the *E*, *V* efficient set. The conditions are that the DM always prefers more to less of the measure of consequences *x*, and is universally not risk preferring with respect to the level of *x*. Additional requirements for the rule to be exact are that: (i) the outcome distribution is normal; or (ii) the DM's utility function is quadratic. Since normal distributions are the exception rather than the rule in decision analysis and since a quadratic utility function implies the unlikely characteristic that absolute risk aversion increases with level of payoff, the *E*, *V* efficiency criterion is usually best regarded as an approximate rule only. The advantage of the *E*, *V* approach, however, is that only information on means and variances of the outcome distributions is needed in order to permit at least a partial ordering of alternatives. This advantage largely explains the popularity of the approach.

To apply the *E*, *V* efficiency criterion to the example relating to choice of crop rotations, the adviser sets out the mean and variance of each alternative (Table 7.1) in two-dimensional *E*, *V* space, such as Fig. 7.1. Assuming that the farmer is risk averse, the relevant *E*, *V* indifference or iso-utility curves (in this case linear) will slope upwards, as illustrated in the figure. Three imaginary indifference lines for utility levels $U_1 < U_2 < U_3$ are shown. The more risk averse the farmer is, the steeper these curves will be.

Inspection of Fig. 7.1 reveals that, for the degree of risk aversion expressed in the indifference curves, crop rotation K is the best alternative, as it is located on the highest indifference curve U_3 (i.e. K has the most preferred combination of $E[x]$ = \$418 and $V[x]$ = \$6519). Alternative H, for instance, with the relatively high overall mean of \$398, but also by far the highest variance of \$25,901, is far from being the most preferred alternative for the indifference lines shown.

Since it is assumed for the purposes of this chapter that the degree of risk aversion of the farmer is not known, we cannot in fact identify the decision alternative that gives the highest expected utility. Instead, we can apply the following rule for *E*, *V* efficiency: an alternative is in the *E*, *V* efficient set if there is no other alternative that lies in its 'north-western' quadrant. As can be visually deduced from Fig. 7.1,

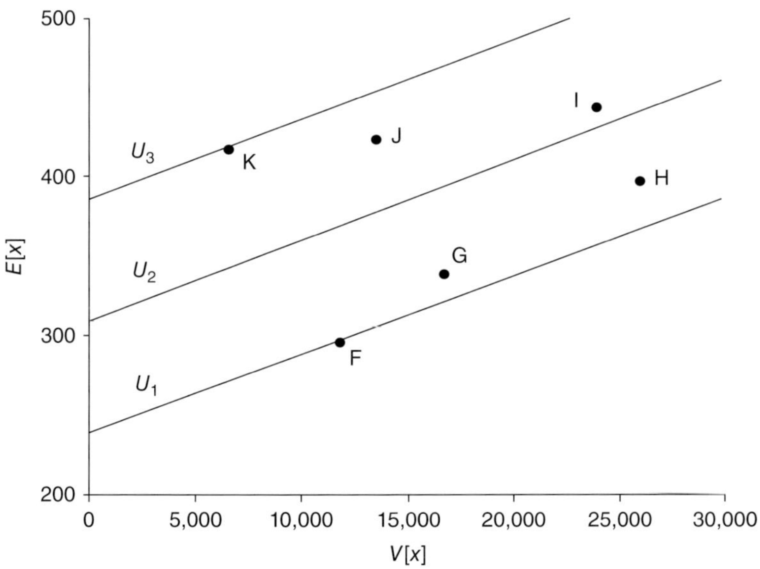

Fig. 7.1. Six crop rotations (F–K) in *E*,*V* space and three indifference curves (U_1, U_2 and U_3) representing risk aversion.

applying this rule for all possible pairwise comparisons of alternatives reveals that alternatives I, J and K are not dominated by any other alternatives, so these three comprise the *E,V* efficient set. It would have to be left to the farmer to choose between them, unless more information could be elicited about the farmer's attitude to risk.

Mean–standard deviation analysis

A variant of *E,V* efficiency analysis that some may find easier to follow is mean–standard deviation efficiency. Recall that standard deviation is the positive square root of the variance. Clearly, *E,V* analysis can be equivalently presented in mean–standard deviation or *E,S* space.

In terms of *E,S efficiency*, as for *E,V* efficiency, any alternative is dominated if there is another alternative that lies to the north-west of it on the *E,S* graph. More formally, those risky prospects that form the *E,S* efficient set are the ones for which there is no other prospect with the same or higher mean and the same or lower standard deviation, with at least one strict inequality. The *E,S* and the *E,V* efficient sets are identical for the same set of alternatives.

Portfolio analysis

One common application of *E,V* efficiency analysis is to make decisions about a mix of risky prospects, such as an investment portfolio or a farm plan (Brealey *et al.*, 2013). The mean and variance of any mix involving q_i (units or proportions) of prospect i are given by:

$$E = \Sigma_i q_i e_i, \text{ and } V = \Sigma_i \Sigma_j \text{cov}_{ij} q_i q_j \tag{7.3}$$

where e_i is the expected return of prospect i, and cov_{ij} is the covariance of returns of prospects i and j (and the variance of returns when $i = j$).

With one or more constraints on the q_i, such as a budget limit for an investment portfolio, or land, labour and capital constraints on a farm plan (and if the possible levels of the components of the portfolio are continuous), the set of possible mixtures of prospects forms a convex set in *E,V* space, as illustrated in Fig. 7.2. A risk-averse DM will have indifference curves with positive gradients such as U_1, U_2 and U_3, as illustrated in Fig. 7.2, again depicted as straight lines. Therefore, the optimal mix will be a point on the north-western frontier of the set, such as *C*. The frontier comprises the *E,V* efficient set and can be generated by quadratic programming (see Chapter 9, this volume). It differs from the efficient set in applications such as that illustrated in Fig. 7.1 in that it contains an infinite number of points representing different decisions about the combination of portfolio components.

In the event that a 'riskless' asset (with a positive *E* and zero *V*) is available that can be held in any positive (saving) or negative (borrowing) amounts, the efficient frontier takes on a different character. The effect is best envisaged in an *E,S* graph (otherwise similar to Fig. 7.2) in which the efficient frontier is now defined by a straight line drawn from the riskless mean point on the *E* axis and tangent to the previous risk-efficient frontier. In this case, the optimal portfolio among the original set is fixed at this point of

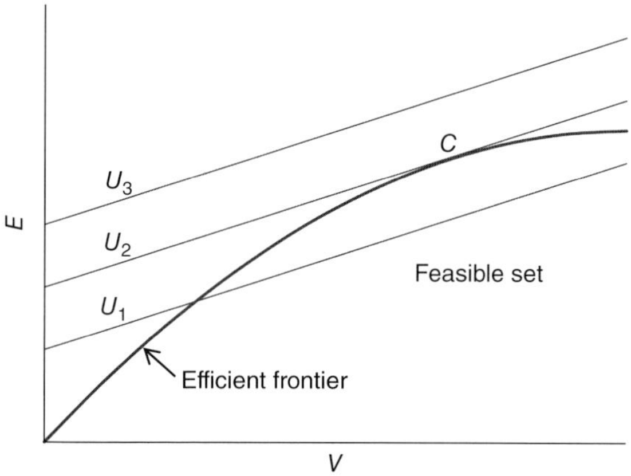

Fig. 7.2. The feasible set, efficient frontier and utility-maximizing point (*C*) in a portfolio selection model.

tangency and is not determined by the degree of risk aversion of the DM. Risk aversion only enters the analysis in determining how much of the riskless asset to borrow or invest in. This principle is known as the *Separation Theorem* since the optimal portfolio choice is separate from the DM's attitude to risk. There is an extensive literature in financial economics exploiting this theorem, but since we regard the postulated existence of such a riskless asset as a mere theoretical nicety, we give no more attention to the theorem here. The important lesson for practical risk analysis is that the set of alternatives considered for inclusion in any portfolio should be as complete as is practicable. For example, in farm planning, holding positive or negative amounts of financial instruments – such as borrowing, off-farm investments, futures contracts, and insurance – should be considered along with the on-farm risky prospects such as crops and livestock.

Portfolio analysis in an E,V (or E,S) framework is a widely used, and sometimes abused, method of decision analysis. It is too often overlooked that the E,V rule is an approximation of expected utility maximization unless the rather demanding conditions noted earlier are satisfied. Where direct maximization of expected utility is possible, it is usually to be preferred to the E,V approximation, as discussed in Chapter 9 in the context of whole-farm planning. However, the convenience of E,V analysis, and experience that it often works just about as well as more theoretically correct forms of analysis, means that it is likely to remain in the tool-kit of agricultural economists for some time to come.

Stochastic Efficiency Methods

As illustrated below, there are different forms of stochastic efficiency analysis that vary with regard to the assumptions made about the nature of the relevant utility function and the risk attitudes implied. Unlike E,V efficiency and other moment-based methods, these methods are based firmly on the notion of direct expected utility maximization. Alternative risky prospects are compared, at least in principle, in terms of full distributions of outcomes, not just in terms of moments, even though, operationally, it is usual to work with a 'sufficient number' of fractile values. The cost of this more rigorous assessment is some increase in conceptual complexity of the required calculations.

We deal first with *stochastic dominance* methods, which require pairwise comparison of alternatives. As a result, the potential number of such comparisons rises rapidly with the number of alternatives.

Moreover, since the comparisons must be made at every specified point along the distributions, the computational task quickly becomes burdensome unless efficient computer routines are used. Fortunately, suitable software packages for stochastic dominance analysis are available (e.g. Goh *et al.*, 1989; Richardson *et al.*, 2001). However, the method of stochastic efficiency analysis called SERF, described later in the chapter, avoids or reduces most of the computational problems with dominance analysis.

In the sub-sections that follow, forms of efficiency analysis are described in an order that, in the main, involves progressively stronger assumptions about risk preferences, therefore potentially producing progressive reductions in the size of the efficient set.

First-degree stochastic dominance

With *first-degree stochastic dominance* (FSD), the restriction on the utility function is simply that the DM has positive marginal utility for the performance measure (i.e. more is preferred to less).[1] Then, given two alternatives A and B, each with a probability distribution of outcomes x defined by CDFs $F_A(x)$ and $F_B(x)$, respectively, A dominates B in the first-degree sense if:

$$F_A(x) \leq F_B(x) \text{ for all } x \qquad (7.4)$$

with at least one strong inequality. Graphically this means that the CDF of A must always lie below and to the right of the CDF of B. If two CDFs cross, then neither dominates the other in the first-degree sense. Note that this stochastic dominance criterion, like others to be described below, works by the identification of the dominated alternatives, such that those that are not dominated represent the efficient set. Dominance analysis requires the pairwise comparison of the distributions of all the alternatives being considered with the proviso that, once an alternative has been found to be dominated by another, the dominated one can be dismissed from all further consideration.

We now return to the crop rotation problem, for which the CDF data are presented in the lower part of Table 7.1 above. Because these data came from a stochastic simulation, it would have been easy to obtain more fractiles to make the analysis more precise. However, as explained before, to permit readers to replicate the calculations if they wish, we have confined this and subsequent stochastic dominance analysis to only the 11 fractile values shown in Table 7.1. Using these data, Fig. 7.3 portrays the CDFs of the four alternatives F, G, H and J represented as linear segments joining the given fractile values. Only four of the six CDFs are shown in the figure to illustrate FSD without making the graph too complicated.

As can be seen in the figure, $F_G(x) < F_F(x)$ for all x and so alternative F cannot be a member of the efficient set under the FSD rule. Also $F_H(x) < F_G(x)$ for all x so G cannot be a member of the first-degree efficient set. Although J dominates F and G in the FSD sense, we do not need to know this to eliminate F and G from further consideration. As noted above, once an alternative is found to be dominated by one other, it can be eliminated as a candidate for inclusion in the efficient set. If we compare H and J, however, we see that the CDFs cross, so there is no dominance in terms of FSD; for lower values of x, $F_J(x)$ is the better of the two, but for higher ones $F_H(x)$ is better. Systematic pairwise comparisons of all the crop rotations reveals that the efficient set by the FSD rule is H, I, J and K.

[1] If performance is in terms of costs (or some other 'bad'), all the same methods of stochastic efficiency analysis can be applied by treating the outcomes as negative benefits.

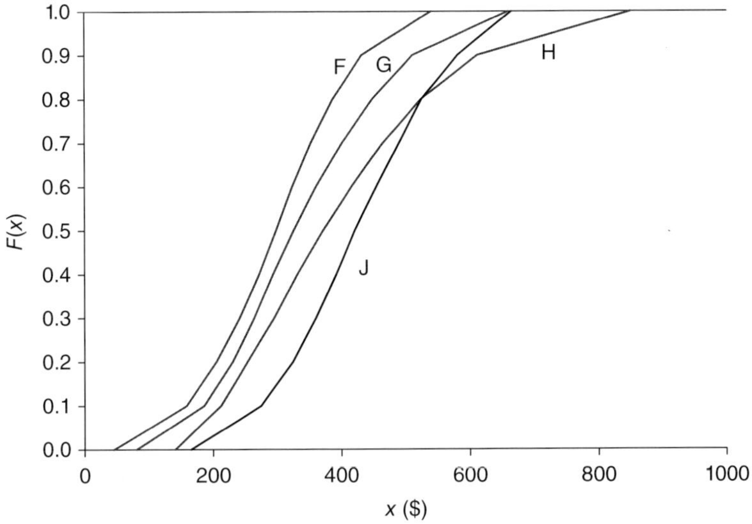

Fig. 7.3. Illustration of first-degree stochastic dominance (FSD) applied to four of the alternatives in the cropping problem (alternatives H and J dominate F and G, there is no FSD between H and J).

It should come as no surprise that four of the six alternatives survived this comparison since only one weak behavioural assumption has been invoked. Our empirical experience is, however, that managing to eliminate one-third of the alternatives merely with FSD is a relatively high level of first-stage shrinkage to the efficient set. Usually, a significant shrinkage has to rely on the use of additional behavioural assumptions, which we deal with below.

Second-degree stochastic dominance

The additional restriction on the utility function necessary for *second-degree stochastic dominance* (SSD) is that the DM must be risk averse for all values of x, and thus have a utility function of positive but decreasing slope (i.e. $U^{(1)}(x) > 0$ and $U^{(2)}(x) < 0$). With SSD, A is preferred to B if:

$$\int_{-\infty}^{x^*} F_A(x)dx \leq \int_{-\infty}^{x^*} F_B(x)dx, \text{ for all values of } x^* \tag{7.5}$$

with at least one strong inequality. Hence, under this criterion, distributions of outcomes are compared based on areas under their CDFs. SSD requires that the curve of the cumulative area under the CDF for the dominant alternative lies everywhere below and to the right of the corresponding curve for the dominated alternative. SSD has more discriminatory power than FSD and the efficient set under SSD is a subset of that under FSD.

The application of SSD to the crop rotation problem is illustrated in Fig. 7.4 where the CDFs of three alternatives (H, J and K) are shown. (Logically, this analysis should be confined to the FSD set, but the principles of SSD are more clearly illustrated using these chosen alternatives.) We could do the analysis by plotting the areas under the CDFs, as indicated in Eqn 7.5. However, in this case it is possible to judge SSD by eye, using only the CDFs, and the basis of the method is more easily understood by inspecting the CDFs. (Of course, in practice the analysis is done numerically using appropriate software.)

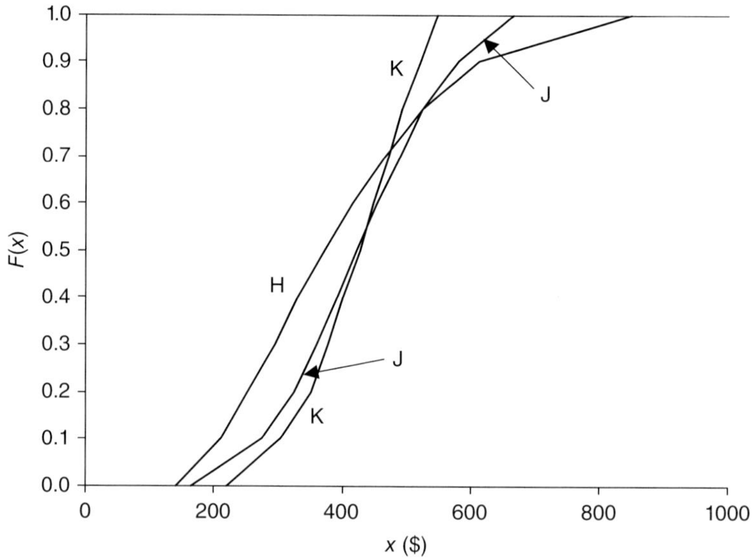

Fig. 7.4. Illustration of second-degree stochastic dominance (SSD) applied to three of the alternatives in the cropping problem (alternatives J and K dominate H, there is no dominance between J and K).

First we check whether there is SSD between alternatives H and J. Since the minimum of alternative H is less than the minimum for alternative J, H cannot dominate J, but J can dominate H in the sense of SSD. As just described, we have to compare the areas below the two CDFs. The area below $F_J(x)$ is smaller than the area below $F_H(x)$ for all values for x. The difference between these two areas is reflected by the area between the two CDFs. This latter area is divided into two parts, as can be seen in the figure. For fractiles between 0 and 0.8 (read from the vertical axis) $F_J(x)$ lies to the right of $F_H(x)$ and this portion of the area between the curves is bigger than the area for fractiles between 0.8 and 1 where $F_H(x)$ lies to the right of $F_J(x)$. It follows that J dominates H in the sense of SSD. Comparing J and K reveals that the area below $F_K(x)$ is smaller for small values for x (until about $430 or fractiles below 0.55; see Fig. 7.4), but for higher values the area below $F_J(x)$ is much smaller. In other words, the lower area between K and J is much smaller than the upper area between them in Fig. 7.4, indicating the cumulative area (accumulating from the lower tails) under K is not everywhere less than that under J, so that there is no SSD between J and K.

In the further comparisons that for brevity are not reported here in graphical (or tabular) terms, only H is eliminated from the FSD efficient set. Hence the SSD efficient set is I, J and K. This example therefore only hints at the typical power of going to the SSD step, which commonly eliminates more options than does the first-degree dominance analysis.

Some other more discriminatory dominance criteria

In some cases application of FSD and SSD may not be able to discriminate between alternatives sufficiently, in the sense that it is judged that there are still too many choice alternatives in the efficient set. In that case, other more discriminating criteria that have been developed may be considered.

A *third-degree stochastic dominance* criterion (TSD) exists. It depends on the same behavioural assumptions as SSD but with the new assumption that the coefficient of absolute risk aversion is decreasing with income or wealth. Experience suggesting that the additional discriminating power of TSD over SSD is often slight (Anderson, 1974a,b, 1975; Anderson *et al.*, 1977, Chapter 9), makes it relatively less useful than alternative methods based on more restrictive ranges of risk aversion or on using probabilistic mixtures of prospects. We therefore give no further consideration to TSD or to forms of dominance based on assumptions about still higher derivatives of the utility function.

There also exists a technique called *convex stochastic dominance* that is based on the formation of *convex combinations* of the CDFs (Fishburn, 1974a,b; Cochran *et al.*, 1985). It can be shown that an alternative is dominated if there is some convex combination formed from the other alternatives that satisfies the relevant dominance rule. However, although this extension of stochastic dominance analysis adds to the discrimination power of the one-pair-at-a-time methods, it is somewhat difficult to implement and appears not to have been widely and successfully used. We therefore have chosen not to discuss it further here.

Stochastic efficiency with stronger assumptions about risk aversion

It is possible to devise a range of more discriminating forms of stochastic dominance analysis that depend on more demanding assumptions about the risk attitudes of the DM or group of DMs. One approach that has been widely used is *stochastic dominance with respect to a function* (SDRF), also known as *generalized stochastic dominance*. The method has stronger discriminatory power than FSD and SSD, achieved through the introduction of bounds on the absolute risk-aversion coefficient within a second-degree stochastic dominance analysis (Meyer, 1977a,b). Hence the analysis applies to DMs who have a degree of risk aversion that falls within specified bounds. Eliciting the bounds on their risk-aversion coefficients from DMs may be simpler than eliciting complete utility functions. Alternatively, it may be possible to make a guess about the plausible range of risk aversion in any reasonably homogeneous target group.

In applying SDRF to the crop rotation problem, therefore, the first step is to set bounds on the absolute risk-aversion coefficient. In fact, bounds already exist with the other efficiency criteria, and so the SDRF approach amounts to tightening the bounds. If r_a is the absolute risk-aversion coefficient for an assumed approximately constant level of wealth, then with FSD the bounds are $-\infty < r_a < +\infty$, and with SSD they are $0 < r_a < +\infty$. For SDRF the bounds are reduced to $r_1 \leq r_a \leq r_2$, where r_1 is a non-negative number and r_2 a positive number. As should be evident, the narrower the bounds set on risk aversion, the more powerful the rule.

While specialist computer programs can be developed to implement a number of variants of generalized stochastic dominance analysis, we have chosen to confine this treatment to a simpler and perhaps better way of doing much the same thing. As Richardson *et al.* (2001) illustrate, and as expounded by Hardaker *et al.* (2004), a simpler option is to compare the alternative risky prospects in terms of certainty equivalents (CEs) over the range of risk aversion of interest. The latter authors have named this method *stochastic efficiency with respect to a function* (SERF).

For each risky alternative and for a chosen form of the utility function, the SEU hypothesis means that utility can be calculated depending on the degree of risk aversion r and stochastic outcome of x as:

$$U(x,r) = \int U(x,r)\, f(x)dx \qquad (7.6)$$

Then U is calculated for selected values of r in the range r_1 to r_2. The CEs for each of these values of U are found by:

$$CE\ (x,r) = U^{-1}(x,r) \tag{7.7}$$

where U^{-1} is the inverse form of the utility function.

The general rule for SERF analysis for the given assumptions is that the efficient set contains only those alternatives that have the highest (or equal highest) CE for some value of r in the relevant range.

This method requires the choice of a particular form for the utility function and associated measure of risk aversion. Usually, for partial analyses, the familiar negative exponential function, exhibiting CARA, will generally serve well, implying that the degree of risk aversion is defined by a range of the absolute risk aversion coefficient r_a. Where outcomes are measured in terms of terminal wealth, however, the CRRA function will often be more appropriate.

The SERF method was applied to the crop rotation example as follows. First, a decision was made to use the negative exponential utility function for this partial analysis, requiring the choice of a relevant range in the coefficient of absolute risk aversion. For the illustrative example the range chosen was between $r_1 = 0$ and $r_2 = 0.006$. Then the expected utility of each alternative was calculated in Microsoft Excel for several values of the risk aversion coefficient to span the chosen range of r. Again, as for the FSD and SSD analyses, for consistency we have confined the calculations to only those fractile values given in Table 7.1 and have assumed a linearly segmented CDF between the fractile values. Under these assumptions and for the CARA utility function, expected utility, as defined in Eqn 7.6, can be approximated using the formula:

$$U(x,r_a) = \Sigma_i (F_{i+1} - F_i)\,[1 - \{\exp(-r_a x_i) - \exp(-r_a x_{i+1})\}/r_a(x_{i+1} - x_i)],\ \ r_1 \le r_a \le r_2 \tag{7.8}$$

Then for each alternative rotation and for each value of the risk-aversion coefficient, the CE was calculated from:

$$CE = -\ln\{1 - U(x,r_a)\}/r_a \tag{7.9}$$

The CEs obtained were then plotted against the value of the risk-aversion coefficient for each rotation, as shown in Fig. 7.5.

The graphical presentation in Fig. 7.5 shows that alternative I has the highest CE for values of r_a between 0 and 0.0033, whereas alternative K is superior for r_a values above 0.0033 but less than the upper limit of 0.006.

It is easy to see from the graph which alternatives form the efficient set. It is also possible to read off from the graph the point of changeover, and this might be useful in formulating recommendations to a group of farmer clients. Those more risk averse than the crossover value of r_a should choose K rather than I, and vice versa for those less risk averse than the crossover value.

The way SERF results are depicted clearly reveals that the more that is known or presumed to narrow the relevant range of risk aversion, the more the efficient set can be reduced to a small number of alternatives, perhaps just one. With some thought, it is often possible to limit the range more than has been commonly assumed in past studies (Hardaker and Lien, 2010).

The SERF method requires no special software. It can readily be implemented in worksheet software such as Excel. Usually it suffices to compute the expected utility of each alternative for a relatively small

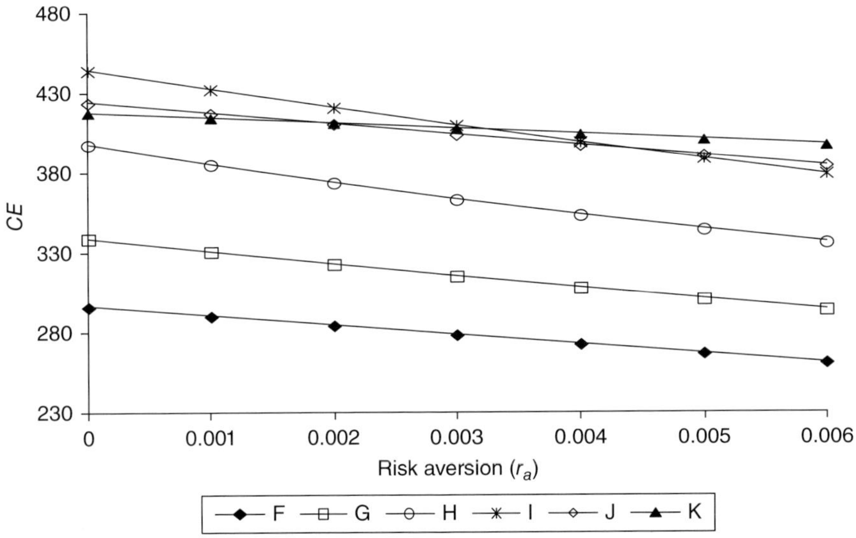

Fig. 7.5. Illustration of stochastic efficiency with respect to a function (SERF), showing certainty equivalents (CEs) for crop rotations (F–K) across a range of values of r_a.

number of values of r_a in the relevant range, especially when using a graphical presentation of results. (More values can be inserted as required if it is important to zero-in on crossover points.) Comparisons can be made directly in terms of utility values but we prefer the use of CEs, computed from the inverse utility function, mainly because they are more readily interpreted.

When many alternatives are to be compared, the graphical presentation illustrated will not be immediately applicable, but it is a simple matter to identify within the worksheet format those alternatives that form the efficient set. These few can then be plotted if required.

Concluding Comment

Decision analysis can clearly only go so far in the absence of good knowledge of the preferences of the DMs on whose behalf the work is being done, and accordingly we give somewhat more emphasis in this book to methods where it is presumed that access to the DM is sufficiently direct as to permit application of some of the methods exposited in other chapters. Often, however, there will be incomplete information about the exact risk attitude of the DM or DMs. In such cases, the methods of efficiency analysis, particularly SERF, provide an approach that is consistent with the SEU hypothesis and that narrows down the range of choice to an efficient set. Moreover, the simple principle of exploring the impact on risky choice of varying the assumed degree of risk aversion within a plausible range can be applied to almost any of the forms of analysis described in this book. That should surely quieten, if not silence, those critics of decision analysis who see problems in utility function elicitation as a main stumbling block.

Selected Additional Reading

Stochastic dominance analysis was launched in the late 1960s with the pioneering work of Hadar and Russell (1969) and Hanoch and Levy (1969). It was soon picked up by financial market analysts but it was seemingly not applied to agricultural problems until the work of Hardaker and Tanago (1973) and Anderson (1974a,b). It may have been popularized by Anderson *et al.* (1977) and Meyer (1977a,b) but adoption was slow, and applications by US agricultural economists, for instance, did not become widespread until the 1980s (e.g. King and Robison, 1984; Cochran *et al.*, 1985; Raskin and Cochran, 1986; McCamley and Kliebenstein, 1987; McCarl, 1988a,b). For a summary of the field see Levy (1992) or Chavas (2004, Chapter 5).

Stochastic efficiency analysis in terms of CEs (SERF) is outlined by Richardson *et al.* (2001), who also provide software for those who prefer not to program the required (rather straightforward) calculations themselves. The rationale for SERF is provided by Hardaker *et al.* (2004).

References

Anderson, J.R. (1974a) Risk efficiency in the interpretation of agricultural production research. *Review of Marketing and Agricultural Economics* 42, 131–184.

Anderson, J.R. (1974b) Sparse data, estimational reliability and risk-efficient decisions. *American Journal of Agricultural Economics* 56, 564–572.

Anderson, J.R. (1975) Programming for efficient planning against non-normal risk. *Australian Journal of Agricultural Economics* 19, 94–107.

Anderson, J.R., Dillon, J.L. and Hardaker, J.B. (1977) *Agricultural Decision Analysis*. Iowa State University Press, Ames, Iowa.

Brealey, R.A., Myers, S.C. and Allen, F. (2013) *Principles of Corporate Finance*, 11th edn. McGraw-Hill/Irwin, Boston, Massachusetts.

Chavas, J.-P. (2004) *Risk Analysis in Theory and Practice*. Elsevier Academic Press, San Diego, California.

Cochran, M.J., Robison, L.J. and Lodwick, W. (1985) Improving efficiency of stochastic dominance techniques using convex sets. *American Journal of Agricultural Economics* 67, 289–295.

Fishburn, P.C. (1974a) Convex stochastic dominance with continuous distributions. *Journal of Economic Theory* 7, 143–158.

Fishburn, P.C. (1974b) Convex stochastic dominance with finite consequence sets. *Theory and Decision* 5, 119–137.

Goh, S., Shih, C.-C., Cochran, M.J. and Raskin, R. (1989) A generalized stochastic dominance program for the IBM PC. *Southern Journal of Agricultural Economics* 21, 175–182.

Hadar, J. and Russell, W.R. (1969) Rules for ordering uncertain prospects. *American Economic Review* 59, 25–34.

Hanoch, G. and Levy, H. (1969) Efficiency analysis of choices involving risk. *Review of Economic Studies* 36, 335–345.

Hardaker, J.B. and Lien, G. (2010) Stochastic efficiency analysis with risk aversion bounds: a comment. *Australian Journal of Agricultural and Resource Economics* 54, 279–283.

Hardaker, J.B. and Tanago, A.G. (1973) Assessment of the output of a stochastic decision model. *Australian Journal of Agricultural Economics* 17, 170–178.

Hardaker, J.B., Richardson, J.W., Lien, G. and Schumann, K.D. (2004) Stochastic efficiency analysis with risk aversion bounds: a simplified approach. *Australian Journal of Agricultural and Resource Economics* 48, 253–270.

King, R.P. and Robison, L.J. (1984) Risk efficiency models. In: Barry, P.J. (ed.) (1984) *Risk Management in Agriculture*. Iowa State University Press, Ames, Iowa, pp. 68–81.

Levy, H. (1992) Stochastic dominance and expected utility: survey and analysis. *Management Science* 38, 555–583.

McCamley, F. and Kliebenstein, J.B. (1987) Identifying the set of SSD-efficient mixtures of risky alternatives. *Western Journal of Agricultural Economics* 12, 86–94.

McCarl, B.A. (1988a) Generalized stochastic dominance: an empirical examination. *Southern Journal of Agricultural Economics* 22, 49–55.

McCarl, B.A. (1988b) Preferences among risky prospects under constant risk aversion. *Southern Journal of Agricultural Economics* 20, 24–33.

Meyer, J. (1977a) Choice among distributions. *Journal of Economic Theory* 14, 326–336.

Meyer, J. (1977b) Second degree stochastic dominance with respect to a function. *International Economic Review* 18, 477–487.

Raskin, R. and Cochran, M.J. (1986) Interpretation and transformations of scale for the Pratt-Arrow absolute risk aversion coefficient: implications for stochastic dominance. *Western Journal of Agricultural Economics* 11, 204–210.

Richardson, J.W., Schumann, K.D. and Feldman, P. (2001) *Simetar: Simulation and Econometrics to Analyze Risk*. Department of Agricultural Economics, Texas A&M University, College Station, Texas.

8 The State-contingent Approach to Decision Analysis

Introduction

Until relatively recently the analysis of production under uncertainty in agriculture had been dominated by the use of *stochastic production functions* and related methods. First proposed by Sandmo (1971) and refined by Just and Pope (1978) (JP), a stochastic production function can be specified to accommodate both increasing and decreasing output variance in inputs. The single-output JP production function has the general form:

$$y = g(x) + u = g(x) + h(x)^{0.5}\varepsilon \tag{8.1}$$

where $g(\cdot)$ is the *mean function* (or deterministic component of production), $h(\cdot)$ is the *variance function* that captures the relationship between input use and output variation, and ε is an index of exogenous production shocks with zero mean and variance σ_ε^2. This formulation allows inputs x to influence mean output $E(y)$ and variance of output $V(y)$ independently, since:

$$E(y) = g(x) \text{ and } V(y) = h(x)\,\sigma_\varepsilon^2 \tag{8.2}$$

Applying prices to inputs and outputs converts the above two functions into functions of expected net revenue and variance of net revenue, both in terms of input levels x. (For simplicity, we ignore the complication that uncertainty about output prices often also needs to be accommodated.) Then it is possible to find the level of inputs to maximize expected utility expressed via the approximate indirect utility function:

$$E(u) \approx E(y) - 0.5r_a V(y) \tag{8.3}$$

where r_a is the coefficient of absolute risk aversion.

Perhaps partly because of the many studies along these lines, risk in agricultural production tended to be viewed as a sort of 'friction' to the efficient allocation of resources. It has generally been treated as if it is beyond the scope of conventional production economics, requiring a different form of analysis (typically E,V analysis) to deal with it.

Yet back in the 1950s, the state-contingent (SC) approach was pioneered by Debreu (1952) and Arrow and Debreu (1954). They showed that, if uncertainty is represented by a set of possible states of nature, production uncertainty can be brought within the ambit of conventional production theory, with output in different states treated in much the same way as for different types of output. After languishing for many years, at least in agricultural decision analysis, the SC approach was revived in an important book by Chambers and Quiggin (2000). Their treatment of the approach was theoretically rigorous, perhaps putting it beyond the reach of many practitioners of decision analysis. Fortunately, however, in an agricultural context, clarifying papers by Quiggin and Chambers (2006) and Rasmussen (2011), among others, have largely dispersed the mists so that a number of agricultural applications have appeared. The aim below is to provide a simple explanation and illustration of SC analysis.

In the context of this approach, the word 'contingent' means 'depending upon'. As indicated, central to the approach is the notion of consequences of risky choices being analysed as depending upon which of the possible uncertain states of the world eventuates. Of course, there is nothing new in this notion. Decision trees, payoff tables and other forms of analysis discussed in previous chapters account for risky consequences contingent upon which state eventuates. Nevertheless, as noted above, and as we shall demonstrate, analysing risky choices specifically in terms of SC outcomes has some theoretical advantages that may make it a superior approach to the alternatives in some situations.

Basics of the Approach

The SC approach is best explained with the aid of an example. A farmer, operating on light, sandy soil in a location where rainfall is somewhat low and unreliable, is trying to decide on a sowing rate for a field of wheat that is to be sown soon. If the season turns out to have moderate to high rainfall, a relatively high sowing rate produces the best yield. But if it is a dry season, competition between the many wheat seedlings means that few will grow to full maturity and the yield from a high sowing rate will be less, in such a season, than if fewer seeds had been sown.

To keep things simple at the start, only two sowing rates are considered – Low (a_1) and High (a_2). Similarly, only two possible states of seasonal conditions are considered – Wet (S_1) or Dry (S_2). The corresponding probabilities are $P(S_1) = 0.7$ and $P(S_2) = 0.3$. When the costs and return of each action for each state have been budgeted out, and the resulting net revenues multiplied up to the field scale of 50 ha, the payoff table shown in Table 8.1 is derived.

Note that this is a genuine decision problem since the optimal choice of low or high sowing rate depends on the type of season. If the season is wet, the best choice is a high sowing rate, but if the season turns out to be dry, a low rate is best, as expected. And, of course, the farmer does not know for sure which type of season will eventuate.

If the farmer is indifferent to risk (risk neutral), a choice could be make on the basis of expected money values (EMVs), as calculated in the table. These two values indicate a small advantage for a high sowing rate (a_2) over a low rate (a_1). However, the spread of outcomes is higher for the higher rate, as the table shows, so a risk-averse DM might prefer a_1 over a_2.

The same problem can be represented as a decision tree, as shown in Fig. 8.1.

Table 8.1. Payoff table for wheat-sowing decision.

States		Probabilities	Sowing rate	
			Low	High
		$P(S_i)$	a_1	a_2
S_1	Wet	0.7	\$8250	\$9000
S_2	Dry	0.3	\$4875	\$3525
	EMV[a]		\$7237.5	\$7357.5

[a]EMV, expected money value.

Analysing this tree by 'averaging out and folding back', maximizing EMV, of course leads to the same prescription as that derived from the payoff table.

Now see what happens when we represent this same problem in state space in Fig. 8.2.

In this figure the axes z_1 and z_2 represent the payoffs in dollars if the season is wet (S_1) or the season is dry (S_2), respectively. Points a_1 and a_2, each represented by a cross ×, are the two decision options of low or high sowing rates, respectively. Thus, for example, a_1 is plotted to show that the net return would be \$8250 on the z_1 axis should S_1 occur and \$4875 on the z_2 axis should S_2 occur, and similarly for a_2.

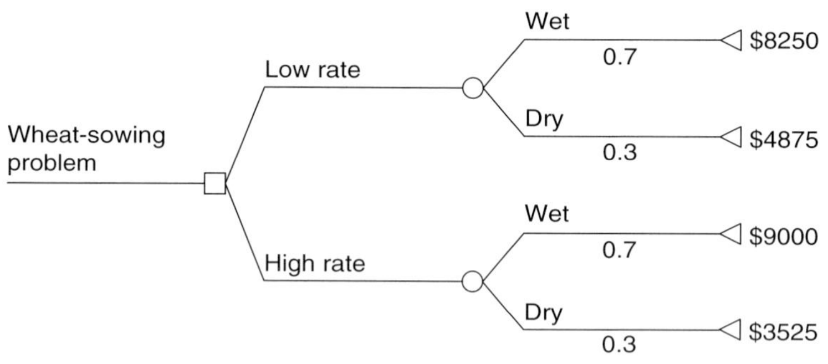

Fig. 8.1. Decision tree for sowing rate problem with two rates and two states.

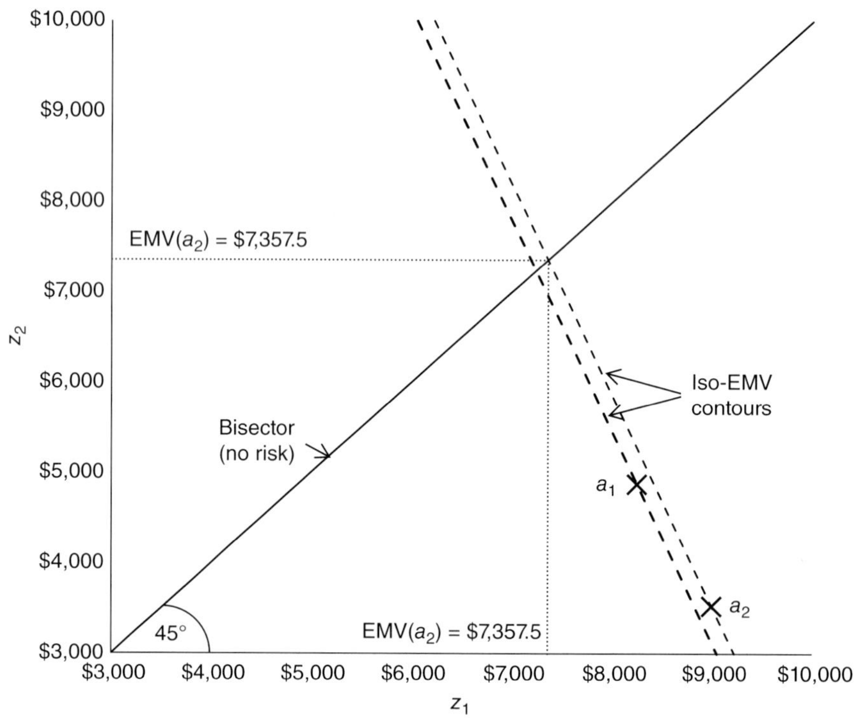

Fig. 8.2. Representation of the wheat-sowing rate choice in state space, maximizing expected money value.

We now add to the figure some contour lines passing through points of the same expected money values. It can be shown that these iso-EMV contours are parallel straight lines of gradient $-P(S_1)/P(S_2)$, which in this case is −0.7/0.3 = −2.333.

The value of the EMV increases in a north-easterly direction (towards the top right-hand corner). Two of these contours are shown in the figure, one passing through a_1 and one passing through a_2. Since the contour for a_2 lies above and to the right of that for a_1, we may conclude that a_2 will be preferred to a_1 by a risk-indifferent DM, as we already established. For such a person, these contours can be seen as indifference curves familiar in a range of economic settings.

Included in the figure is the bisector or 45° line. This is a line passing through points of the same payoff for the two states. Hence, points along this line represent 'risk-free' options. Of course, there may be no real choice option in a given decision problem that is risk free, or, if there is, such an option may not be optimal. For example, a 'do nothing' option may be risk free, but will seldom be a preferred option in the presence of opportunities to make money in some or all states. It is nevertheless instructive to track along the EMV contours to the points where they cross the bisector. As illustrated in the figure for point a_2, it is then possible to read off the EMV value of this option on either of the axes. In the case of a_2, the value is $7357.5, as established earlier.

Accounting for Risk Aversion

The next step is to consider what happens if the DM is risk averse. To illustrate, we assume that the farmer's attitude to risk is captured by a negative exponential utility function with a coefficient of absolute risk aversion r_a equal to 0.00008. Applying this utility function to the payoffs in the farmer's decision problem in the form of a payoff table gives the results shown in Table 8.2.

As can be seen, with the introduction of this degree of risk aversion, the optimal choice shifts from a_2 under indifference to risk to a_1 with risk aversion. The shift is illustrated graphically in the SC representation by replacing the iso-EMV contours with iso-utility contours or risk-averse indifference curves, as shown in Fig. 8.3. With risk aversion, these contours are convex to the origin and, as with EMV contours, preference increases to the north-east. In Fig. 8.3 only one such iso-utility contour is plotted, passing through the new optimal point a_1. This time, the point of intersection of the contour with the bisector gives the value of the certainty equivalent on both axes, i.e. CE = $7138.5.

Table 8.2. Payoff table for wheat-sowing decision in utility terms.

States[a]		Probabilities	Sowing rate	
			Low	High
		$P(S_i)$	a_1	a_2
S_1	Wet	0.7	0.48315	0.51325
S_2	Dry	0.3	0.32294	0.24573
	EU		0.43509	0.43299
	CE		$7138.5	$7092.3

[a]CE, certainty equivalent; EU, expected utility.

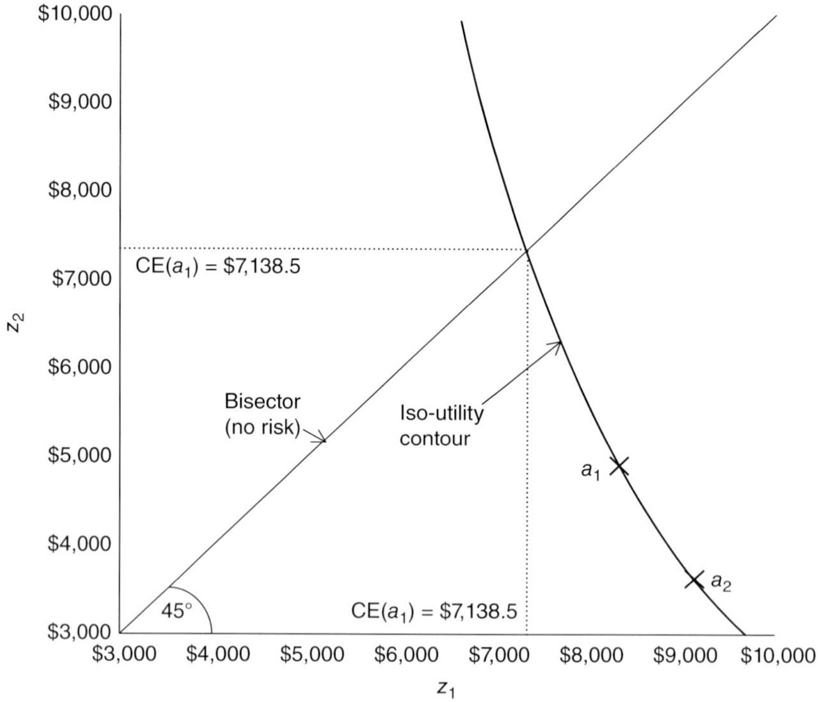

Fig. 8.3. Representation of the wheat-sowing rate choice in state space, maximizing expected utility.

Extending to a Continuous Decision Variable

The above example is confined to discrete levels of input whereas, of course, the sowing rate can be varied continuously, meaning that we need to determine the optimal level.

The first step in this process is to estimate SC production functions, measuring the impact on yield of different sowing rates contingent upon the state of nature that eventuates. At least in principle, this might be possible for several states but, in order to show the process graphically, we'll continue to work with just the two states.

We assume that the following SC production functions showing the effect of sowing rate on yield in the two states, wet and dry season, respectively, are:

$$y_1 = -1.6 + 0.085x - 0.0005x^2 \tag{8.4}$$

$$y_2 = -1.7 + 0.087x - 0.0006x^2 \tag{8.5}$$

where y_i is yield per hectare in tonnes in state i and x is sowing rate in kilograms per hectare. For a wet year $i = 1$ and $i = 2$ for a dry year.

These functions are illustrated in Fig. 8.4. They might have been obtained by regression using data from field trials over a number of years, from farm survey data or, conceivably, as subjective assessments based on the farmer's or adviser's experience.

We assume that the price of wheat, p_w = \$150/t, the price of seed, p_x = \$0.15/kg, other variable costs, v = \$110/ha and the area sown, a = 50 ha. With all these assumptions it is now possible to convert the two production functions into net revenue functions for each of the two states and to transcribe the results into state space, as illustrated in Fig. 8.5.

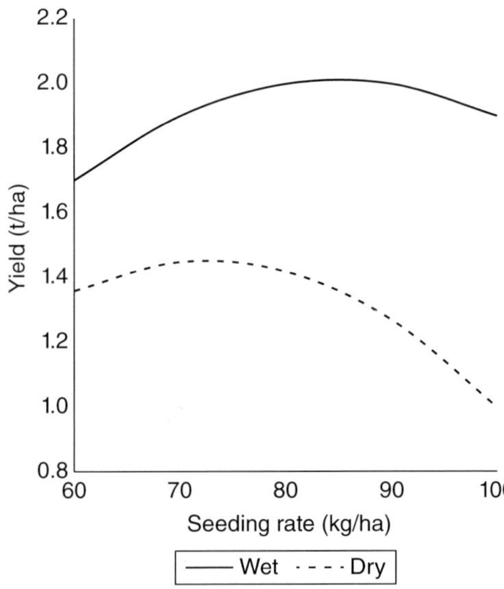

Fig. 8.4. State-contingent production functions for wheat yield against sowing rate on sandy soils.

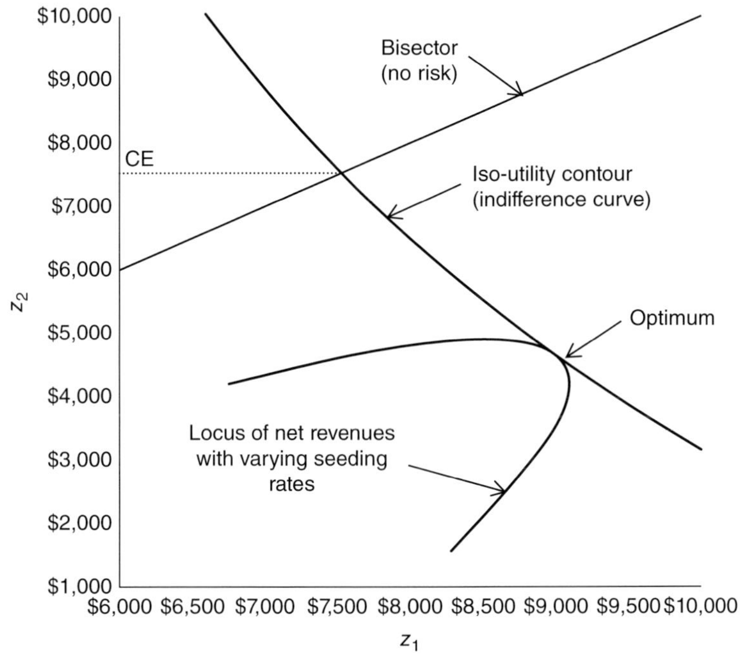

Fig. 8.5. Representation of optimal choice of sowing rate in state space.

Note that, for clarity of the graphs, the two axes in this figure do not have the same scales, so that the bisector is not at 45°. However, the bisector is drawn passing through the risk-free points of equal outcome in the two states.

The locus of net revenue values shown in the figure may be thought of as analogous to a production possibility set of two outputs. Then using the same utility function as above, utility indifference contours of the shape shown can be formulated and the one furthest to the north-east, just tangential to the possibility set of net revenues from different sowing rates, is drawn. The point of tangency indicates optimality. The CE of this decision can be obtained as before on either axis at the intersection of the optimal indifference curve and the bisector. In this case, the CE is \$7529.4, compared with \$7138.5 when only two discrete sowing rates were considered.

This graphical presentation is provided to aid understanding of what's going on and of the relationship with economic theory. The parallel with consumer choice theory in which optimal choice between two goods with an expenditure constraint can be represented using a very similar graphical format.

It is useful to note that, risk preference apart, only the portion of the locus of net revenues in Fig. 8.5 between the point where z_1 is maximized and the point where z_2 is maximized can contain the optimum. This part of the locus therefore represents the *efficient set*, somewhat akin to an E,V efficient set. Moreover, this set of points is efficient for any kind of non-risk preferring utility function, not just a function consistent with the expected utility hypothesis.

In the SC approach to production the optimal choices for a range of formulations can be found by deriving the appropriate marginal conditions – see the references cited earlier or Rasmussen (2011). For the simple example above, finding the optimal solution requires finding the value of the decision variable (sowing rate) that maximizes the chosen objective function. We need to maximize:

$$E[U(x)] = P(S_1)U(z_1) + P(S_2)U(z_2) \tag{8.6}$$

with

$$z_i = a\{p_w y_i(x) - p_x x - v\} \tag{8.7}$$

and

$$U(z_i) = 1 - \exp(-r_a z_i),\ i = 1,2 \tag{8.8}$$

where z_i is net revenue in state i and other variables are as defined above.

The relevant optimization may be done by differentiation or numerically. In the above example, the optimum of $x = 78.8$ kg/ha was readily found using the Solver option in Excel before the utility indifference curve could be drawn in Fig. 8.5.

In reality we might expect to have more than one decision variable. So, for example, the SC production functions for wheat might include variables for soil moisture at sowing, measured levels of key plant nutrients in the seedbed, nutrients applied as fertilizer, etc. We next to extend the above example to illustrate what happens with more than one variable.

Example with More than One Input

To illustrate the case with more than one input variable included in an SC production function, we now extend this example to include two input variables: (i) sowing rate; and (ii) application rate of nitrogen (N) fertilizer at sowing time. It is to be expected that N application will be more effective in a wet year than in a dry one, meaning that the optimal rate and associated yield will be higher under wet conditions than dry. It is also assumed that there will be a small positive interaction between N application and sowing rate in a wet season but that the opposite effect can be expected in a dry year.

The quadratic production functions adopted to reflect these assumptions are:

$$y_1 = -1.83 + 0.0838x_1 + 0.008x_2 - 0.0005x_1^2 - 0.00004x_2^2 + 0.00005x_1x_2 \tag{8.9}$$

$$y_2 = -1.87 + 0.088x_1 + 0.0082x_2 - 0.0006x_1^2 - 0.00004x_2^2 - 0.00005x_1x_2 \tag{8.10}$$

where y_1 and y_2 are yields in wet and dry years, respectively, x_1 is sowing rate in kilograms per hectare and x_2 is N application rate in kilograms of N per hectare. All other variables take the same values as for the one-input case discussed earlier.

The analysis was performed as before. The SC production functions were transformed into net revenue functions using the price and other assumptions. This time the generation of an SC efficient possibility set is more complex because of the two variables. While the formula for this set could be obtained using calculus, for Fig. 8.6 it was produced by maximizing EMV for a full range of probabilities of the two states. Then the point on this efficient set that maximized expected utility was found by solving the values of x_1 and x_2 that maximize:

$$E[U(x_1, x_2)] = P(S_1)U(z_1) + P(S_2)U(z_2) \tag{8.11}$$

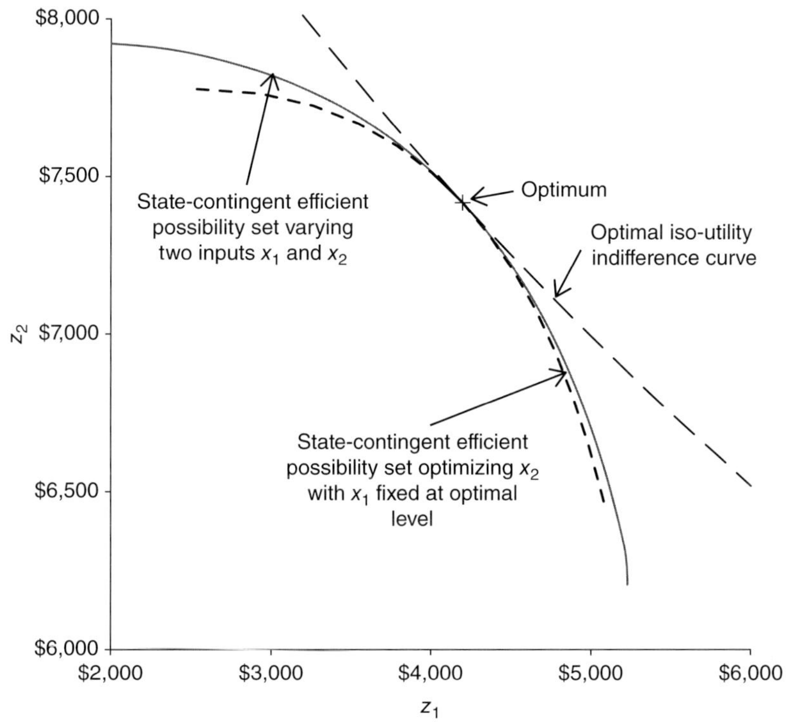

Fig. 8.6. Net revenue possibility set in state space with two input variables.

with

$$z_i = A\{p_y g_i(x_1, x_2) - p_{x1}x_1 - p_{x2}x_2 - v\} \quad (8.12)$$

and

$$U(z_i) = 1 - \exp(-r_a z_i),\ i = 1,2 \quad (8.13)$$

where $g_i(.)$ is the SC production function for yield in state i, px_1 is the price of seed and p_{x2} is the price of N, which is \$1.00/kg, and other variables are as defined earlier.

To generate the optimal iso-utility curve shown in Fig. 8.6 it is first necessary to perform the above optimization, then to find other values of, say, z_2 for given values of z_1 that would give the same expected utility for the given probabilities. Finally, in this figure, we repeated the optimization but holding x_1 constant at the optimal value and optimizing x_2 for different probability combinations. The point of this exercise was merely to emphasize that, with more than one input variable there are interior feasible points within the efficient frontier.

It is important to emphasize that the graphical presentations of the analysis of optimal production in an SC framework are for didactic purposes only. The aim is only to show the similarities of the SC analysis with conventional production economics. In fact, to develop the graphs shown, it was first necessary to find the optimal solutions, and if that is all that matters, which will often be the case in real applications, no graphical representation is needed.

Types of Input Variables

In considering the types of input variables that might be used in developing SC production functions, it is useful first to make a distinction between fixed and variable inputs. At the moment the production choice is being made, there are likely to be a number of measurable variables that describe the current state of the system but that now cannot be altered. In the sowing rate example, current soil moisture level and soil nutrients carried over from the previous crop are two such examples. Similarly, any relevant previous investments, such as the installation of an irrigation system, represent fixed inputs. Such fixed inputs may enter the SC production functions but obviously cannot be changed as part of the process of optimizing input and output levels. Only the variable inputs are relevant in this regard.

Chambers and Quiggin (2000) defined three types of variable inputs that need to be differentiated in the SC approach:

1. state-general inputs that affect output in two or many, possibly all, states of nature;
2. state-specific inputs that, as the name suggests, affect output in only one state of nature; and
3. state-allocable inputs that can be allocated differentially *ex ante* across the states.

Rasmussen (2011) describes the different procedures to be applied to optimize the input use in each of these cases.

The first two input types are reasonably self-evident. For example, the sowing rate of wheat in the example given above is a state-general input. The application of a fungicide to the seed to protect against rot in a wet season would be a state-specific input, offering negligible benefit in a dry season. State-allocable inputs are harder to get to grips with, since they must be allocated before it is known what the actual state will be. An example other authors have given is the division of available labour between measures to drain water away should the season prove to be wet, versus preparations to irrigate the crop should the season be dry.

Of course, in reality, the unfolding of the uncertainty might happen over time, providing opportunities for the DM to take actions to mitigate potentially bad outcomes or to exploit good ones more fully. For example, if our wheat farmer observed good rainfall early in the season, he might elect to apply a top-dressing of fertilizer that would not be warranted if it remained dry. Similarly, later in the season it might become obvious that conditions are so dry as to make it unlikely that the grain yield would be worth harvesting. The farmer could then decide to bring in livestock to graze off the wheat crop. Accounting for such evolving circumstances and opportunities is generally outside the scope of simple production function analyses (Dillon and Anderson, 1990, Sections 6.9 and 8.2). Approaches described in Chapter 11, this volume, may be more relevant.

Relationship between State-contingent and *E*,*V* Analysis

Note that, at each point in the SC production set, we know the SC payoffs and their associated probabilities. Hence, we can calculate the mean and variance of each point and plot these in *E*,*V* space. Then, for the sowing rate example, we get something very like the results for a JP's stochastic production function approach, and we can find the optimum using the indirect utility function:

$$CE = E[U] \approx E - 0.5r_a V \tag{8.14}$$

as shown in Fig. 8.7.

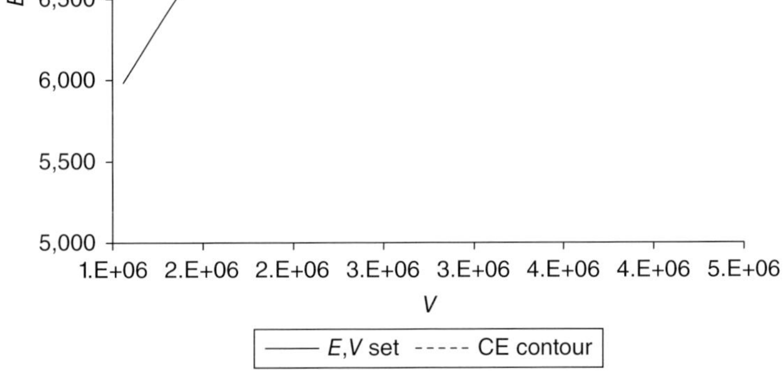

Fig. 8.7. State-contingent results for sowing rate problem expressed in *E*,*V* space.

For this example, the optimum is the same for both forms of analysis. Similarly, if we had the stochastic production function that led to this figure (which we do not have), it could be represented in SC space with two states using the fact that any pair of E, V values can be represented by two equally likely states as $E + SD$ and $E - SD$. Making this transformation from the data for Fig. 8.7 again makes almost no appreciable difference to the optimum, for this particular example.

Commentary

A main appeal of the SC approach to the analysis of production under uncertainty is that it reveals that risk can be accommodated within the ambit of existing production economics theory. It makes it clear that risk should not be regarded as a limitation of that theory. The method also has appeal as providing at least potentially a more satisfactory way of solving risky choices about production. Moreover, the scope is wider than just the analysis of a production function problem, as illustrated here. The same methods work with cost functions, profit functions, etc. Indeed, as should be apparent, almost any problem in decision analysis can be framed in an SC context, though not all may be best solved that way, for reasons to be discussed shortly.

Note that the SC approach is not confined to formulations with only two states. In theory, any number of states is possible, but obviously graphical representation of such problems is not possible and they must be handled analytically. In reality, however, there may be data limitations or other issues that will prevent the analyses being extended to multiple states. And there are other difficulties in application. How are states to be identified? Ideally, the states should be chosen based on the values of uncontrolled variables affecting production. Yet seldom will all such variables have been measured. Nor may there be reliable information on how such variables, individually and jointly, affect production.

There is a view that just two or three states are likely to be 'enough' to provide a good representation of most decision problems (J. Quiggin, 2006, personal communication). One justification for this notion comes from the experience that E, V approximations of expected utility are often found to give results quite close to those obtained using the full distribution of risky outcomes. Moreover, as we have shown, any E, V analysis, including a stochastic production function analysis, can be converted to an equivalent SC formation in two states, and vice versa. We might expect that, if the distribution of outcomes in the production process under study is reasonably symmetric, an SC analysis with only two states is likely to give a close approximation to the true (but generally unknown) optimal solution, and the same can probably be said for a stochastic production function analysis in terms of E and V. The more skewed the distributions of outcomes, the more likely it is that at least three states would be needed of an SC analysis, and the less satisfactory an E, V analysis via a stochastic production function would be.

In practice, of course, it may be hard to judge how many states would be needed in a particular SC analysis to give an answer that is 'good enough'. With only two or three states, the estimated SC functions must span a range of situations, meaning that they are in effect stochastic functions with non-zero error terms. Yet the approach, as it is currently presented, makes no provision for such uncertainty. The approximations inherent in analyses such as illustrated in this chapter are a reminder that some degree of approximation is inevitable in the practice of whatever analytic approach is taken.

At the time of writing, it seems that the relative merits of the SC approach versus the stochastic production function approach need more investigation. In the meantime, it seems likely that the SC approach might perform better than the stochastic production function approach: (i) when there is a clear basis for identifying a sufficient number of differentiated states; (ii) when there is a sound basis for estimating contingent functions for these states; and especially (iii) when the distributions of outcomes are highly asymmetric. Experience suggests that these criteria will be met rather rarely. In particular, data deficiencies seem likely to limit the use of SC analysis of production in agriculture, at least until experimental or other data-generating methods are adapted to provide the extra data needed. On the other hand, the conceptual framework of the SC, as illustrated by the figures in this chapter, offer a different and perhaps more insightful way of understanding the nature of risky choice. For this reason if for no other, we might expect to see more applications of SC analysis in agriculture in future.

Selected Additional Reading

See the references cited in this chapter. Some empirical examples of SC analysis of agricultural production are provided by O'Donnell and Griffiths (2006), Chavas (2008) and Nauges *et al.* (2011).

References

Arrow, K. and Debreu, G. (1954) Existence of an equilibrium for a competitive economy. *Econometrica* 22, 265–290.

Chambers, R.G. and Quiggin, J. (2000) *Uncertainty, Production, Choice and Agency: the State-Contingent Approach*. Cambridge University Press, New York.

Chavas, J.-P. (2008) A cost approach to economic analysis under state-contingent production uncertainty. *American Journal of Agricultural Economics* 90, 435–466.

Debreu, G. (1952) A social equilibrium existence theorem. *Proceedings of the National Academy of Sciences, USA* 38, 886–893.

Dillon, J.L. and Anderson, J.R. (1990) *The Analysis of Response in Crop and Livestock Production*, 3rd edn. Pergamon, Oxford.

Just, R.E. and Pope, R.D. (1978) Stochastic specification of production functions and economic implications. *Journal of Econometrics* 7, 67–86.

Nagues, C., O'Donnell, C.J. and Quiggin, J. (2011) Uncertainty and technical efficiency in Finish agriculture: a state-contingent approach. *European Review of Agricultural Economics* 38, 449–467.

O'Donnell, C.J. and Griffiths, W.E. (2006) Estimating state-contingent production functions. *American Journal of Agricultural Economics* 88, 249–266.

Quiggin, J. and Chambers, R.G. (2006) The state-contingent approach to production under uncertainty. *Australian Journal of Agricultural and Resource Economics* 50, 153–169.

Rasmussen, S. (2011) *Optimisation of Production under Uncertainty: the State-Contingent Approach*. Springer, Heidelberg.

Sandmo, A. (1971) On the theory of the competitive firm under price uncertainty. *American Economic Review* 61, 65–73.

9 Risk and Mathematical Programming Models

A Brief Introduction to Mathematical Programming

Mathematical programming (MP) is the term used to describe a family of optimization methods that may be familiar to many readers. Therefore, only a brief introduction to these useful methods is provided here and readers needing to learn more by way of background are referred to one of the many introductory texts on the topic, such as Williams (2013).

Almost all optimization problems, including planning problems accounting for risk and uncertainty, can be expressed as the optimization of some objective function subject to a set of constraints. MP has been developed just for such problems. Unfortunately, however, many real-world planning problems are complex, especially when accounting for risk and uncertainty. The result is that, while most risk-planning problems can be formulated as MP models, at least conceptually, not all can be reliably solved using available algorithms. Sometimes it may not be possible to be sure that a generated solution is the true optimum. The models for some problems may grow so large that the computing task may be beyond the capacity of the computer or software being used.

The simplest form of MP, and the first for which a solution routine was developed and applied, is *linear programming* (LP). LP models are specified in terms of maximizing a linear objective function subject to a set of linear constraints. For example, a planning problem might be to find the mix of activities on a farm to maximize expected profit, subject to constraints imposed by the limited availability of resources of various kinds. A simple example of the application of LP to just such a planning problem is illustrated later in this chapter.

The main and obvious limitation of LP is that the real world is seldom linear. With a little ingenuity, however, many real problems can be approximated reasonably well by an LP model. Moreover, reliable computer software to solve LP models is widely available.

Because of the ready availability of LP software, it is not necessary for users to know how to perform the massive amount of arithmetic often needed to find a solution. However, it is useful to have at least a broad idea of the nature of the underlying optimization problem and how it is solved.

The linear constraints, taken together, define a *convex set* containing an infinite number of feasible solutions. Each solution in this set is defined in terms of the levels of the decision variables. In three dimensions, such a set would look somewhat like a faceted gemstone, with exterior facets, edges and corners. In LP, each facet is defined by one linear constraint. In reality, with n variables, the set is defined in n-dimensional space, but this is a good deal harder to imagine, so we'll stick to three dimensions for the analogy!

The linear objective can be thought of as a plane or slice passing through the set. Every point on this plane has the same value of the objective function. Of course, only those points on the plane within the convex set defined by the constraints are feasible. The plane moves parallel to itself according to the total value of the objective function. Conceptually, the aim is to move this plane in the direction that improves the value of the objective until the position is found where it is still in contact with the feasible set while taking its optimal value. This optimum necessarily includes corner points and will often be a single corner point.

This latter observation is important because the LP solution method, called the *simplex method*, involves a search that moves from corner point to corner point of the feasible set. The optimum is found using an algorithm that 'crawls' round the surface of the convex set from corner to corner, always in the direction of some improvement in the objective function, until no more progress can be made. Given that the feasible set is convex and that the objective function is linear, it is certain that this point is the optimum.

This solution method can be likened to climbing a mountain in a mist by always heading up hill. When the terrain all about you slopes down, you can assume that you are at the top, but only if the hill is convex. Otherwise, you might have reached a minor summit but would not realize it unless the mist cleared.

In addition to the strong assumptions of linearity, LP models (and most other MP models) are usually said to be deterministic, meaning that all the coefficients in such a model are treated as known constants. This might appear to make LP very poorly suited to risk analysis. In fact, however, considerable effort and ingenuity have been applied to the development of a wide range of model formulations within the basic LP framework to allow for reasonable representation of risk and risk aversion. Some of these are better than others and some limitations remain, as explained in the rest of this chapter. Most involve making discrete approximations of what in reality are often continuous probability distributions.

An obvious problem with LP for risk analysis is the linear objective function, since risk aversion implies optimization in terms of a generally non-linear utility function. Fortunately, reasonably reliable methods exist to solve MP models with a well-behaved non-linear objective function and linear constraints. In this context, a well-behaved objective function is one that is convex to the feasible set, which is usually the case when maximizing expected utility under risk aversion. The relatively wide availability and ease of use of software to solve MP models with a non-linear objective have made many of the earlier LP approximations obsolete.

Non-linearity in the constraints is more difficult to deal with in MP models because it often amounts to a violation of the convexity requirement for the set of feasible solutions. That can mean that only a local optimum may be found, as in the hill-climbing analogy. The same problem can occur if a non-linear objective function is not well behaved. While there are some ways to improve the chances of finding the global optimum in these cases, such as repeating the search starting at different points, there is always a risk that solutions found for such models will not include the true optimum. Whether that matters will depend on circumstances, but sometimes a solution that is appreciably better than the present one may be good enough, even though there is no way to be sure that it is the very best.

A Basic Example

The following quite basic example illustrates the bare essentials of a non-linear MP model, although its very simplicity reveals little about the power of the method to deal with much more complex optimization problems. The example is an extension of the case already described in Chapters 2 and 6 of the dairy farmer who is trying to decide whether to insure against losses from a possible outbreak of foot-and-mouth disease. We now assume that the farmer not only has a choice of whether or not to insure, but can also decide what level of cover to buy. If the decision is to buy less than 100% cover, the premium paid is directly proportional to the level of cover chosen. Similarly, should a disease outbreak occur, the indemnity payments will be scaled according to the proportion of the risk that was insured. The farmer is assumed to be risk averse and all other assumptions are as in Fig. 6.9 in Chapter 6. The problem is reformulated as in the worksheet of Fig. 9.1.

Solver is an optional add-in for Microsoft Excel that allows an objective to be maximized, minimized or set to a fixed value subject to constraints specified by the user. In this case we set Solver to maximize the contents of cell E14, containing the CE from the decision, by varying cell C5, the proportion of cover purchased. We specify that the optimization is subject to the constraints that C5 must be greater than or equal to zero and less than or equal to 100%. Note that any trial starting value in cell C5 can be entered; in this case 100% has been chosen. These details are entered into the Solver window that pops up when the option is activated within Excel. Then clicking the 'Solve' button produces the solution quickly.

The solution found by Solver in this case sets cell C5 to 79%, leading to a CE of $492.8k.

Of course, we could have found the same answer for this simple example in a number of other ways, such as trial and error in Excel or simple calculus. Nevertheless, the example does illustrate the bare bones of MP. A non-linear objective function was optimized subject to a set of two constraints. In this case, only one cell representing a single decision variable was used. In most MP models there will be an array of such adjustable decision variables. Similarly, there will usually be a large number of constraints defined for the decision variables. MP models with hundreds of adjustable variables and constraints are quite common in farm planning applications.

Solver for Excel could be used to solve most, if not all, of the examples in the remainder of this chapter. However, for large problems the standard version of Solver provided with Excel may not be sufficiently powerful and other specialist MP software may be needed. Some options are listed in the Appendix of this volume.

	A	B	C	D	E
1	**Solver analysis for dairy farmer's insurance problem**				
2					
3	*Varying percentage cover bought to maximize CE*				
4					
5	Cover bought (%)		100%	Utility function:	$U = 1 - \exp(-r_a W)$
6	Premium/unit % cover		7.245	$r_a =$	0.003658
7			(Money amounts in $ thousands)		
8					
9			Probabilities	Terminal wealth	Utility
10	No outbreak		0.94	492.8	0.8351
11	Outbreak	Bans	0.03	492.8	0.8351
12		Slaughter	0.03	492.8	0.8351
13				EU	0.8351
14				CE	492.8

Fig. 9.1. Excel worksheet for Solver application to optimize the proportion of dairy farmer's insurance cover. CE, certainty equivalent; EU, expected utility (both of wealth).

Mathematical Programming Approaches to Whole-farm System Planning under Risk

Farms, along with other kinds of agricultural businesses, are often best thought of in a systems context. Taking this view makes it clear that a decision relating to one part of the business will often affect other parts. Therefore, in modelling a decision about the operation of such a business, it may be wise to cast the analysis in a whole-farm (or whole-business) context, rather than in a partial way. This is particularly true for many risk analyses for which stochastic dependencies between risky prospects may make partial evaluations unreliable. For the same reasons, in planning a farm (or other) business, accounting for risk, it makes sense to include in the analysis any opportunities to extend the range of alternatives considered.

Farms (again as for other businesses) are constrained systems, in the sense that what can be done is limited by the available amounts of resources such as land, labour and capital in its various forms (e.g. buildings, machinery, livestock), and by the restrictions imposed from outside the farm, such as quotas, environmental limitations, marketing restrictions, etc. So the whole-farm planning problem can be seen as one of maximizing some objective function that reflects the goals of the farmer, subject to a set of constraints. As indicated, MP methods are well adapted for such problems. MP models of whole-farm systems are relatively simple to construct and solve. They permit production, financing, risk-sharing and consumption components of the system to be accommodated as appropriate, and have considerable flexibility in use.

Most of the early MP modelling studies of farm systems took little or no account of risk. Since farming is inherently a risky business, it seems likely that many of these analyses will have overlooked an essential component of the farm planning task. These days, ways to accommodate risk in MP models for farm planning are well developed and reasonably easy to use. Most of these methods, described in this chapter, can also accommodate components representing the use of financial products such as futures hedging, buying insurance or trading in weather derivatives (if available), allowing a comprehensive representation of the planning problem in a risky world.

The impact of risk and uncertainty on planning any farm is likely to be pervasive and complex. It is usually impossible to contemplate accounting for all sources and impacts of uncertainty. Rather, some simplification will be necessary. Thus, the modelling of any risky farming system must start with an assessment of the main ways in which uncertainty impacts on that system. An outline decision tree can provide a good means of capturing in a diagram the principal kinds of decision that can be made and the main sources of uncertainty impinging on those decisions and their consequences.

In developing MP models of risky farming systems it is necessary to distinguish cases with *embedded risk* from those with *non-embedded risk*. These two cases are represented using outline decision trees in Fig. 9.2. As indicated in the figure, embedded risk occurs when some decisions depend on both earlier decisions and on the outcomes of some uncertain events.

An example of the first case, of non-embedded risk, is an arable farm where the farmer has to decide at the start of the cropping season what crops to grow. At that time, crop yields and prices are taken to be still uncertain and become known only after harvest. However, once the crops are sown, there are no more important decisions to be made. An example of embedded risk, as illustrated in the lower part of the figure, is a mixed farm with both crops and grazing livestock in an area of uncertain rainfall. Crop and pasture areas and numbers of livestock kept must be decided at the start of the farming year. As before, yields of crops and prices of products are uncertain, but the amount of forage for the grazing stock is also uncertain. This means that the farmer must adopt a strategy to make best use of any forage surpluses, or to meet

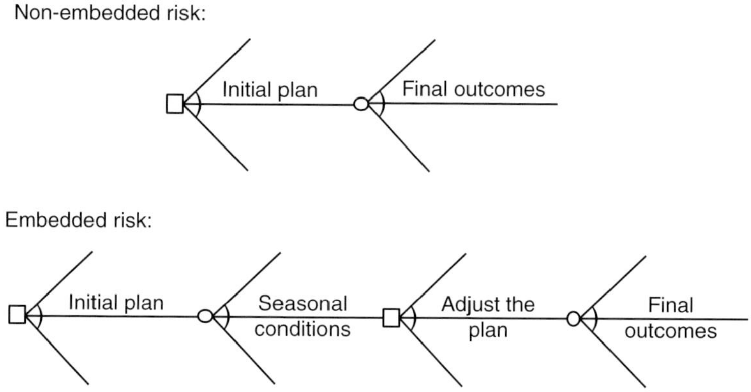

Fig. 9.2. Outline decision trees showing non-embedded and embedded risk.

shortfalls in feed availability for the grazing animals, perhaps including using crops originally intended for harvest and sale for stock feed.

In reality, most real systems have embedded risk but, because it is somewhat difficult to handle, it may be assumed away in MP modelling whenever such an assumption is judged not to be too far removed from reality. We call MP models for non-embedded risk *risk-programming* models; and we call those accounting for embedded risk *stochastic programming* models.

Risk Programming

Linear risk programming

LP is the most widely applied MP method used for farm planning. The method has also been routinely applied in other planning problems in agriculture, such as the formulation of least-cost diets for farm animals. In the context of whole-farm planning accounting for risk, LP may be used to maximize expected profit subject to the farm resource constraints and other restrictions. The notation for a problem of n activities and m constraints is as follows:

$$\text{maximize } E = c\,x - f \tag{9.1}$$

subject to

$$A\,x \leq b \text{ and } x \geq 0$$

where E is expected profit, c is a 1 by n vector of activity expected net revenues, x is an n by 1 vector of activity levels, f is fixed or overhead costs, A is an m by n matrix of technical coefficients, b is an m by 1 vector of resource stocks, and 0 is an n by 1 vector of zeroes. ('Net revenue' is the term used for the measure of activity net profitability per unit level. For most activities, but not necessarily all, it will be the same as the more familiar notion of activity gross margin.) Since the fixed costs make no difference to the optimal solution for this linear model, they can be omitted from the model formulation. Note, however, that the level of fixed costs can affect the solution in models set up to maximize expected utility under risk aversion.

For readers unfamiliar with matrix notation, some brief interpretation of the above may be helpful. In Eqn 9.1 the product of the vectors c and x is simply the total net revenue obtained by multiplying the level of each activity in x by the corresponding expected net revenue per unit in c and summing the results. Deducting fixed costs f from the total gives the expected farm income E. The constraints $A\,x \leq b$ is matrix shorthand to say that, for each constraint represented by the rows of the matrix A, the sum of the products of each activity level in vector x and the corresponding resource requirements per unit in that row of A must be no more than the corresponding amount of that resource available, as indicated by the relevant entry in the vector b. (It is also possible to set some constraints to have the left-hand side greater than some amount specified, such as a minimum farm area set aside for conservation. Similarly, it is possible to set some strict equalities.) Last, the non-negativity restrictions on x simply mean that negative levels of activities are not permitted. This convenient representation of MP problems in matrix algebra may be clearer later when we show the relationship between the parts of an example LP model in tabular format and the various components defined above.

We now define $c = p\,C$ (9.2)

where c is the 1 by n vector of activity expected net revenues (as before) and p is a 1 by s vector of state probabilities and C is an s by n states of nature matrix of activity net revenues per unit level by state (row) and activity (column). The matrix C may be based on adjusted historical data or may be wholly or partially subjective. The matrix specifies a number of possible future sets of outcomes of activity per unit net revenues that might occur in the planning period. While such states of nature are often treated as equiprobable, they can be weighted with probabilities according to assessments of how likely it is that each state will be somewhat representative of the future real outcome in the period when the farm plan will be implemented. The matrix shows not only the uncertainty about individual activity net revenues but also embodies a measure of the stochastic dependency between activity returns (see Chapter 4, this volume).

An example for purposes of illustration is provided in Fig. 9.3.

Note that the sample size is small with only six states. Small samples are common in such states of nature matrices for farm planning because they are typically based on historical records of the farmer, which

	A	B	C	D	E	F	G	H	I	J
1	**Subjectively adjusted states of nature matrix[a]**									
2										
3		State	Prob	Potatoes	S Beet	Onions	W Wheat	Sp Wheat	Sp Barley	Grass
4		1	0.15	744.78	2041.57	−3618.50	1109.04	931.26	877.31	1239.33
5		2	0.18	2509.94	2274.27	−1213.83	1120.23	937.30	699.13	1095.13
6		3	0.20	3982.65	2728.56	12898.73	1147.26	997.02	697.77	1412.63
7		4	0.08	3854.27	2570.80	7261.60	1138.07	938.24	831.54	1372.82
8		5	0.29	2820.44	2465.65	2799.56	1165.81	955.54	955.04	1556.50
9		6	0.10	1566.94	2559.81	−2302.82	1194.21	990.27	1005.25	1362.39
10	Calculated means			2643	2438	2981	1146	959	841	1363
11	Calculated SDs			1070	220	5800	26	25	120	163
12										
13	[a] Based on historical data, adjusted for inflation, detrended and with subjective probabilities									
14	assigned to states to provide 'best guesses' of possible outcomes in the planning period.									

Fig. 9.3. States of nature matrix of risk-programming models.

seldom extend over many years for all activities of interest. Moreover, even if a long sequence of such records is available, they can seldom be regarded as representing possible future outcomes due to changes in production methods and market conditions. Nor can area averages of net revenues be used since, while they may exist for longer periods than individual farm records, such averages tend to 'iron out' a significant part of the variation and hence of the uncertainty likely to be experienced on a single farm. Methods described in Chapter 4 can be used to make the best use of such sparse data as may be available from a particular farm, coupled with subjective judgements, for example, by smoothing CDFs through the individual activity returns and by capturing stochastic dependency via an appropriate copula (Lien *et al.*, 2009).

Suppose that the states of nature matrix shown in Fig. 9.3 relates to a crop farm in The Netherlands for which the plan that maximizes expected profit is to be sought. The farm has a total arable area of 45 ha. The farmer can choose from the seven different crops listed. Figure 9.3 also shows the calculation of expected per unit net revenues for the crops. These values will be used in the MP model.

The farm work is normally carried out by the farmer and by farm family members, but the farmer also has the option to hire additional labour in the peak labour demand months of May, August, September and October. The maximum farm labour hours available from the permanent workers and the maximum additional hours that can be hired are given in Table 9.1.

Technical input–output coefficients for seasonal labour requirements of the crops in these busy 6 months are known and are assumed to be fixed, but are not listed here to save space (but they can be found in Fig. 9.4 below). Total fixed costs are $40,000.

Other constraints that the farmer must take into account are restrictions on the areas of individual crops owing to rotational considerations or to marketing limits, as listed in Table 9.2.

Table 9.1. Availability and costs of labour for the example problem.

Month	Maximum own labour (h)	Maximum hired labour (h)	Cost of hired labour ($/h)
May	290	60	30
June	230	Nil	–
July	220	Nil	–
August	240	120	10
September	240	120	10
October	300	60	30

Table 9.2. Area restrictions of crops for the example problem.

Crop	Maximum area (ha)
Potatoes	9.00
Sugarbeet	11.25
Onions	11.25
Cereals	25.00
Winter wheat	22.50
Spring wheat	11.25

	A	B	C	D	E	F	G	H	I	J	K	L	M	N	O	P
1	**Example LP Model**															
2			Activities:	Potatoes	Sugar	Onions	Winter	Spring	Spring	Grass-	Hire May	Hire Aug	Hire Sept	Hire Oct	Rel	Right-
3					beet		wheat	wheat	barley	land	labour	labour	labour	labour		hand
4			Units:	ha	ha	ha	ha	ha	ha	ha	h	h	h	h		side
5				POTS	SB	ONIONS	W WH	SP WH	SP BAR	GRASS	LABMAY	LABAUG	LABSEP	LABOCT		
6	Net revenue	$		2643	2438	2981	1146	959	841	1363	–30	–10	–10	–30		
7	Land	ha	LAND	1	1	1	1	1	1	1					<=	45
8	Cereals max	ha	MX CER				1	1	1	1					<=	25
9	Potatoes max	ha	MX POT	1											<=	9
10	Sugar beet max	ha	MX SB		1										<=	11.25
11	Onions max	ha	MX ON			1									<=	11.25
12	W wheat max	ha	MX WW				1								<=	22.5
13	Sp wheat max	ha	MX SW					1							<=	11.25
14	May labour	h	MAYLAB	6	15	17	5	1.5		1	–1				<=	290
15	June labour	h	JUNLAB	8	10		5	1	1						<=	230
16	July labour	h	JULLAB		7	3		1		4					<=	220
17	Aug labour	h	AUGLAB	5			15	1	5	6		–1			<=	240
18	Sept labour	h	SEPLAB	20		7		6	1	4			–1		<=	240
19	Oct labour	h	OCTLAB	15	10	19	8	3	2					–1	<=	300
20	Max May lab hire	h	MXHMAY								1				<=	60
21	Max Aug lab hire	h	MXHAUG									1			<=	120
22	Max Sep lab hire	h	MXHSEP										1		<=	120
23	Max Oct lab hire	h	MXHOCT											1	<=	60

Fig. 9.4. Linear programming (LP) tableau for the example problem.

The representation of this example farm in LP format, called a *tableau*, is shown in Fig. 9.4. This tableau represents an LP model set up to maximize the sum of the activity expected net revenues, subject to the constraints described above. Such a model would be appropriate for cases when a farmer is effectively indifferent to risk and is therefore prepared to base the choice of farm plan on maximization of expected income. It is also implicitly assumed that embedded risk is of little importance for the analysed problem. Note that, since the model represented here is for a plan for the next cropping year, and most things in the plan can be changed in the future, the income being measured falls into the category of transitory income, as described in Chapter 5, this volume. Hence, as also described in that chapter, we can expect the somewhat affluent Dutch farmer for whom the analysis is done to be only moderately averse to risk in transitory income, so that assuming no risk aversion may be not too wide of the mark.

The tableau is set up in Fig. 9.4 with the activities as columns D to N. The objective function values, representing the vector c, are in row 6. The input–output coefficients, comprising the matrix A are in D7:N23. These numbers show the amount of each constraint or intermediate quantity consumed (or produced if the entry is negative) per unit level of each activity. The vector of resource stocks b is in column P. Not shown are the solution vector x and the value of the objective function E which are found by the solution procedure. The non-negativity restrictions on the x variables are also handled by the solution procedure. Note the inclusion of abbreviated names for activities and constraints which will appear in the output file to facilitate interpretation of the solution.

This information in the tableau can now be processed to find the optimal solution using suitable LP software. Options include Solver from Frontline Systems Inc. and What's*Best*! from Lindo Systems Inc., both of which work as add-ins to spreadsheet software such as Microsoft Excel. Other packages accept spreadsheet files as data files but some programs require a special data entry format. Surveys of available software are published from time to time in *Informs Online* (http://www.orms-today.org/).

The results obtained for the example LP problem using the basic version of Solver supplied with Excel indicate a farm plan containing the activities and levels listed in Table 9.3.

Evidently, a risk-neutral farmer should plant the maximum areas of potatoes and sugarbeet, along with nearly 6 ha of onions, devoting the remainder of the farm to grassland. Shortages of farm labour in May, September and October are satisfied by hiring additional labour, in the latter month to the limit of labour hiring that was set. The total value of expected net revenue is nearly \$90,600 which, with fixed costs of \$40,000, leaves the farm family with an expected net income from farming of some \$50,600.

Table 9.3. Linear programming (LP) solution for the example problem.

Activity	Level
Potatoes	9 ha
Sugarbeet	11.25 ha
Onions	5.92 ha
Grassland	18.83 ha
Hire labour in May	52.2 h
Hire labour in September	56.78 h
Hire labour in October	60 h
Total expected net revenue	\$90,594

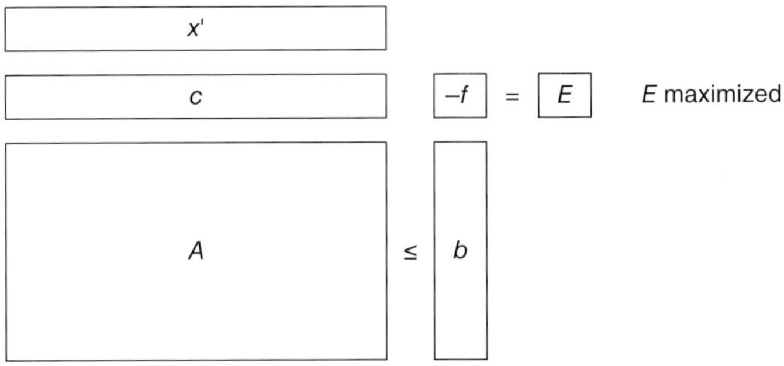

Fig. 9.5. Outline of an LP tableau.

In addition to the *primal solution* given above, the so-called *dual solution* can also be generated, providing information about the marginal value products of the limiting resources and the marginal opportunity costs of the excluded activities. However, since the focus here is on risk programming, these additional features, described fully in most standard texts on LP, will not be dwelt upon.

To clarify the relationship between the general description of the LP problem in matrix algebra and the example, the tableau in Fig. 9.4 can be re-cast in general terms as shown in Fig. 9.5. The matrix notation applied to the various parts of the tableau is shown in this figure.

The solution, which is of course unknown at the time the tableau is formulated, is represented by the row vector x' of adjustable variables representing activity levels. (Note that in matrix notation a prime (′) indicates the transposition of a column vector to a row, or vice versa. In this case, x itself was defined as a column vector.) The vector of objective function coefficients (expected net revenues per unit level of each activity) is shown as the row vector c. As explained above, expected income, which is to be maximized, is calculated by multiplying each activity level in x' by the corresponding expected net revenue per unit in c and summing. In this case we have also shown for completeness the deduction of the fixed costs f in calculating E. The matrix of input–output coefficients A shows the use of each resource or constraint per unit level of each activity, while the vector of right-hand-side coefficients, indicating resource stocks or accounting balances, is shown as the column b. For each constraint (row of A) the sum of the products of the per unit entries in A and the activity levels in x' must be less than or equal to the corresponding maxima in b.

As illustrated by the above example, in the ordinary LP formulation of a farm planning problem, non-embedded risk can be at least partly accommodated by the use of expected activity net revenues calculated across possible states of nature. However, such a linear risk-programming model does not account for any non-neutral risk attitude of the farmer. Models that overcome this latter deficiency are described next.

Quadratic risk programming

Quadratic risk programming (QRP) can be used to generate the set of farm plans lying on the E,V efficient frontier (Freund, 1956). The notion of E,V efficiency was discussed in Chapter 7, this volume.

The QRP model may be formulated in a number of ways. One option would be to maximize CE as defined in Eqn 5.22 (Chapter 5) and reintroduced here:

$$CE \approx E - 0.5 r_a V \quad (9.3)$$

where E is expected income, r_a is the absolute risk-aversion coefficient (assumed constant) and V is variance of income. This will work if we know r_a. However, if all we know is that the DM is risk averse, it would be appropriate to generate the whole E,V efficient frontier, as illustrated in Fig. 7.2 (Chapter 7). That can be done by minimizing V while varying E over its feasible range, or by maximizing E while varying V over its feasible range. Both will give the identical set of solutions, as would maximizing Eqn 9.3 while varying r_a from 0 to infinity. Here we use the option:

$$\text{minimize } V = x'Qx \tag{9.4}$$

subject to

$$Ax \leq b$$

$E = cx - f$, with E varied over the feasible range

and $x \geq 0$

where Q is an n by n variance–covariance matrix for activity net revenues per unit. The advantage of this formulation is that the non-linear component is confined to the objective function, so that the feasible set remains convex and problems in finding the optimal solution are minimized.

The above algebra can be represented in tableau format as shown in Fig. 9.6.

In applying this model to the example problem, the components c, A and b in Fig. 9.6 are identical to the corresponding sections of Fig. 9.4 as outlined in Fig. 9.5. The variance–covariance matrix Q is calculated from $Q = D'(PD)$ where $D = C - uc$, i.e. an s by n matrix of deviations of activity net revenues from respective means, with u an s by 1 vector of ones, and P an s by s matrix of the probabilities of the s states along the main diagonal and zeroes elsewhere.

For the example problem, we need to use the states of nature table (Table 9.3) that shows activity net revenues per unit, along with the associated probabilities. These data have been transcribed to Fig. 9.7, which shows the calculations made to obtain the matrix Q.

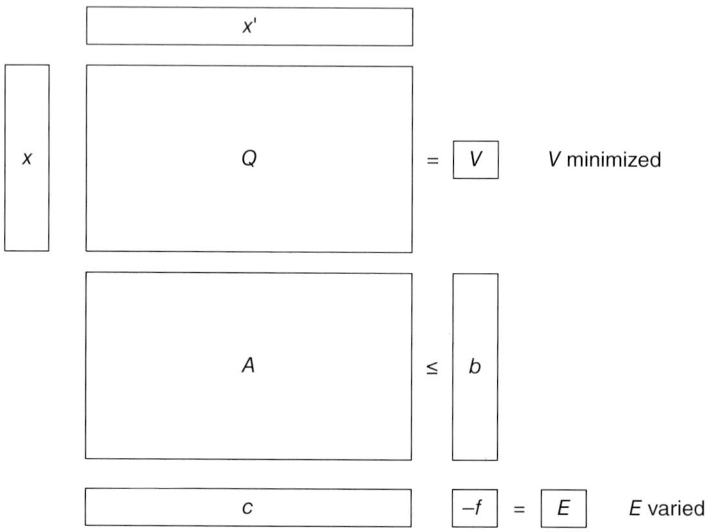

Fig. 9.6. Outline of a quadratic risk-programming tableau.

	A	B	C	D	E	F	G	H
1	**Calculation of deviations matrix and covariance matrix from states of nature matrix of GMs matrix C**							
2	Using Excel array formulae, as indicated							
3	**GMs**							
4		744.78	2041.57	−3618.50	1109.04	931.26	877.31	1239.33
5		2509.94	2274.27	−1213.83	1120.23	937.30	699.13	1095.13
6	C =	3982.65	2728.56	12898.73	1147.26	997.02	697.77	1412.63
7	(6 by 7)	3854.27	2570.80	7261.60	1138.07	938.24	831.54	1372.82
8		2820.44	2465.65	2799.56	1165.81	955.54	955.04	1556.50
9		1566.94	2559.81	−2302.82	1194.21	990.27	1005.25	1362.39
10	**Probs**							
11	p =	0.15	0.18	0.20	0.08	0.29	0.10	
12	(1 by 6)							
13	**E[GM]**	(=MMULT(B11:G11;B4:H9))						
14	c =	2643	2438	2981	1146	959	841	1363
15	(1 by 7)							
16	u' =	1.00	1.00	1.00	1.00	1.00	1.00	
17	(1 by 6)							
18	**Devns**	(=B4:H9-MMULT(TRANSPOSE(B16:G16),B14:H14))						
19		−1898.22	−396.43	−6599.50	−36.96	−27.74	36.31	−123.67
20		−133.06	−163.73	−4194.83	−25.77	−21.70	−141.87	−267.87
21		1339.65	290.56	9917.73	1.26	38.02	−143.23	49.63
22		1211.27	132.80	4280.60	−7.93	−20.76	−9.46	9.82
23		177.44	27.65	−181.44	19.81	−3.46	114.04	193.50
24		−1076.06	121.81	−5283.82	48.21	31.27	164.25	−0.61
25	**P matrix**							
26		0.15	0	0	0	0	0	
27		0	0.18	0	0	0	0	
28	P =	0	0	0.20	0	0	0	
29	(6 by 6)	0	0	0	0.08	0	0	
30		0	0	0	0	0.29	0	
31		0	0	0	0	0	0.10	
32								
33	**Cov(GM)**	(=MMULT(TRANSPOSE(B19:H24),MMULT(B26:G31;B19:H24)))						
34		1144900	195831	5610847	6544	13050	−58039	65899
35		195831	48400	1072059	3693	4631	−3487	19781
36		5610847	1072059	33640000	29325	95808	−308950	416627
37	$Q = D'(PD)$ =	6544	3693	29325	676	408	1873	3043
38	(7 by 7)	13050	4631	95808	408	625	−271	1726
39		−58039	−3487	−308950	1873	−271	14400	11127
40		65899	19781	416627	3043	1726	11127	26569
41								
42	**SDs**	1070.0	220.0	5800.0	26.0	25.0	120.0	163.0

Fig. 9.7. Calculation of variance–covariance matrix for the example problem.

Solving this model requires access to software capable of solving non-linear problems. For this example, the model was solved using GAMS (General Algebraic Modeling System from GAMS Development Corporation) (Rosenthal, 2014) that incorporates several powerful non-linear programming packages. There are also several other software packages that will do the same job, including Lingo from Lindo Systems Inc.

Note that this simple example problem does not include activities for sharing risk off farm, such as purchase of crop insurance or use of derivative markets. Ideally, such opportunities need to be accommodated in the model if the farmer is risk averse and if the solution is to indicate how risk is to be best managed both on and off farm. Such additional opportunities have been omitted to keep the example as simple as possible. The same qualification applies to other models presented later in this chapter.

Some of the *E,V* efficient change-of-basis solutions to this problem are listed in Table 9.4. (Change-of-basis solutions can be thought of as the corner points of the convex set, as described earlier.) Only the solutions that use all the land are presented on the grounds that it is unrealistic to leave land idle. Intermediate solutions between these change-of-basis solutions can be found by linear interpolation for all features except the variance. Anderson *et al.* (1977, p. 202) provide the formula to interpolate the variance. However, for practical purposes it is more satisfactory to solve the problem for a number of points on the *E,V* efficient set to obtain a good indication of the solutions likely to be of interest to the DM.

The *E,V* efficient set for this problem is illustrated in Fig. 9.8. The figure illustrates the typical result that, starting from the solution maximizing expected income at the top right of the set, it is possible at first to achieve a significant reduction in variance for the sacrifice of relatively little expected income. Later, however, the curve falls more steeply, indicating that a greater sacrifice of expected income is needed to reduce variance further. Evidently, provided that the results were presented in an understandable fashion, such as in terms of the expected value and a range corresponding to, say, the 0.05 and 0.95 fractiles, a risk-averse DM could decide where on the frontier to locate. Then it would be a simple task to identify the corresponding farm plan from the MP results. Alternatively, if the DM's utility function is known, the point of maximum expected utility can be calculated – again see Anderson *et al.* (1977, Chapter 7) for an explanation of how this can be done.

As explained in Chapter 5, this volume, the assumptions necessary to validate the use of QRP for farm planning under risk are that the farmer's utility function is quadratic or the distribution of total net revenue is normal. Quadratic utility functions are not increasing at all points and also imply increasing absolute risk aversion. Both these properties are generally regarded as unacceptable. The distribution of total net revenue varies from case to case and is unlikely to be normal – in agriculture, returns from individual activities are often skewed, although, at least for a mixed farming system, the distribution of total net revenue may be approximately normal in so far as it is influenced by many small and unrelated

Table 9.4. Change-of-basis solutions for the quadratic risk programming (QRP) formulation of the example problem (variance is in $\$^{2}*10^{9}$).

	Change-of-basis solutions					
Item	**1**	**2**	**3**	**4**	**5**	**6**
Mean ($)	50,594	48,895	48,777	44,559	44,279	43,367
Variance ($\$^{2}10^{9}$)	2.1917	0.8404	0.8060	0.1952	0.1889	0.1760
Crops (ha)						
Potatoes	9.00	9.00	9.00	8.99	8.75	8.75
Sugarbeet	11.25	11.25	11.25	11.25	11.25	11.25
Onions	5.92	2.76	2.66	0.00	0.00	0.00
Winter wheat	0.00	0.00	0.00	0.00	0.00	5.14
Grassland	18.83	21.99	22.09	24.76	25.00	19.86
Hire labour (h)						
May	52	2	0	0	0	0
September	57	47	47	39	35	14
October	60	0	0	0	0	0

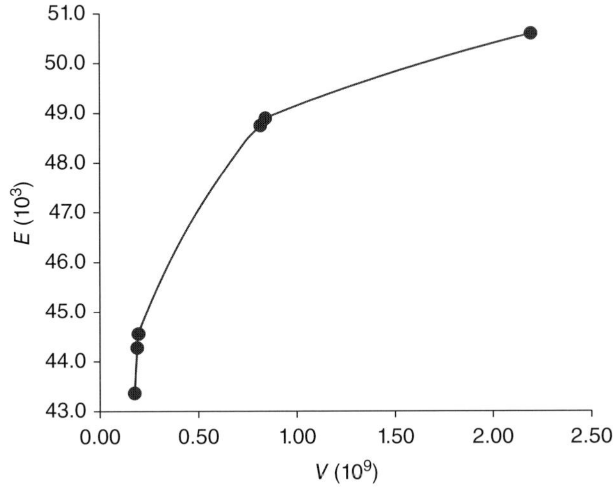

Fig. 9.8. *E,V* efficient frontier for the quadratic risk programming (QRP) formulation of the example problem.

random events. If the distribution of total net revenue is indeed close to normal, then QRP will generate a set of solutions that will be second-degree stochastically efficient. Usually it will be a matter of making a judgement as to whether the departures from a normal distribution in a given case matter to the DM – conventional tests of adequacy of a normal approximation are based on statistical criteria rather than on the DM's preferences and so are not necessarily appropriate.

Other linear risk-programming models

When MP software to handle non-linear objective functions was less available and less reliable, attempts to find LP approximations to the QRP formulation were made. *MOTAD programming*, developed by Hazell (1971), has been the most widely used of these approximations.

In MOTAD programming the variance constraint of the QRP model above is replaced with a constraint on the mean absolute deviation of net income. The advantage is that mean absolute deviation can be obtained as a linear expression, therefore requiring only LP to find the solutions.

The application of the MOTAD approach entails use of the same technical input–output tableau *A* as for the LP and QRP models, but additional constraints are added for the calculation of total absolute deviations for each of a number of discrete states of nature. These total deviations are averaged using the probabilities of the states (nearly always implicitly and perhaps unrealistically assumed to be equi-probable in published examples) to calculate the mean absolute deviation, *M*. The model is then solved with *M* set to an arbitrarily high value, which is then progressively reduced until no further solutions of interest are found.

The MOTAD formulation generates the *E,M* efficient frontier that approximates the *E,V* frontier. Since the latter is normally an approximation to the set of utility-maximizing solutions for varying degrees of risk aversion, it is possible in theory that MOTAD models could produce solutions somewhat less attractive to the DM than the utility maximizing optimum.

Target MOTAD programming, as developed by Tauer (1983), is related to MOTAD in that it entails a constraint on income deviations, this time from a target level of income. In other words, the efficient set of solutions is obtained by maximizing expected income for a range of values of the mean deviation from the target income. The main advantage is that the solutions are second-degree stochastically dominant (regardless of the distribution of income), and so are efficient for risk-averse DMs.

The disadvantage of Target MOTAD is that there is usually no good a priori reason to set any particular value of target income. That means that the model is usually solved maximizing E for a relatively large number of combinations of the target income and deviations therefrom, making the results rather extensive, and therefore more difficult to interpret.

Mean–Gini programming is yet another linear risk-programming formulation that also has the advantage of generating solutions that are always second-degree stochastically efficient (Okenev and Dillon, 1988). It requires the construction of a matrix of net revenue differences for all the activities and all possible discrete pairs of states, together with a vector of probabilities of these pairs.

An advantage of the mean–Gini model over Target MOTAD, which, as noted, also gives second-degree stochastically efficient solutions, is that the former is a two-parameter model while Target MOTAD involves three parameters. A limitation of the mean–Gini approach, however, is that some stochastically efficient solutions that would be preferred by strongly risk-averse DMs might be excluded from the efficient set. Also, the tableau for mean–Gini programming is much larger than that for Target MOTAD.

We have chosen not to fully specify nor illustrate these linear approximations to risk programming because we believe that the need for them has passed with the ready availability of reliable non-linear algorithms. The models based on direct utility maximization, described below, use exactly the same data as all three of the linear approximations mentioned above, and yet lead directly to a set of solutions fully consistent with the SEU hypothesis.

Direct maximization of expected utility

In view of the general availability of non-linear programming software, it is straightforward to set up a risk-programming model to maximize expected utility (Lambert and McCarl, 1985). The model is of the form:

$$\text{maximize } E[U] = p\ U(z) \qquad (9.5)$$

subject to

$$A\,x \leq b$$

$$C\,x - I\,z = u\,f$$

$$\text{and } x \geq 0$$

where $E[U]$ is expected utility, z is an s by 1 vector of net incomes by state, and $U(z)$ is an s by 1 vector of utilities of net income by state.

Because the utility function will be monotonic and concave for a risk averter, a good non-linear algorithm will find the global optimum.

From the decision analysis perspective, the method of direct utility maximization is clearly superior to others discussed above, since it fits exactly with the SEU hypothesis. However, it can be applied only when there is an individual DM whose utility function is available. When this is not the case, as when there are many DMs such as some group of farmers for whom advice is being formulated, it would be desirable to develop an efficient set of farm plans following something akin to the SERF rule (stochastic efficiency with respect to a function), discussed in Chapter 7, this volume. This can be achieved using *utility-efficient (UE) programming* (Patten *et al.*, 1988). Since UE programming is a simple extension of direct maximization of expected utility, only the latter is described and illustrated below.

Utility-efficient (UE) programming

UE programming takes the form:

$$\text{maximize } E[U] = p\ U(z, r),\ r \text{ varied} \quad (9.6)$$

subject to

$$A\,x \le b$$

$$C\,x - I\,z = u\,f$$

$$\text{and } x \ge 0$$

where the utility function is defined for a measure of risk aversion r that is varied. In most applications, r has represented the coefficient of absolute risk aversion, r_a. However, as for SERF, any appropriate form of utility function can be used. It is then possible, using modern software, to generate solutions for a range of values of r with little trouble.

The layout of an UE programming tableau is as shown in Fig. 9.9.

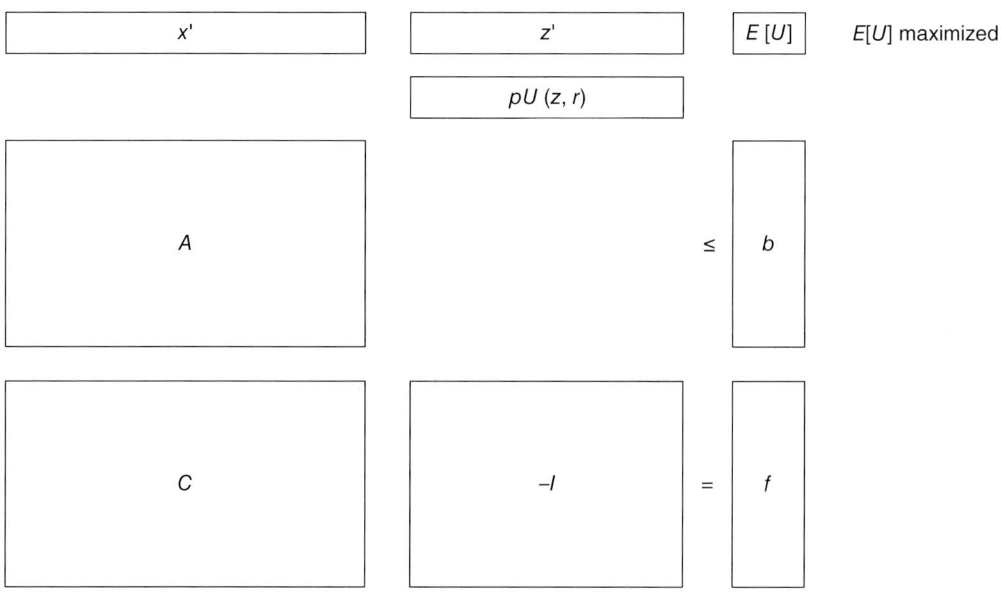

Fig. 9.9. Outline of tableau for utility-efficient (UE) programming.

The general form of objective function above can be adapted to a number of types of utility function. One approach is to use the negative exponential function in the form:

$$U = 1 - \exp[-\{(1-a)\ r_{min} + a\,r_{max}\}\ z],\ a \text{ varied between zero and 1} \qquad (9.7)$$

which gives a coefficient of absolute risk aversion between r_{min} when a is zero and r_{max} when a is 1.0. Similarly to the SERF rule, this method can be expected to generate the set of solutions that are stochastically efficient for all DMs whose coefficient of absolute risk aversion is in the relevant range.

This latter approach was used for the example problem. The A matrix of Fig. 9.9 was taken from the LP model for this problem in Fig. 9.4. The LP tableau was extended to include the matrix of activity net revenues by state, C (from Fig. 9.3), as indicated in Fig. 9.9. In addition, six additional activities were added to measure the income of the current solution for each state of nature. A GAMS input file was written to convert these incomes into utilities using the negative exponential function chosen, defined for a relatively large number of values of a in the range 0 to 1. The range of risk aversion used was from $r_{min} = 4 \times 10^{-6}$ to $r_{max} = 10^{-5}$. (These values were chosen accounting for the fact noted above that this is an annual planning model so it is aversion to transitory income that needs to be reflected.) The utility values for each state were then multiplied by the corresponding probabilities to calculate expected utility, which was set as the maximand in the GAMS input file.

Results for selected levels of risk aversion are shown in Table 9.5. The solutions for this simple problem are very similar to those generated by QRP, and also match results, not shown, from the linear risk-programming models described above. For larger and more realistic models, such close congruence is not necessarily to be expected.

The solutions to the UE programming model can be plotted on a graph as cumulative mass functions (CMFs), as shown for three contrasting levels of risk aversion in Fig. 9.10. This form of presentation emphasizes the link between UE programming and stochastic efficiency analysis. Indeed, the efficient solutions found define the SERF set (see Chapter 7).

Table 9.5. Results of the utility-efficient (UE) programming formulation of the example problem for selected levels of risk aversion.

	Absolute risk aversion coefficient r_a*1000					
Item	**0.0013**	**0.0022**	**0.0025**	**0.0028**	**0.0031**	**0.0034**
CE ($)[a]	49,209	48,214	47,960	47,776	47,596	47,489
Crops (ha)						
Potatoes	9.00	9.00	9.00	9.00	9.00	9.00
Sugarbeet	11.25	11.25	11.25	11.25	11.25	11.25
Onions	5.92	5.46	4.54	3.81	3.22	2.76
Grassland	18.33	19.29	20.21	20.94	21.52	21.99
Hire labour (h)						
May	52	45	30	18	9	2
September	57	55	53	50	48	47
October	60	51	34	20	9	0

[a]CE, certainty equivalent.

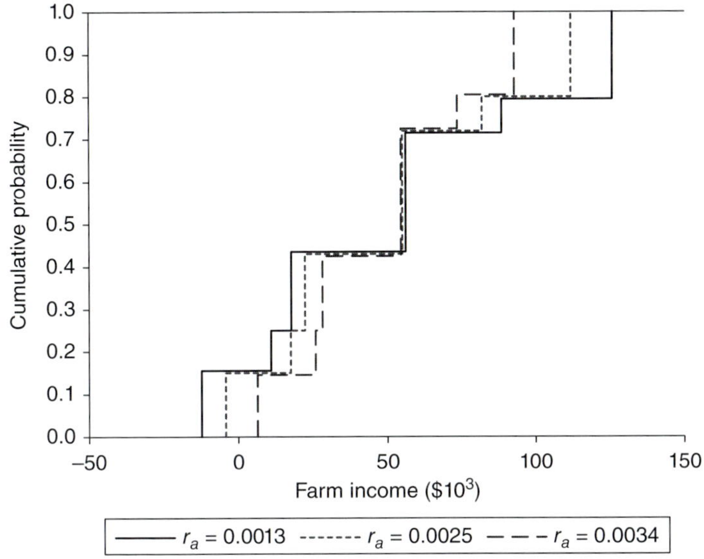

Fig. 9.10. Cumulative mass functions (CMFs) for UE programming solutions for three levels of risk aversion.

As noted earlier, risk-programming models for farm planning can usually be readily extended to include opportunities for the DM to make use of risk-sharing instruments such as futures contracts and crop yield or revenue insurance. Usually such opportunities will enter the models as additional activities. For example, to include the option to buy wheat yield insurance for the case farm, the existing wheat activity could simply be duplicated with the new activity net revenue recalculated for each state of the nature in matrix C to reflect the insurance premium paid and any indemnities received in years of low yields. Because of the obvious stochastic dependency between returns from on-farm production activities and the costs and returns of related insurance or hedging options, it will be sensible to evaluate such opportunities in the context of a whole-farm planning model, not in a partial budgeting context as is often done.

Stochastic Programming

As discussed at the start of this chapter, many farm planning problems entail choices at various stages with risky events embedded between stages. Consequently, some later decisions depend on both earlier decisions and the outcomes of earlier uncertain events. In principle, MP can be extended to solve such problems. In all but a few special cases, the best approach is to use what is known as *discrete stochastic programming* (DSP) (Cocks, 1968; Rae, 1971). DSP is a form of state-contingent analysis (see Chapter 8). Any decision tree can be represented as an equivalent DSP model, with obvious advantages for decision analysis. In practice, however, there are limitations to what is possible, even with DSP.

When represented as decision trees, all such problems have a tendency to explode into 'bushy messes', meaning that the problem grows to have too many branches to be drawn easily or at all, and may be hard or impossible to solve if specified in all its detail. Another name for this phenomenon is 'the curse of dimensionality'.

It is apparent that the size of a decision problem increases with each of:

1. the number of stages, represented by decision–event sequences;
2. the number of choice options at each decision; and
3. the number of possible outcomes at each event.

Forks with many branches early in the tree obviously have a more dramatic effect on the overall size of the tree than do forks towards the end of the tree.

MP representation of such decision problems with embedded risk can easily cope with any number of choice options for decisions – indeed, MP handles situations with an infinite number of choice options more easily than when the range of options is finite. But the MP approach runs into the familiar dimensionality problems when there are many stages and many possible outcomes for each uncertain event. To solve such problems by MP, at least approximately, it will usually be necessary to simplify the problem by restricting the number of stages to two or three, and by limiting the number of possible events at each stage to just a few. Otherwise, as will be seen when the approach is explained, an MP tableau may quickly grow to a size that is difficult to handle and, depending on the available computing facilities, too large to solve.

While such abridgement of a decision problem may appear to be too far removed from reality, there are three mitigating considerations:

1. Computing capacity rises year by year as more powerful computers with better software become available. As a result, what today may seem to be an unacceptably large MP tableau may easily be handled within just a few years.
2. Modern software such as GAMS also helps to manage the data handling task for very large problems. Such problems often have repeated sub-matrices that can be manipulated as such in the software, rather than element by element, so reducing the risk of errors.
3. In a multi-stage decision problem, the modelling of the problem need only be just good enough to get the first-stage decisions right since, after these have been implemented, a revised version of the problem, entailing less approximation of its later parts, can be formulated and solved to determine the second-stage decision, and so on.

The DSP approach, although messy in terms of notation, is conceptually simple. The discrete treatment is consistent with much of the spirit of simpler decision analyses, notably decision trees. The same kinds of maximands can be used as for risk programming.

A DSP model for a simple two-stage problem may be formulated as:

$$\text{maximize } E[U] = p_t\, U(z_{2t}) \tag{9.8}$$

subject to

$$A_1 x_1 \leq b_1$$

$$-\mathrm{L}_{1t} x_1 + A_{2t} x_{2t} \leq b_{2t}$$

$$C_{2t} x_{2t} - I_{2t}\, z_{2t} = f_{2t}$$

$$\text{and } x_1, x_{2t} \geq 0,\ t = 1, \ldots, s$$

where subscripts 1 and 2 indicate first and second stages, respectively, the subscript t indicates the state of nature; p_t are 1 by s vectors of joint probabilities of activity net revenue outcomes given that state of nature t has occurred; and L_{1t} is a set of s matrices linking first- and second-stage activities. Thus, in this formulation it is assumed that some initial decisions are made (x_1), then one of two states of nature ensues (t = 1 or 2), when some second-stage decision must be made (x_{2t}), conditioned by the first-stage decisions and the state of the world. The returns from these activities are subject to further uncertainty, defined by the matrices of activity net revenues C_{2t} and associated probabilities p_t. The objective function is the maximization of expected utility. If the further uncertainty in activity returns is assumed away, so that the matrices C_{2t} have only one row, further simplification of the matrix is possible, as illustrated in an example below.

The above formulation is set out in diagrammatic form in Fig. 9.11.

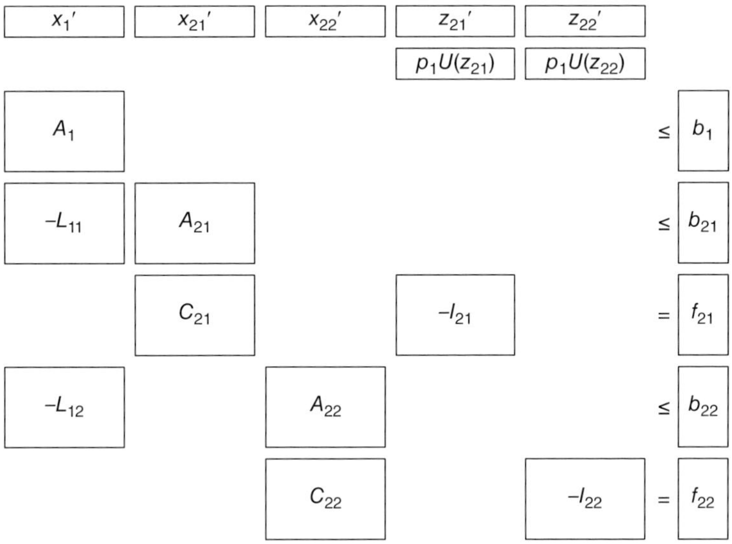

Fig. 9.11. Outline of tableau for discrete stochastic programming (DSP).

In this rather ugly figure, the first column of sub-matrices represents the first-stage activities and transfers, respectively. The second and third columns contain the sub-matrices corresponding to the second-stage activities for the two possible second-stage states 1 and 2, respectively. Columns 4 and 5 have sub-matrices where net income is measured based on s possible net revenue outcomes for each of the two states. The income measures collected in these sub-matrices go into the non-linear part of the solution procedure to be converted to expected utility. The first row of the table shows the activity levels, determined in the solution procedure; the second row is included to indicate the use of a non-linear objective function; in a GAMS formulation, this part will be programmed in the command file, and not included in the data matrix. The third row represents the constraints on the first-stage activities. Next come two pairs of rows, the first of each pair containing sub-matrices linking first- and second-stage decisions for each state, and the second of each pair providing the accounting of the final uncertain payoffs, transferring the net incomes for each of s possible activity net revenue outcomes for each state to the income activities for transformation into expected utility.

Despite the ungainly and extensive nature of the matrix, a major advantage of DSP is that the sequential nature of the decision problem can be represented in the model. Consequently, risks in both the constraints and the input–output coefficient can be modelled. Both are often important sources of risk in farming. Whether or not a farmer is risk averse, the downside risk that is embedded in most farming systems, discussed in Chapter 1, can be captured, at least in approximate fashion, in DSP. If risk aversion is important, it is usually relatively trivial to extend the DSP model formulation to account for it.

For purposes of illustration, we present a quite simple DSP model. A dairy farmer is planning production for next year in the face of a rigid milk quota and uncertain production per cow. The quota year starts in April, at which time the farmer expects to have 67 cows and a quota of 500 t. Milk supplied in excess of the quota attracts a penalty, making it essentially valueless. The farmer has the option to lease quota in or out at that time. Another option is to reduce the number of cows by selling some. These decisions have to be made against the possibility that milk yield per cow may be high, medium or low, depending mainly on seasonal conditions.

Later in the year in January, 3 months before the end of the quota year, it is possible to reassess the situation. At that time, the type of season and the level of milk production to date will be known. It is then possible to sell more cows, to bring up to ten calving heifers into the herd to boost production, or to lease quota in or out.

Of course, the relationship between production and quota could be assessed at several stages through the year, not just in January. However, to keep the example simple to illustrate the DSP method, these other possible decision stages have been ignored. Restricting the problem to just two stages may or may not be judged to capture the essence of the real planning problem. Furthermore, it is assumed for simplicity that the market prices for selling cows and the costs of bringing heifers into the milking herd perfectly reflect the true economic value of these animals to the farmer. For example, selling a cow would result in a return of \$1000 per head, but at the same time the total economic value of the herd decreases by \$1000, with no net effect on terminal wealth. Table 9.6 summarizes some other background information.

The problem is first formulated as an LP problem, maximizing expected money value. The tableau is shown in Fig. 9.12.

The relationship between this formulation and the outline tableau in Fig. 9.11 needs some comment. The first-stage activities in Fig. 9.11 are in columns B to E of Fig. 9.12. There are three states of nature represented, not two as in Fig. 9.11. The second-stage activities for these states are in columns F to I, J to N and O to Q, respectively. The sub-matrix A_1 is merely the cell range B8:E8. (Some of the cells of this and other sub-matrices are blank.) The three L_{1t} sub-matrices are in B11:E12, B14:E16 and B18:E20, respectively. Similarly, the three A_{2t} sub-matrices are in F11:I12, J14:N16 and O18:Q20, respectively. Because no uncertainty in the final payoffs is assumed, the two sections of Fig. 9.11 reflecting the income accounting are not needed, since the payoffs and their corresponding probabilities can be attached directly to the activities, as shown in Fig. 9.12, cells B4:Q5. This arrangement permits

Table 9.6. Background information on the discrete stochastic programming (DSP) milk production example.

	Milk production		
	High	**Normal**	**Low**
Probability	0.3	0.4	0.3
Milk production per cow[a] (kg/stage 1)	6000	5625	5250
Milk production per cow[a] (kg/stage 2)	2000	1875	1750
Leasing in quotas (\$/1000 kg)	400	300	–
Leasing out quotas (\$/1000 kg)	–	250	200

[a]Cows and heifers are assumed to have the same milk production.

	A	B	C	D	E	F	G	H	I	J	K	L	M	N	O	P	Q	R	S
1	**DSP programming model – milk production under quota**																		
2		Stage 1				Stage 2 – High production				Stage 2 – Medium production					Stage 2 – Low prodn			Rel	RHS
3		Sell cows	Keep cows	Lease quota in	Lease quota out	Sell cows	Keep cows	Lease quota in	Lease quota out	Sell cows	Keep cows	Insert heifers	Lease quota in	Lease quota out	Keep cows	Insert heifers	Lease quota out		
4	NR/unit	0	0	–300	250	2400	3200	–400	350	2250	3000	750	–300	250	2800	700	200		
5	Probs	1	1	1	1	0.3	0.3	0.3	0.3	0.4	0.4	0.4	0.4	0.4	0.3	0.3	0.3		
6	E[NR]	0	0	–300	250	720	960	–120	105	900	1200	300	–120	100	840	210	60		
7	**Stage 1**																		
8	Cows	1	1															<=	67
9	**Stage 2**																		
10	*High:*																		
11	Cows		–1			1	1											<=	0
12	Quota		6	–1	1		2	–1	1									<=	500
13	*Normal:*																		
14	Cows		–1							1	1							<=	0
15	Heifers											1						<=	10
16	Quota		5.625	–1	1						1.875	1.875	–1	1				<=	500
17	*Low:*																		
18	Cows		–1												1			<=	0
19	Heifers		0													1		<=	10
20	Quota		5.25	–1	1										1.75	1.75	1	<=	500

Fig. 9.12. LP tableau for DSP formulation of the milk quota problem. NR, net revenue; RHS, right-hand side.

Table 9.7. LP solution for the DSP milk production example.

Activities	Units	Levels
Stage 1		
Keep cows	head	67
Lease in quota	t	21.25
Stage 2		
High production:		
Sell cows	head	7.4
Keep cows	head	59.6
Normal production:		
Keep cows	head	67
Add heifers	head	10
Low production:		
Keep cows	head	67
Add heifers	head	10
Lease out quota	t	34.75

the expected net revenue per unit activity level, which is the objective function for the LP formulation, to be calculated in cells B6:Q6.

The LP solution to the problem is as shown in Table 9.7. The expected net revenue is $200,040. It may be noted that there is a problem of a non-integer solution for the numbers of cows kept and sold in a high-production year, but the rounding error entailed may not be serious enough to worry about. Otherwise, an integer programming routine would have to be applied to get a solution in terms of whole numbers of animals.

Evidently the farmer should retain all the existing 67 cows and should immediately lease in additional quota to match the production expected in a normal season from this number of cows plus the ten heifers that can be added to the milking herd in January. If production is above this normal level, the heifers should not be brought into the herd and excess cows should be sold to keep within quota. On the other hand, if production is low, the heifers should be brought into the herd and that part of the quota that still cannot be filled should be leased out.

The model can also be formulated for direct expected utility maximization using a non-linear solution procedure, as illustrated above in relation to UE programming. In this case, setting the coefficient of absolute risk aversion to range between 10^{-6} and 10^{-5} produced solutions identical to that in Table 9.7.

Concluding Comment

Given the extra complexity of accounting for risk in MP models for farm planning, it is important to consider in each case whether risk really matters. Sometimes it is clear that downside risk is important and so should be properly represented in the planning model. Too often, models are built with seemingly cavalier assumptions of certainty. Analysts would do well to review the major sources of uncertainty in any system of interest, noting how this uncertainty could affect system performance.

In much previous work, risk-programming methods have been used to account for risk aversion, with apparently no thought for the treatment of embedded risk and its impact, particularly any downside risk impacts. The increased power of computers means that DSP models of considerable size can now be solved easily, while modern software has eased the data-handling burden on the analyst in formulating large models.

If risk aversion is to be taken into account, the analyst needs to give careful consideration to selecting the model that is best suited to expected utility maximization. The widespread use of MOTAD models, for example, when direct utility maximization or UE programming would seem to be more appropriate, suggests that the choice of method is often based on computational convenience rather than on theoretical and empirical suitability.

In summary, the challenge in accounting for risk in farm system modelling using MP lies in: (i) carefully identifying the major sources of risk and their impact; (ii) representing these adequately in an MP model; and (iii) setting the chosen model up to generate the smallest appropriate set of risk-efficient solutions, ideally comprising just the utility-maximizing one.

Selected Additional Reading

As mentioned at the start of this chapter, there are many good texts on MP. Because methods and software evolve quite rapidly, a recent book such as Williams (2013) is a good starting point for readers wanting to learn more about these methods and their application. The discussion of the application of MP to farm planning problems under uncertainty above draws heavily on Anderson *et al.* (1977, Chapter 7) and on Hardaker *et al.* (1991). There is a large and growing literature on the application of these methods in agriculture and resource management that will be revealed in any careful literature search. A particularly useful source for would-be practitioners of MP in agriculture who elect to use the GAMS software is McCarl and Spreen (1997), while Hazell and Norton (1986), Rae (1994) and Moss (2010, Chapter 6) provide useful guidance to the construction of MP models in agriculture. Arriaza and Gómez-Limon (2003) offer some evidence on the importance of including risk considerations when trying to model the responses of farmers to policy changes.

References

Anderson, J.R., Dillon, J.L. and Hardaker, J.B. (1977) *Agricultural Decision Analysis*. Iowa State University Press, Ames, Iowa.

Arriaza, M. and Gómez-Limon, J.A. (2003) Comparative performance of selected mathematical programming models. *Agricultural Systems* 77, 155–171.

Cocks, K.D. (1968) Discrete stochastic programming. *Management Science* 15, 72–81.

Freund, R.J. (1956) The introduction of risk into a programming model. *Econometrica* 24, 253–261.

Hardaker, J.B., Pandey, S. and Patten, L.H. (1991) Farm planning under uncertainty: a review of alternative programming models. *Review of Marketing and Agricultural Economics* 59, 9–22.

Hazell, P.B.R. (1971) A linear alternative to quadratic and semivariance programming for farm planning under uncertainty. *American Journal of Agricultural Economics* 53, 53–62.

Hazell, P.B.R. and Norton, R.D. (1986) *Mathematical Programming for Economic Analysis in Agriculture.* Macmillan, New York. Available at: http://www.ifpri.org/sites/default/files/publications/mathprogall.pdf (accessed 1 June 2014).

Lambert, D.K. and McCarl, B.A. (1985) Risk modeling using direct solution of nonlinear approximations of the utility function. *American Journal of Agricultural Economics* 67, 846–852.

Lien, G., Hardaker, J.B., van Asseldonk, M.A.P.M. and Richardson, J.W. (2009) Risk programming and sparse data: how to get more reliable results. *Agricultural Systems* 101, 42–48.

McCarl, B.A. and Spreen, T.H. (1997) *Applied Mathematical Programming Using Algebraic Systems.* Texas A&M University, College Station, Texas.

Moss, C.B. (2010) *Risk, Uncertainty and the Agricultural Firm.* World Scientific Publishing, Singapore.

Okenev, J. and Dillon, J.L. (1988) A linear programming algorithm for determining mean-Gini efficient farm plans. *Agricultural Economics* 2, 273–285.

Patten, L.H., Hardaker, J.B. and Pannell, D.J. (1988) Utility-efficient programming for whole-farm planning. *Australian Journal of Agricultural Economics* 32, 88–97.

Rae, A.N. (1971) Stochastic programming, utility, and sequential decision problems in farm management. *American Journal of Agricultural Economics* 53, 448–460.

Rae, A.N. (1994) *Agricultural Management Economics: Activity Analysis and Decision Making.* CAB International, Wallingford, UK.

Rosenthal, R.E. (2014) *GAMS – A User's Guide.* GAMS Development Corporation, Washington, DC. Available at: http://www.gams.com/dd/docs/bigdocs/GAMSUsersGuide.pdf (accessed 1 June 2014).

Tauer, L.W. (1983) Target MOTAD. *American Journal of Agricultural Economics* 65, 608–610.

Williams, H.P. (2013) *Model Building in Mathematical Programming*, 5th edn. Wiley, Chichester, UK.

10 Decision Analysis with Multiple Objectives

Introduction with Some Examples

In earlier chapters we have defined utility functions that indirectly embody an important trade-off: expected monetary return versus variance. Such a utility function represents a preference model for choice that captures the DM's attitude to expected return and variance. Obtaining high returns and reducing exposure to variability are usually two conflicting objectives in decision making. We have shown in Chapters 5 and 7 how to model the preference trade-off between these objectives. In many situations, however, the action chosen depends on how each possible choice meets several objectives, as the following examples show.

A dairy farmer has become concerned about some long-term negative impacts of the current system of milk production on the farm and is therefore considering changing this system. The current production system is a high-input/high-output system. Large amounts of resources are used per cow to produce a high milk yield. In the short term, this system gives the farmer a good income and a high status in the local community. However, because of its intensive nature, it may cause some environmental problems in the future, as well as some problems with cow health and welfare. In thinking about changing the production system, the dairy farmer might consider diverse possible objectives such as the following: (i) maximizing current farm income; (ii) maximizing farm income in the future; (iii) minimizing environmental damage; (iv) maximizing animal health and welfare; (v) achieving a high status in the local community; and (vi) minimizing the workload.

The owner of a large and expanding flower-growing business engaged in breeding, multiplying and processing tulips, wants to hire a new manager. When evaluating potential candidates, the owner wants to satisfy several objectives. These include minimizing the salary and benefits that must be paid to attract a good candidate and maximizing relevant years of experience. In addition, the candidate should have a wide range of technical skills, be a good people manager and have the ability to work without supervision.

The government needs to act in response to an outbreak of a contagious cattle disease. Because the disease may cause a fatal illness in humans who eat products made from infected animals, public health may be in danger. Furthermore, it is difficult to identify infected cattle because the incubation time may be as long as 5 or even 10 years before they show signs of the disease, and there is as yet no reliable diagnostic test that can be used on live animals. The most effective, albeit costly, way to get rid of the disease is to kill and destroy all potentially infected animals. This method would mean killing many animals that are not in fact infected, which could be considered a brutal violation of animal rights. Another consideration is that consumer prices for milk and beef are likely to rise if a large number of animals were to be taken out of production. Policy makers need to balance such objectives as: (i) food safety; (ii) consumer confidence; (iii) effects on food prices; (iv) incomes of affected farmers; (v) the financial cost for the government; and (vi) animal welfare.

A large agribusiness enterprise is planning to establish a production and marketing chain for fresh fish. The chain involves several stages from breeding, multiplying and on-farm fish production, processing and packing, and the transportation of the fresh fish from the farm to retailers. It is important to satisfy all consumers' needs. Consumers like cheap, fresh and healthy fish produced using animal-friendly and environmentally responsible methods. So, the enterprise has to develop a chain that meets a variety of objectives: (i) low-cost fish; (ii) high chain profit; (iii) minimal stocks and frequent transport between the stages (for fresh fish); and (iv) animal-friendly and environmentally responsible production.

These examples illustrate the need for some form of multi-attribute analysis. Below, we set out the components and methods that permit such decision problems to be tackled in a way consistent with, but building on, the approach taken in earlier chapters.

Objectives and Attributes

The terms 'objectives' and 'attributes' are frequently used interchangeably, but it is desirable to make a distinction between them. We take *objectives* – in some studies called *criteria* – to refer to the directions in which the DM wants to go. For example, a person looking for a better job may want to earn a high income without having to travel too far to work. In this case, 'high income' and 'short travel distance' are objectives.

Attributes relate to the properties of each alternative. Examples are 'starting salary of a job' and 'distance from home'. The performance of an alternative relative to a particular objective is reflected in the attained level of the relevant attribute (or attributes). Although ideally there is a one to one correspondence between objectives and attributes, in reality the degree of attainment of some objectives may require the specification of more than one attribute. For example, scores of attributes have been put forward in the attempt to measure the objective of sustainability of farming systems.

A numerical value describing the attained level of a given attribute we call the *attribute measure*. So, in terms of the above example, in looking for a better job, one may have the objective earning a 'high income', reflected partly by the attribute 'starting net income per year before tax'. The measure of this attribute for a particular job may be \$80,000. Not all objectives may have quantitative attributes but it may still be possible to specify different levels of attainment of an attribute that can at least be ranked in order of preference, if not quantified. With some limitations, such ranked levels can be treated as measures.

The aim of the DM is presumed to be to find the particular choice alternative that provides the best attainment of the objectives (i.e. the alternative with the most preferred combination of attribute measures).

Structuring the Decision Problem

The essence of *multi-objective decision analysis* is to break down a complicated decision into smaller pieces that can be dealt with more easily, followed by recombination of the parts to reach a preferred choice. The dissection of the problem typically involves the following five steps:

1. identify alternatives;
2. identify objectives and attributes;

3. quantify (or specify) attribute measures;
4. quantify preferences; and
5. rank alternatives.

These steps are explained later in this section, but first an example is introduced. The example is then carried through the chapter to illustrate the several techniques to be described.

Introduction of a worked example

The example relates to a multi-objective policy decision with respect to choosing a site for a large-scale animal manure processing plant. A senior policy adviser has been instructed to prepare a recommendation for the government. A multi-objective decision analysis is undertaken to help the adviser make the best possible recommendation.

Because of high livestock densities in the immediate region for which the policy adviser has responsibility, an excess of manure is produced by the farm animals. To spread all this over the agricultural land would cause groundwater contamination and other pollution problems. The government has therefore three policy options: (i) reduce the number of animals (i.e. stocking rate per hectare); (ii) move the excess manure to other regions to be used on arable land; or (iii) process the excess manure in a large-scale factory and convert it into dry fertilizer that can be used outside agriculture or that can be profitably exported to other countries with a fertilizer shortage. The policy adviser decides to explore the latter option, and is committed to finding a suitable site for the construction of a new large-scale manure processing plant. Excess manure from livestock farms from the whole region will then be collected and brought to this plant to be processed. A desirable site will be of reasonable cost, of sufficient area for all needed facilities, far from areas that might be affected by undesirable odours and possible ammonia emission, yet not so out of the way as to add excessively to costs of vehicle travel for staff and for trucks carting manure in and fertilizer out. A short distance between the site and the farms is also important to limit damage to road pavements by the heavy traffic and to minimize the risk of road accidents, not only as a safety matter but also to avoid effluent spills. The factory should also be located to fit in with regional development needs and priorities.

It is clear that, even in this simplified example, all the objectives cannot be optimized at the same time. The site that is eventually selected will be a compromise between the objectives. The way this compromise may best be worked out is examined in subsequent sections.

Identify alternatives

Alternatives are the choices that have to be ranked in order to come to a decision. Alternatives are seldom easily defined because there are usually many possible alternatives – such as makes and models of tractors that a farmer might buy – and the DM may need to do a preliminary subjective screening to reduce the set of alternatives to a manageable number. Such screening might start with (but not be confined to) consideration of which desirable attributes or attribute measures are vital, and which undesirable attributes or attribute measures are unacceptable. Alternatives that failed to meet one or other of these criteria would then be excluded from further consideration. After as many alternatives as possible have been culled, considerable creativity may be needed to specify the alternatives in sufficient detail to allow systematic comparisons.

In the site-selection example introduced above, we assume that the policy maker has settled on five alternative sites for further consideration. They are denoted as 'North', 'South', 'East', 'West' and 'Central', describing their location within the area.

Identify objectives and attributes

Objectives are the considerations that influence the desirability of the alternatives. What objectives need to be considered? The answer to this question can often be clarified by constructing a so-called *value tree* to express the objectives and attributes judged to be relevant in comparing the alternatives. A value tree begins with the fundamental objectives or main branches. Each fundamental objective is then expanded with further branches describing more specific objectives.

Completing the value tree requires identification of attributes that reflect the detailed objectives. This means that, at the end of each branch, there must be an operational way to judge the extent to which an alternative attains that objective, as reflected in the attribute measure. Finding operational attribute measures may sometimes be difficult. An attribute measure is operational if it is possible to explain to someone else what information is needed and why, so that, if appropriate, that person can be expected to provide the information.

According to Clemen and Reilly (2014), a value tree should meet the following criteria:

1. It should be complete – it should include all relevant aspects of the DM's objectives.
2. It should at the same time be as small as possible and therefore it should not contain redundant elements.
3. The attributes at the end of the branches should be operational.
4. It should be decomposable as far as possible, meaning that the DM should be able to think about each attribute individually without having to consider others.

Avoiding redundancy and maximizing decomposability means that attributes that overlap to some extent should be avoided. For example, it would probably be unsatisfactory to include in the same value tree the attributes 'total costs' (to be minimized) and 'profit' (to be maximized).

In the site-selection example, the policy adviser must find a balance between such objectives as how to: (i) maximize general site conditions (with the most important attribute 'Site isolation'); (ii) maximize access for manure trucks (attribute 'Distance to farms'); and (iii) minimize the total costs of processing the excess manure (with attribute 'Net costs' per year). The value tree, including the objectives and attributes considered to be relevant by the policy adviser, is presented in Fig. 10.1.

Quantify attribute measures

Attribute measures describe the alternatives in terms of their performance in contributing to each attribute. The set of measures across all attributes ideally fully describes each alternative for ranking purposes. Attribute measures can be natural, such as the cost of an investment, or constructed, such as a five-point scale

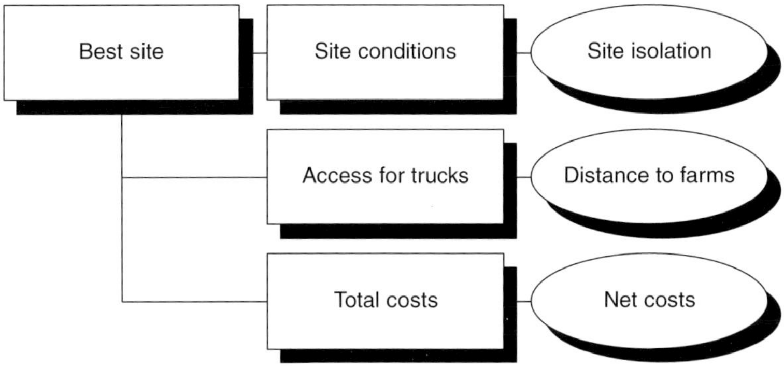

Fig. 10.1. Value tree for the site-selection example.

that describes the social impacts of an investment. An attribute measure may be a point estimate, which is a single number equal to or representing a specific level of an attribute for a given alternative. It may, alternatively, be a probability distribution. When a DM does not know with reasonable certainty the measure of an attribute for some alternative, it may be best described by a probability distribution. If risk aversion is considered unimportant, it will usually be sufficient to summarize such a probability distribution by its expected value, which may well be different from the 'best estimate' or 'most likely value' all too often used.

If uncertainty in more than one attribute measure is recognized, and if there is stochastic dependency between them, then in principle, the full joint distribution of outcomes is required. Procedures for and difficulties in specifying joint probability distributions were discussed in Chapter 4, this volume. The difficulties of joint specification are avoided in the fortunate eventuality that attributes can be chosen for which an assumption of stochastic independence is reasonable. Then only the marginal distributions of the uncertain attribute measures need be specified.

If all attributes and attribute measures have been defined, then the alternatives should be checked for *dominance*. An alternative dominates another if it is at least as preferred as the other in terms of all attributes, and strictly preferred in terms of at least one attribute. Dominated alternatives can be dismissed from further consideration. Applying this rule in a pairwise fashion to all the alternatives may allow the set of candidates for evaluation to be reduced. This reduced set of alternatives may be called the *efficient set*. Obviously, it is likely that more dominated alternatives will be found when the number of alternatives to be considered is relatively large, yet, in such cases, the efficient set may still be large.

Only in rare cases is it likely that the efficient set will contain just one member, meaning that the choice problem has been solved. Nevertheless, it is good practice to check for dominance before doing further analysis, in order to avoid performing unnecessary calculations on dominated options. Moreover, even when strict dominance is not present, a subjective judgement may be made at this stage to eliminate less promising alternatives in order to simplify the analytical task.

The attribute measures for the site-selection example as assessed by the policy adviser are summarized in Table 10.1. As can be seen in the table, some attributes are measured with concrete dimensions, such as 'Distance to farms' in kilometres. Other attributes are measured on a scale ranging from 1 to 5, with higher attribute measures being preferred to lower. For example, the site isolation of North is considered to be very poor (attribute measure of 1), because there are other businesses in the immediate environment of the site. In contrast, the site isolation of West is judged to be excellent (attribute measure of 5) owing to its relative isolation.

Table 10.1. Attribute measures for the site-selection example.

	Attribute measure		
Name of site	**Distance to farms (km)**	**Net costs (millions $)**	**Site isolation**[a]
North	150	0.5	1
East	70	1.0	5
South	50	1.2	4
West	120	1.8	5
Central	100	2.0	1

[a]Site isolation is rated on a scale ranging from 1 (very poor) to 5 (excellent).

To simplify the explanation we assume for now that all these attribute measures are known with reasonable certainty so that there is no need to account for uncertainty by specifying probability distributions. This assumption is relaxed in the section 'Multi-attribute Utility Analysis under Uncertainty' at the end of this chapter to illustrate the more likely case where at least some attribute measures are uncertain.

In our site-selection example under assumed certainty there are two dominated sites. West and Central are both dominated by East since all the attribute measures for East are equally or more preferred to those of West and Central. West and Central can therefore be discarded from further analysis, provided that uncertainty about the attribute levels can be ignored.

Quantify preferences

Just as in the single-attribute case, reflecting the preferences of a DM for alternative consequences is a matter of defining an appropriate utility function. In the multi-attribute case, suppose there are n different attributes to be considered in making a particular decision, then the total multi-attribute utility U may be assumed to be some function of the individual attribute measures x_i that can be written as:

$$U = U(x_1, x_2, \ldots, x_n) \tag{10.1}$$

However, considerable practical problems arise in eliciting such a multi-attribute function because of the need to consider and reflect in the function the DM's preferences for all the possible combinations of attribute measures. Assumptions, sometimes very strong ones, must be made about the nature of the multi-attribute utility function in order to make it possible to elicit and specify an appropriate function, as discussed below.

Once the multi-attribute utility function has been determined, the overall utility of each of the alternatives can be calculated by entering the corresponding attribute measures into this utility function.

Rank alternatives

Once the overall utilities of all alternatives have been determined, the DM is able to rank the alternatives. The alternative that ranks highest (i.e. that with the highest overall utility) should be chosen.

As for single-attribute analysis, the limitations following from the arbitrary scales used for this type of analysis mean that the magnitudes of differences in overall utilities among alternatives are not readily interpreted. Furthermore, comparisons of overall utilities from one analysis to another, or between individuals, are generally meaningless.

Basic Concepts in Multi-attribute Utility

The notion of a multi-attribute utility function

Theoretically, there is virtually no difference between a multi-attribute utility function and a single- attribute utility function as described in Chapter 5. Both are functions that assign single utility values to consequences in a way that reflects the preferences and attitudes to risk of the DM for the set of possible outcomes.

To understand what is implied in a multi-attribute function such as Eqn 10.1, it is useful to examine the properties we would normally expect the function to exhibit.

Indifference curves and optimality

Under assumed certainty, a multi-attribute utility function with only two attributes can be represented graphically using indifference curves. An *indifference curve* is a set of combinations of two attribute measures among which the DM is indifferent (i.e. that the DM gets the same level of utility for all combinations of attribute measures in the set). For this reason, indifference curves are sometimes also called *iso-utility curves* or *iso-preference curves*. A two-attribute utility function can be characterized by a family of indifference curves corresponding to increasing levels of utility. For 'normal' goods, with more preferred to less, utility will increase to the north-east. Such indifference curves have already been discussed in Chapter 8 for the case where the attribute measures were outcomes contingent on one or other of two states of nature.

The slope of the indifference curve at a particular point reflects the utility trade-off between the attributes, such as costs and distance in our site-selection example. This gradient is called the *marginal rate of substitution*. A special case of marginal substitution occurs when the substitution rate at point (x_1, y_1) does not depend on the attribute measures x_1 of attribute X and y_1 of attribute Y. That means that there is a constant rate of substitution and therefore the corresponding indifference curve is linear. However, in most cases, the marginal rate of substitution at (x_1, y_1) is not constant, but depends on the measures x_1 and y_1. If, at point (x_1, y_1), the DM is willing to give up λ units of X for 1 unit of Y, then the marginal substitution rate of X for Y at point (x_1, y_1) is λ. For example, a person who is very hungry has a whole loaf of bread but almost no milk would be willing to give up several slices of bread for one cup of milk. But as the amount of milk acquired increases, the number of slices of bread the person would trade for still more milk will decrease. So, λ can be interpreted as the amount of X the DM is just willing to 'pay' for a unit of Y at the given levels of x_1 of attribute X and y_1 of attribute Y.

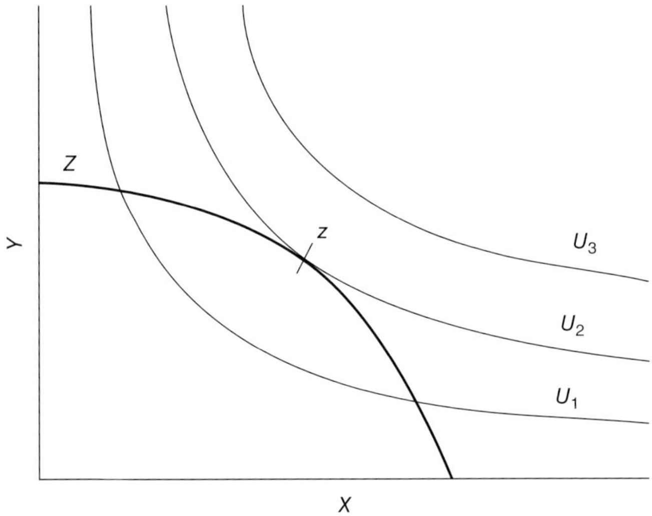

Fig. 10.2. Optimal choice and indifference curves with Z being the north-eastern frontier of the attribute measures of the set of all feasible combinations of alternatives.

The marginal rate of substitution at (x_1, y_1) is the negative reciprocal of the slope of the indifference curve at (x_1, y_1). If X and Y are both desirable, such indifference curves can usually be presented in a two-dimensional evaluation space as curves convex to the origin that asymptotically approach the X- and the Y-axis. The mathematics underlying non-constant substitution is beyond the scope of this book but can be found in accessible form in Keeney and Raiffa (1976, Chapter 3).

If the DM's preferences can be represented in the form of a set of indifference curves, the optimal choice can also be represented graphically. Assuming that X and Y are normal goods – the more of each attribute the better (cf. our site-selection example, where this is not the case), given any fixed level of the other attribute – the general case of this maximization problem is depicted in Fig. 10.2. Again under assumed certainty, if the north-east boundary of the set of possible combinations of alternatives is the locus shown, the optimal alternative z is that alternative from the set of feasible alternatives Z that is on the highest attainable indifference curve.

Simplifying assumptions to make analyses possible

The task in multi-attribute utility analysis is to find a way of eliciting the preferences of a DM in order to be able to locate alternatives in utility space, as illustrated theoretically in Fig. 10.2. Moreover, especially for risk analysis of multi-attribute decision problems, it will be desirable to know the formula of the utility function. If there are many attributes to consider, the elicitation task may seem daunting. However, decision analysis has been introduced in this book as the 'art of the possible'. This statement is especially true with respect to multi-objective decision making. In cases of complex multi-objective analysis, simplifications may have to be made. In this section, we discuss some basic assumptions that are commonly made about multi-attribute preferences in order to make analysis more straightforward. In so far as the assumptions may be strictly inapplicable in many cases, the need to rely on these assumptions can be considered to be a limitation. On the other hand, these assumptions are generally less demanding than those made, usually implicitly, in many of the more ad hoc approaches to multi-attribute decision making.

Preferential independence and utility independence

In order to be able to decompose the general multi-attribute utility function with n attributes (Eqn 10.1) into a convenient functional form of the n individual attributes, we have to make, and ideally verify, two assumptions about the nature of the DM's preferences for the underlying attributes. These are mutually *preferential independence* and *utility independence.*

An attribute X is said to be *preferentially independent* of another attribute Y if preferences for levels of attribute X do not depend on the level of attribute Y. For example, let Y be the time to completion of building the new manure processing plant, and let X be the total investment. Suppose the policy adviser prefers a time of 6 months to one of 12 months when the total investment is $100 million. Then, if Y is preferentially independent of X, the DM will also prefer the shorter building time to the longer one, no matter what the total investment is. If the policy adviser also prefers lower total investment no matter what the building time, it implies that X is preferentially independent of Y. Then the two attributes are said to be mutually preferentially independent.

Cases of breakdown of preferential independence are not uncommon. For example, an Australian grazier may prefer riding a horse rather than a motor bike when mustering sheep, but the reverse when mustering cattle. In multi-objective decision analysis it may be possible to overcome such breakdowns of the preferential independence assumption by defining the attributes of the alternatives differently. In our above example, we might specify one attribute of mustering stock (sheep or cattle) with the type of transport defined differently for sheep and cattle.

Attribute X is *utility independent* of attribute Y if preferences for uncertain choices (such as lotteries) involving different levels of attribute X do not depend on the level of attribute Y. Imagine assessing a certainty equivalent for a lottery involving only outcomes in X. If the DM's CE for the X lottery is the same no matter what the level of Y, then X is utility independent of Y. If Y is also utility independent of X, the two attributes are mutually utility independent.

The assumption of mutual utility independence is stronger than that of mutual preferential independence since it requires that the CE of a risky prospect for each attribute is independent of the level of all other attributes. In other words, the concept of utility independence can be viewed as an extension to risky choice of the concept of preferential independence. Hence, the assumption is not required for analysis under assumed certainty. An example of breakdown of the assumption of utility independence is the case of a grazier who is more averse to the risk of drought if running stud animals rather than normal commercial stock. Full mutual utility independence is probably the exception rather than the rule, yet the assumption is commonly made (often implicitly) since to do otherwise would make the analysis too difficult. Again, the more obvious exceptions can sometimes be avoided by appropriate choice of attributes or attribute measures.

A procedure to verify the conditions for both types of independence is described in Keeney and Raiffa (1976, pp. 299–301). Such verification appears to be rarely attempted in practice.

If mutual preferential and utility independence exist, it is possible to define the multi-attribute utility function in the general form:

$$U(x_1, x_2, \ldots, x_n) = U\{u_1(x_1), u_2(x_2), \ldots, u_n(x_n)\} \tag{10.2}$$

Elicitation of a multi-attribute function of this form can proceed in two stages. First, attribute utility functions $u_i(x_i)$ are derived for each attribute measure in turn, then these individual attribute utilities are combined in some way into a total utility value.

The quasi-separable utility function

Of the operational forms of Eqn 10.2, perhaps the most flexible that has found practical application is the so-called *quasi-separable utility function*. If the attribute utility functions $u_i(x_i)$ are scaled from zero to one, and if U is also scaled from zero to one, the function U is either of the additive form:

$$U(x_1, x_2, \ldots, x_n) = \Sigma_i\, k_i u_i(x_i) \text{ (for } i = 1, 2, \ldots, n) \qquad (10.3)$$

or of the multiplicative form:

$$U(x_1, x_2, \ldots, x_n) = \{\Pi_i\, (K k_i\, u_i(x_i) + 1) - 1\}/K \quad \text{(for } i = 1, 2, \ldots, n) \qquad (10.4)$$

where k_i is a *scaling factor* between zero and one for $u_i(x_i)$ for all i, and K is another scaling constant, the value of which depends on the values k_i. K can be interpreted as an indicator of the effect of interaction between the attributes on total utility. So if $\Sigma_i\, k_i = 1$, then U takes the additive form, as displayed above, and $K = 0$ meaning there is no interaction between the attributes. In contrast, if $\Sigma_i\, k_i \neq 1$, then $K \neq 0$ and U takes the multiplicative form. If K is greater than 0, then the attributes interact destructively so that a low utility for one attribute can result in a low overall utility U. When K is less than 0, the attributes interact constructively so that a high individual attribute utility results in a high overall utility U. These implications follow from the specific quasi-separable form of the utility function U. For more details and background including the derivation of K from the k_i values in the multiplicative case, see Keeney (1974) or Anderson *et al.* (1977, p. 86).

The essence is that, instead of trying to assess the n-dimensional utility function $U(x_1, x_2, \ldots, x_n)$ directly, a very difficult task for DMs of normal introspective capacity, it is only necessary to assess n one-dimensional utility functions $u_i(x_i)$ and the n scaling factors k_i. In other words, by this method the multi-attribute utility function can be decomposed into component parts each of which is more readily handled.

Approximations of the full quasi-separable model

The full approach to multi-objective decision analysis based on a quasi-separable utility function may not be judged to be applicable for one of a number of reasons. First, quite inappropriately, some analysts may be 'scared off' by the rather ugly algebra, particularly for the multiplicative form of the utility function. More rationally, the full assessment of (generally non-linear) attribute utilities and of attribute weights may be judged to be too demanding of the introspective capacity of the DM, or the DM may be unavailable to supply the required assessments. In such cases, one or both of the following further simplifying assumptions may be made:

1. The attribute utility functions may be treated as linear.
2. The attribute weights may be assumed to sum to one, so that the form of the overall utility function is taken to be additive, not multiplicative.

The crudest operational form of Eqn 10.2, found in some applications, is based on the assumptions that all the attribute utility functions $u_i(x_i)$ are linear, so that $u_i(x_i) = x_i$, and that the total utility function U is a simple weighted sum of the attribute measures, that is:

$$U(x_1, x_2, \ldots, x_n) = \Sigma_i a_i x_i \quad (10.5)$$

where U is the multi-attribute utility function, x_i is the attribute measure corresponding to the i-th attribute and the a_i are weighting constants. This is the form of function often adopted, seldom with any explicit consideration of its limitations, in many so-called *multi-criteria analyses*. The attribute measures x_i are best normalized to lie in the range zero to one, but this is not always done. Without normalization, the derivation and interpretation of the a_i constants may be problematic for reasons discussed later.

The model of Eqn 10.5 implies linear indifference curves, which is unlikely to be realistic over a wide range of attribute measures, but may be a reasonable approximation over a relatively narrow range in some cases. The model is used implicitly in cases where it is judged to be appropriate to 'cost out' all attributes but one in order to convert them all to a single money value, typically using constant prices. With some modifications in some applications, that is the method commonly used in cost–benefit analysis. Applied to the site-selection example, this model would be appropriate if the policy adviser were willing to trade off the attribute of distance to farms against net cost at a rate of, say, \$5000/km.

The measurement of a linear multi-attribute utility function involves determining a number of such trade-offs. If there are n attributes, then $n - 1$ trade-offs need to be evaluated. Unless n is very large, the measurement process is not very tedious once the strong assumption of linearity has been accepted.

A slightly more general additive model than that in Eqn 10.5 allows for non-linearity in the attribute utilities. It takes the form:

$$U(x_1, x_2, \ldots, x_n) = \Sigma_i a_i u_i(x_i) \quad (10.6)$$

where u_i is a function of the attribute measure x_i corresponding to the i-th attribute and a_i are attribute utility weights, usually scaled to sum to 1.0. Again, it will also be desirable to scale the attribute utilities from zero to one over the range from the worst to the best values of the individual attribute measures to clarify exactly what the weights attach to. The form of Eqn 10.6 implies that the attributes are assumed to be independent, meaning that the rate of trade-off between any two attributes depends only on the measures of those two attributes and not on measures of other attributes. Moreover, the additive form implies that even the lowest level of any attribute, assigned a utility of zero, can always be compensated for by higher levels of other attributes. Clearly, that is not always reasonable. For example, if one of the essential attributes of a life support system is at zero level, no amount of the others will compensate. By contrast, zero level of any attribute utility in the multiplicative case makes total utility zero, regardless of the other attribute levels.

Assessing Attribute Utilities

The first step after the DM has identified the alternatives, objectives and attributes measures, is to make the individual attribute measures comparable by converting them into utilities. This should be done in such a way that the conversion reflects that person's preferences for these individual attribute measures. Utilities for attribute measures are assessed considering just one attribute at a time and so depend on the assumptions of mutual preferential and utility independence. Comparability across attribute utilities is achieved by scaling the utility values so that the utility of the best possible outcome for any attribute measure is one and the utility of the worst possible outcome is zero.

The definitions of 'worst' and 'best' attribute measures need attention. There are two possible situations. If the analysis is being carried out under assumed certainty, the worst and best attribute measures are the worst and the best *within* the set of alternatives. These are easily found by inspecting the ranges of measures for each attribute across the alternatives being considered. Alternatively, and more realistically, the analysis may require the recognition of uncertainty. Then one or more attribute measures must be described by probability distributions. Hence, the worst and best attribute measures should be interpreted as the worst and best possible attribute measures that the DM can imagine (i.e. the values of x_i corresponding to $F(x_i) = 0.0$ and 1.0, assuming x_i can take continuous values). In this section we address the case of assumed certainty, while in the next major section we focus on dealing with uncertainty.

Eliciting attribute utility functions

The methods of utility function elicitation described in Chapter 5 are directly applicable for assessing individual attribute utility functions. Elicitation of such functions for attributes is generally required for multi-attribute analysis, whether or not attribute measures are uncertain. For attributes that can take any values between worst and best, the ELCE method is usually most appropriate, whereas for attributes that take only discrete levels, either the von Neumann–Morgenstern method, or some form of direct utility assessment method must be used.

In the site selection example, we consider first the two attributes that can take continuous values, namely 'Distance to farms' and 'Net costs'. Starting with 'Distance to farms' the worst and best outcomes of 150 km and 50 km are assigned utility values of zero and one, respectively. Suppose the policy adviser's utility function u_D for 'Distance to farms' is specified using the ELCE method. Using the notation for indifference between risky prospects introduced in Chapter 5, suppose the CEs for 'Distance to farms' are as follows:

$$\text{CE1: } (50\text{ km}, 150\text{ km}; 0.5, 05) \sim (115\text{ km}, 1.0) \tag{10.7}$$

$$\text{CE2: } (50\text{ km}, 115\text{ km}; 0.5, 05) \sim (90\text{ km}, 1.0) \tag{10.8}$$

$$\text{CE3: } (115\text{ km}, 150\text{ km}; 0.5, 05) \sim (135\text{ km}, 1.0) \tag{10.9}$$

With these CEs, the following negative exponential utility function u_D can be estimated $u_D = 1.363 [1 - \exp\{0.0132\ (D - 150)\}]$, where D is the distance in kilometres (calculations are not shown here). The values of u_D can now be calculated as follows:

$u_D(150\text{ km}) = 0.00$ for North

$u_D(70\text{ km}) = 0.89$ for East

$u_D(50\text{ km}) = 1.00$ for South.

The DM's utility function u_C for 'Net costs' is also presumed to have been determined using the ELCE method, resulting in the utility function $u_C = 1.707\ [1 - \exp\{1.259\ (C - 1.2)\}]$ where C is the net cost in millions of dollars (details again not shown here). Using this function, the values for u_C are now found to be:

$u_C(\$0.5 \text{ million/year}) = 1.00$ for North

$u_C(\$1.0 \text{ million/year}) = 0.38$ for East

$u_C(\$1.2 \text{ million/year}) = 0.00$ for South.

Note that use of the ELCE method in this example is not strictly necessary because there are only three levels of the measure of each attribute to consider, two of which are assigned utility values by definition. That leaves only one utility value to be assessed for each continuous attribute. That could be done by the method to be described next with no need to determine the full utility function. However, the function would be needed if there were many alternatives to compare with different attribute levels. Knowledge of the full utility functions is also necessary for risk analysis, as described later in the chapter.

A different approach is needed to elicit the DM's utility function for site isolation since this attribute measure is qualitative and has been defined on a discrete cardinal scale from 1 to 5. Intermediate values have no meaning so the ELCE method is not applicable. Using the von Neumann–Morgenstern method we suppose that the DM reaches the following indifference relationship for scoring site isolation:

$$(4; 1.0) \sim (5, 1; 0.44, 0.66) \quad (10.10)$$

This means that the DM expresses indifference between a score of 4 for certain and a lottery paying scores of 5 or 1 with probabilities of 0.44 and 0.66, respectively. Then, with I representing the scale for site isolation, since $u_I(5) = 1.0$ and $u_I(1) = 0$, $u_I(4) = 0.44$. Hence we have:

$u_I(1) = 0.00$ for North

$u_I(5) = 1.00$ for East

$u_I(4) = 0.44$ for South

We now have all the attribute utilities needed to move to the next step.

Methods for Weighting Attribute Utilities

After standardized utility values of individual attribute measures have been evaluated, the next step is to find appropriate attribute weights that allow these attribute utilities to be amalgamated into estimates of the DM's overall preferences for the alternatives. For this purpose, we shall introduce a method, based on lotteries, that is consistent with the quasi-separable model. The method leads to either an additive or multiplicative form of the utility function, depending on the sum of the attribute scaling factors. Lotteries can be applied both under assumed certainty and under uncertainty.

A number of other methods are to be found in the literature but we do not deal with them here. Most of these alternative methods are based on the assumption of an additive utility function.

Lotteries to trade off attributes

To weight attributes the DM is presented with a choice between two prospects I and II. Prospect I is a lottery with a probability p of getting a hypothetical alternative of the best outcomes for all attribute measures, and a probability $(1 - p)$ of getting a hypothetical alternative which is worst on all attribute measures. Prospect II is a sure outcome of a hypothetical alternative yielding the best outcome for one attribute (say attribute X) and worst outcomes for all others. The task then is to find the probability p_X which makes the DM indifferent between the lottery (prospect I) and the sure thing (prospect II). There must be such probability p_X, because with $p_X = 1.00$ the DM would certainly opt for prospect I, as in that case the outcome would be the 'best for all attributes' for certain. With $p_X = 0.00$ the DM would definitely opt for prospect II, as in that case prospect I yields the worst for all attributes. So, by a trial-and-error process it should be possible to zero in on a probability p_X at which the DM's preference shifts from prospect I to II. The higher the probability p_X, the more important is attribute X, as is explained below. This procedure is repeated with each attribute in turn set to the best measure in prospect II.

The scaling factor k_i for the i-th attribute utility function in the quasi-separable utility function, introduced above, is equal to p_i, as demonstrated below. For convenience, we use the notation of x^+ and x^- to indicate the most and least preferred measures of attribute X, respectively. Suppose there are three relevant attributes X, Y and Z. Then to determine k_X for attribute X using the lottery method we need to elicit p_X, such that:

$$U(x^+, y^-, z^-) = p_X\, U(x^+, y^+, z^+) + (1 - p_X)\, U(x^-, y^-, z^-) \qquad (10.11)$$

The part to the left of the equality sign refers to prospect II in the lottery, and the part to the right to both elements of prospect I. Since scaling of the utility function from zero to one implies $U(x^+, y^+, z^+) = 1$ and $U(x^-, y^-, z^-) = 0$, it follows that $U(x^+, y^-, z^-) = p_X$. Furthermore, again because the attribute utility functions are scaled from zero to one, so that $u(x^+) = 1$ and $u(y^-) = u(z^-) = 0$, it follows that $k_X = p_X$.

Let's return to our site-selection example, with North, East and South as alternatives and 'Distance to farms', 'Net costs' and 'Site isolation' as attributes. To scale the overall utility function U from zero to one we define:

$$U(50 \text{ km}, \$0.5 \text{ million, isolation measure of } 5) = 1.00 \qquad (10.12)$$

$$U(150 \text{ km}, \$1.2 \text{ million, isolation measure of } 1) = 0.00 \qquad (10.13)$$

The decision tree to assess the probability corresponding to the scaling factor for 'Distance to farms' is shown in Fig. 10.3.

Suppose the policy adviser is indifferent between the prospects I and II in Fig. 10.3 when probability p_D equals 0.60. This means that:

$$\begin{aligned}&U(50 \text{ km}, \$1.2 \text{ million, isolation measure of } 1) \sim\\&0.60\ U(50 \text{ km}, \$0.5 \text{ million, isolation measure of } 5) +\\&0.40\ U(150 \text{ km}, \$1.2 \text{ million, isolation measure of } 1)\end{aligned} \qquad (10.14)$$

where ~ means 'is indifferent to'. With the utility values for U(50 km, \$0.5 million, isolation measure of 5) = 1.00 and U(150 km, \$1.2 million, isolation measure of 1) = 0.00, as defined above, it follows that $k_D = p_D = 0.60$ for 'Distance to farms'.

The same approach can be repeated to determine the weights for 'Net costs' and 'Site isolation'. Suppose the probabilities elicited from the policy adviser for indifference in the two further lotteries are $p_C = 0.35$ and

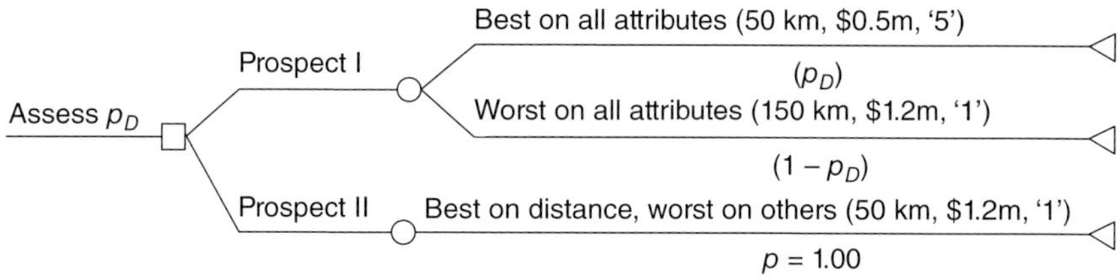

Fig. 10.3. Decision tree using a lottery to assess the attribute scaling factor for 'Distance to farms'.

$p_I = 0.20$. Then $k_C = 0.35$ and $k_I = 0.20$. In this case we have $\Sigma_i k_i = 0.60 + 0.35 + 0.20 = 1.15$, which is not equal to one, and therefore the multiplicative rather than the additive form of the overall utility function applies. K can be determined as follows in this example. As $u_D(d^+) = u_C(c^+) = u_I(i^+) = 1$ due to scaling, and $U(d^+, c^+, i^+) = 1$, then (with d, c and i denoting the attribute measures of distance, costs and isolation, respectively):

$$U(d^+, c^+, i^+) = \{(K k_D u_D(d^+) + 1) \times (K k_C u_C(c^+) + 1) \times (K k_I u_I(i^+) + 1) - 1\}/K \quad (10.15)$$

Entering the known utilities and k_i values gives:

$$1 = \{(0.6\,K + 1) \times (0.35\,K + 1) \times (0.2\,K + 1) - 1\}/K \quad (10.16)$$

Solving this equation gives $K = -0.391$. We now have all the components of the multi-attribute utility function:

$$U(d, c, i) = [\{1 - 0.235\,u_D(d)\} \times \{1 - 0.137\,u_C(c)\} \times \{1 - 0.078\,u_I(i)\} - 1]/-0.391 \quad (10.17)$$

The utility functions $u_D(d)$, $u_C(c)$ and $u_I(i)$ have been determined in the previous section. Using these functions in the multi-attribute utility function of Eqn 10.17, the overall utilities for sites North, East and South are found to be: 0.350 (North), 0.790 (East) and 0.668 (South), respectively. So, East has the highest overall utility and is therefore the most preferred site.

Multi-attribute Utility Analysis under Uncertainty

So far we have assumed that the DM was able to attach single-valued measures to each attribute for each alternative, as in Table 10.1. These attribute measures can be referred to as point estimates. *Point estimates* are single numbers assigned assuming no uncertainty. If the DM wants to recognize the reality of uncertainty, it will be appropriate to replace an uncertain point estimate with a *probabilistic attribute measure*, where a probability distribution is used to represent the uncertainty that the DM wants to include in the analysis. If more than one attribute is uncertain, and there is a stochastic dependency among the uncertain attribute measures, the joint distribution is needed to account for any stochastic dependency. In the case of stochastic dependency between uncertain attribute measures, a purpose-built stochastic simulation model will be needed for the multi-attribute utility analysis. Furthermore, when some attribute measures are stochastic, the analysis of dominance is not straightforward. In cases where stochastic independence between uncertain attribute measures can be assumed, as mentioned above, the methods of stochastic dominance analysis described in Chapter 7, this volume, could be applied. However, it will usually be best to skip the dominance analysis and to go straight to the evaluation in terms of multi-attribute utility.

Before we are able to carry out the multi-attribute utility analysis under uncertainty, we have to re-assess both the utility functions for the (uncertain) attribute measures and the scaling factors for the attributes themselves. Recall that in the corresponding lotteries now we have to define the 'worst' and 'best' attribute measures as the lowest and highest possible measures, respectively, that are relevant to the problem at hand. However, the procedures to be carried out to obtain these utility functions are otherwise the same as presented before – only the numbers in the lotteries are different.

Let's return to the site-selection example. First we need to elicit the policy adviser's probability distributions for 'Distance to farms' and 'Net costs' for all five possible sites. (We have to include all five because the earlier dominance analysis that allowed us to eliminate two sites is no longer valid.) The policy adviser decides that triangular distributions will serve for this purpose and assigns the distributional parameters shown in Table 10.2. It is decided that the two distributions can be treated as stochastically independent. As before, a point estimate is being used for 'Site isolation', also shown in the table.

These estimates include the best possible and worst possible attribute measures for each attribute. From the table it can be seen that the worst and best possible attribute measures for 'Distance to farms' are 180 km and 30 km, respectively. The worst and best possible attribute measures for 'Net costs' are \$3.0 million and \$0.2 million dollars, respectively. With these values the attribute utility functions u_D and u_C can be elicited using the ELCE method. Negative exponential utility functions are fitted and then scaled to run from zero to one between best and worst attribute measures. Suppose they are as follows (where D is distance in kilometres and C net costs in millions of dollars):

$$u_D = 1.3045\,[1 - \exp\{0.0097\,(D - 180)\}] \tag{10.18}$$

$$u_C = 1.2726\,[1 - \exp\{0.5503\,(C - 3.0)\}] \tag{10.19}$$

Utilities of attribute measures of 'Site isolation' are the same as before. This linear attribute utility function is also scaled so that the worst attribute score of 1 in Table 10.2 has a utility of zero and the best score of 5 has a utility of one.

At this stage we could check for any stochastic dominance to eliminate inferior sites. However, because there are only five sites in total and because the calculations are to be done by computer, that step can be omitted.

Table 10.2. Probabilistic attribute measures for the attributes 'Distance to farms' and 'Net costs'.

	Attribute measures		
Name of site	**Distance to farms (km) (*a*, *m*, *b*)[a]**	**Net costs (millions \$) (*a*, *m*, *b*)[a]**	**Site isolation**
North	(125, 145, 180)	(0.2, 0.4, 0.9)	1
East	(50, 60, 100)	(0.8, 1.0, 1.2)	5
South	(30, 45, 75)	(0.6, 1.1, 1.9)	4
West	(70, 115, 175)	(1.3, 1.7, 2.4)	5
Central	(40, 90, 170)	(1.2, 1.8, 3.0)	1

[a]The notation (a, m, b) refers to the lowest, most likely and highest possible values of the triangular probability distributions.

Table 10.3. Expected utility values of the five sites with risky attribute measures.

Name of site	Expected utility	Rank
North	0.4628	5
East	0.8739	1
South	0.8642	2
West	0.6406	3
Central	0.5325	4

The same lottery method as before can be used to find the new scaling factors k_i, using the above worst and best possible attribute measures for 'Distance to farms', 'Net costs' and 'Site isolation'. Suppose they are (close to) $k_D = 0.55$, $k_C = 0.30$ and $k_I = 0.15$. In this case $\Sigma_i\ k_i = 0.55 + 0.30 + 0.15 = 1$, so the additive form of the utility function applies, that is:

$$U(d, c, i) = 0.55\ u_D(d) + 0.30\ u_C(c) + 0.15\ u_I(i) \tag{10.20}$$

where *d*, *c* and *i* denote the attribute measures of distance, costs and isolation, respectively.

Stochastic simulation was used to obtain the final rankings. The number of simulation iterations was set at 5000. The results, which were obtained using @Risk, are as shown in Table 10.3.

East is still the best site to choose: it has the highest overall expected utility of 0.8739. South ranks second with an overall utility of 0.8642. North is the least preferred site with a utility of 0.4628.

Other Forms of Multi-attribute Utility Analysis

Once a multi-attribute utility function has been specified, of whatever form, it can be used in many types of risky decision analysis, in much the same way as a single-attribute utility function can be used. For example, decision trees can be constructed and resolved when the consequences are multi-attributed. Similarly, stochastic simulation models, described in Chapter 6, can be implemented with multi-attribute utility functions. Mathematical programming models, described in Chapter 9, can be extended to accommodate multi-objective planning in a number of ways, including the direct use of a linear or non-linear multi-attribute objective function.

In all the types of risky multi-attribute decision analysis mentioned above, the specification of a multi-attribute utility function is only half the story. It is also necessary to be able to specify the joint distributions of attribute measures if an assumption of stochastic independence is unrealistic. The difficulties likely to be encountered in doing this, and some feasible methods of tackling those difficulties, were described in Chapter 4.

Recall again that we introduced decision analysis as the 'art of the possible', which, as demonstrated in this chapter, applies in particular to multi-objective decision making. As explained, the type of multi-objective analysis described above may be difficult to conduct because DMs may find it too hard to express their beliefs and preferences for all (combinations of) attributes or attribute measures. Then, unavoidably, simplifications have to be made, to arrive at an acceptable approach to the decision problem under investigation. Finding the right compromise between theoretical nicety and practicality is an artistic skill that decision analysts are routinely called upon to exercise. However, we suggest that for some important decision problems, a full multi-attribute utility analysis, albeit perhaps with some short cuts, will be more likely to support good decision making than some of the cruder forms of analysis, which, by the nature of the very strong and implausible assumptions on which they are based, run the risk of providing misleading recommendations.

For whatever reasons, multi-attribute utility theory (MAUT) has not been widely used in agricultural applications. Probably the seeming complexity of the method has discouraged potential analysts. However, the method is not as difficult to apply as may appear to be the case, especially if some short cuts are taken, such as eliciting measures of curvature of the attribute utility functions each with a single, carefully assessed CE. Similarly, the multiplicative form of the MAUT function can be avoided if it is judged to be adequate to use normalized weights to combine the attribute utilities, perhaps assessed by allocating counters to the intervals of the attribute measures. More published work with MAUT would provide a firmer basis for assessing the merit of the method.

Selected Additional Reading

Most of this chapter is based on Keeney and Raiffa (1976). Though not always easy going, this text is essential reading for would-be practitioners of multi-attribute utility analysis. Clemen and Reilly (2014) provide a more readable though somewhat less thorough treatment of the topic. Probably the best known illustrative example of the methods described above is the application by de Neufville and Keeney (1972) relating to the development of an airport for Mexico City. von Winterfeldt and Fischer (1975) provided a thorough survey of state of the art MAUT and estimation at that time. Yntema and Torgerson (1961) showed that the assumption of an additive function generally provides a reasonable approximation in multi-attribute decision problems.

Applications of MAUT to agriculture or natural resource management are hard to find, yet the method has been used in a range of other applications, particularly in health management. Some agricultural examples are Herath *et al.* (1982), Huirne and Hardaker (1998) and van Calker *et al.* (2006). In natural resource management see Delforce and Hardaker (1985) and Ananda and Herath (2005). Kainuma and Tawara (2006) describe an application to lean-and-green supply chain management.

References

Ananda, J. and Herath, G. (2005) Evaluating public risk preferences in forest land-use choices using multi-attribute utility theory. *Ecological Economics* 55, 408–419.

Anderson, J.R., Dillon, J.L. and Hardaker, J.B. (1977) *Agricultural Decision Analysis*. Iowa State University Press, Ames, Iowa.

Clemen, R.T. and Reilly, T. (2014) *Making Hard Decisions with Decision Tools*, 3rd edn. South-Western, Mason, Ohio.

Delforce, R.J. and Hardaker, J.B. (1985) An experiment in multiattribute utility theory. *Australian Journal of Agricultural Economics* 29, 179–198.

de Neufville, R. and Keeney, R.L. (1972) Use of decision analysis in airport development for Mexico City. In: Drake, A.W., Keeney, R.L. and Morse, P.M. (eds) *Analysis of Public Systems*. MIT Press, Cambridge, Massachusetts, pp. 497–519.

Herath, H.M.G., Hardaker, J.B. and Anderson, J.R. (1982) Choice of varieties by Sri Lanka rice farmers: comparing alternative decision models. *American Journal of Agricultural Economics* 64, 87–93.

Huirne, R.B.M. and Hardaker, J.B. (1998) A multi-attribute model to optimise sow replacement decisions. *European Review of Agricultural Economics* 25, 488–505.

Kainuma, Y. and Tawara, N. (2006) A multiple attribute utility theory approach to lean and green supply chain management. *International Journal of Production Economics* 101, 99–108.

Keeney, R.L. (1974) Multiplicative utility functions. *Operations Research* 22, 22–34.

Keeney, R.L. and Raiffa, H. (1976) *Decisions with Multiple Objectives: Preferences and Value Tradeoffs*. Wiley, New York.

van Calker, K.J., Berentsen, P.B.M., Romero, C., Giesen, G.W.T. and Huirne, R.B.M. (2006) Development of an application of a multi-attribute sustainability function for Dutch dairy farming systems. *Ecological Economics* 57, 640–658.

von Winterfeldt, D. and Fischer, G.W. (1975) Multiattribute utility theory: models and assessment procedures. In: Wendt, D. and Vlek, C. (eds) *Utility, Probability, and Human Decision Making*. Reidel, Dordrecht, pp. 47–85.

Yntema, D.B. and Torgerson, W.S. (1961) Mean-computer cooperation in decisions requiring common sense. *IRE Transactions on Human Factors in Electronics HFE-2* 20–26. Reprinted in: Edwards, E. and Tversky, A. (eds) (1967) *Decision Making*. Penguin, Harmondsworth, UK, pp. 300–314.

11 Risky Decision Making and Time

The Time Factor in Decision Analysis

In most of the earlier chapters, decision analysis has been set in the framework of a short time horizon. In reality, however, farm and agribusiness managers make decisions about production, marketing and finance not just for the present but also for the longer run. For these long-term decisions, accounting for the factor 'time' is essential. Some ways of dealing with time in risky decision making are addressed in this chapter.

One important effect of accounting for time is that uncertainty generally increases the further into the future we look. Consequently, the need to account for risk is often greater in decision making for the long run. Unfortunately, accounting for time and for the greater uncertainty that is thereby entailed adds greatly to the complexity of analysis. As we shall explain, there are difficulties in extending the methods of decision analysis to long-run planning. Nevertheless, we believe that ways of considering such decisions that embody at least some key elements of the principles and methods outlined in earlier chapters are likely to lead to better decisions than those emerging from analyses that ignore risk or that seek to accommodate it only in some over-simplified way.

It is usual to distinguish two main types of multi-period decision problems. In the first and seemingly most straightforward, a decision made now has consequences, nearly always unavoidably uncertain, that are experienced for a number of years into the future. Most important decisions in this once-only decision category are *investment decisions* – whether to buy more land, to invest in new and expensive equipment or buildings or, for a government, whether to invest in, say, a new and expensive flood mitigation scheme for a flood-prone river basin.

In the second type of multi-period decision problem, a sequence of decisions must be made through time, these decisions typically being interleaved with the outcomes of uncertain events that impinge on later choices, as well as on the consequences of decisions taken earlier. We shall call decision problems of this second kind *dynamic decision problems*. They are much the same as those that were classified in the mathematical programming (MP) context in Chapter 9 as problems with embedded risk, although most MP analyses with embedded risk deal with relatively short time spans.

In fact, the distinction between investment decisions and dynamic decisions is an artificial one since almost any important decision taken today will have consequences that extend into the future and that require further choices to be taken later as uncertainty unfolds. Treating investment decisions as once-only independent choices leads to some difficulties, as we shall see in relation to real options, discussed below.

With both types of decision, difficulties arise because it is usually impossible to trace out in advance all the future risky outcomes and possible responses. A good analogy here is with planning a move in a game of chess. To work out the best move for any position in the game requires all possible immediate and future moves to be considered along with all possible responses by the opponent. The calculation task to do that even moderately comprehensively in reasonable time is beyond the capacity of all but the largest and fastest computers. What chess players do, with greater or lesser success, is to look far enough into the future so that moves likely to lead to bad outcomes can be eliminated and the one that looks most promising can be chosen. The same principle applies to long-term analysis. It is necessary to analyse future outcomes, and often also future decisions, only to the extent that they impinge substantially on the current decision. Moreover, because the effects on the initial choice of further analysis that has not yet been done cannot be known, a 'stopping rule' for the analysis is inevitably subjective. It is a matter of judgement just how far ahead one has to look to be reasonably sure that decisions taken now are good, or at least not seriously bad.

In the remainder of this chapter we deal first with investment decisions and later consider dynamic decision problems. In so far as dynamic decisions have consequences that extend into the longer run future, the theoretical issues of investment appraisal discussed below also apply to the appraisal of returns from such dynamic decisions.

Investment Appraisal under Uncertainty

Theoretical background

In the imaginary world of no uncertainty and a perfect capital market, investment appraisal would be easy. To see why, suppose a DM has an inter-temporal utility function for income, reflected as net cash flows available for consumption over present and future time periods. Assume that this function is of the quite general form:

$$U = U(c_0, c_1, \ldots, c_T) \tag{11.1}$$

where the subscripts relate to time periods to a finite planning horizon at period T. These net cash flows could equally represent the funds available for distribution to equity holders in a corporate business. At least for now, we impose few conditions on the form of the function $U(.)$ except to assume that more income in any period is always preferred to less and that there is diminishing marginal utility of income, i.e. $\partial U/\partial c_t > 0$ and $\partial^2 U/\partial c_t^2 < 0$ for all t.

Suppose there is a potential investment with cash-flow implications defined by the vector (x_0, x_1, x_2, …, x_T). If it is a typical investment, some of the early net cash flows will be negative and later flows will be positive. How can we determine whether making this investment will increase the DM's utility?

Recall that we assume for now that there is a perfect capital market, which means that the DM can borrow or save any amounts of money at the fixed market rate of interest. These assumptions mean that the compound interest formulae can be applied to reflect the opportunities the DM has to shift money between time periods. In this perfect capital market the DM can convert any pattern of cash flows over time into any other pattern with the same *net present value* (NPV). The NPV is the discounted value of future cash flows less the capital cost in time $t = 0$. Provided the NPV of the investment is positive, the DM can allocate cash to time periods in order to equate marginal utilities of cash for consumption in each period and so maximize the increment in U in Eqn 11.1 available from the investment. It follows that we don't need to know the exact form of the utility function U in order to deduce the following rules for investment appraisal in this imaginary world:

1. Any investment with a positive NPV is potentially utility increasing and is therefore potentially worthwhile.
2. Any investment with a negative NPV must be utility reducing.
3. When comparing alternative investments over the time horizon to period T, the one with the highest NPV will yield the highest potential increment in the DM's utility.

It is this reasoning that underlies the widespread recommendation by economists that NPV is the most appropriate investment criterion. When comparing investments with different time horizons, the corresponding recommendation is to use *equivalent annuity* (EA) as the choice criterion. EA is the NPV averaged over the life of the investment from time $t = 1$ to T using the formula:

$$EA = \frac{NPV\, i(1+i)^T}{(1+i)^T - 1} \tag{11.2}$$

where i is the relevant discount rate and T is the length of the time horizon.[1] For investments of the same duration, NPV and EA will rank alternatives in the same order of profitability.

In practice, the recommendation in favour of NPV (or EA) is often not followed. Instead, the rate of interest that causes NPV to be zero, known as the *internal rate of return* (IRR), is widely used in preference to NPV, mainly on grounds of ease of interpretation, especially for comparing investments of different scale. The two criteria will usually, but not always, rank alternative investments in the same order. Moreover, not only is IRR less sound on theoretical grounds than NPV, but in some particular circumstances there may be more than one value of IRR that makes NPV zero, leading to obvious confusion.

Confronting reality

The above is all very well, but what happens when we move away from the imaginary world to the real one where uncertainty abounds and where capital markets are less than perfect? An imperfect capital market means that borrowing and saving rates of interest diverge and may change over time, and that unlimited borrowing is not feasible. In extreme cases, such as for farmers in remote locations in less developed countries, capital markets may be largely absent. In this real world, devising appropriate methods of investment appraisal gets a good deal more complicated.

Given the uncertain nature of the future, a DM can no longer optimally borrow and save to move money between time periods, even if the capital market works reasonably well. The reason is that it is not possible to know in advance what future income levels will be. If income is generally falling over time, the DM might want to save in the early time periods whereas borrowing early might be appropriate if income is generally rising. But if the trend in income is uncertain, it is impossible to know for sure whether and how much to save or borrow. The best a DM can hope to do is to rationalize decisions about borrowing and saving using the

[1] Note that annuities are conventionally computed from time $t = 1$ to T, not from $t = 0$ to T. If the latter is more appropriate in some applications, the term $(1 + i)^T$ in the denominator of Eqn 11.2 is replaced with $(1 + i)^{T+1}$.

principles of expected utility maximization. Evidently, in the real world, saving decisions, and especially borrowing decisions, are just as much risky decisions as are decisions about what business investments to make.

Unfortunately, making all such decisions to maximize expected utility is easier said than done. The problems in implementing this decision rule are the same as for one-period decision analysis, but magnified by the addition of a long time horizon (Meyer, 1976). Briefly:

1. It is hard to know the nature of the DM's utility function.
2. Capturing the uncertainty in the returns from individual prospects is many times harder when it is recognized that the returns will occur over several time periods.
3. Stochastic dependency between investment alternatives and between returns from the same prospect over time ideally needs to be accounted for, yet it is difficult to assess and handle such dependency.
4. Partly because of stochastic dependency, but also because of problems in specifying and evaluating an inter-temporal utility function, partial analysis of individual investments is usually problematic.

In other words, maximizing expected utility over an extended time horizon is really a multi-stage (dynamic) portfolio selection problem of considerable difficulty – so much so that it is seldom attempted.

We have emphasized that decision analysis is the art of the possible – a principle that applies strongly to investment appraisal. Some 'artistic' simplifications evidently have to be made. We canvass some possibilities below. First, however, we need to look in more detail at the legitimacy of treating investment appraisal choices as once-only (now or never) decisions.

Real options

In recent decades there has been a growing recognition that the simple investment appraisal rules relying on NPV and related measures do not tell the whole story (Dixit and Pindyck, 1994). The reason is that, as conventionally applied, they wrongly isolate each single investment decision from the context in which it is set. In addition to differences in associated risks, every decision both closes and opens doors to future choices. Where investment is concerned, there is one set of future options for action available if a decision is made to invest now, and perhaps another different set available if the immediate investment decision is rejected. An important option not explicitly considered in the conventional approach is whether to postpone the decision for a time until some of the uncertainty is resolved. In other words, at every stage the DM holds the option to act then or to wait and consider acting later. The option to wait has value because it offers the DM more flexibility in later stages, which can be especially important for irreversible decisions. Postponing decisions allows the DM to collect more information and observe outcomes of previously uncertain events. If uncertainty (partially) resolves, the DM can benefit by being able to adjust the original plan. This extra benefit, also called the 'value of flexibility', comes from the option of delaying an action. Of course, delay is not optimal if the immediate action yields better immediate and future net returns than the option value of waiting.

Postponing choice is not the only option to consider. In principle, all options that are influenced by the current choice ought to be considered. *Real option valuation* takes account of the uncertainty of predicted cash flows and of the DM's opportunities to react to changed circumstances. It may be possible to abandon a project or sell some of the assets if the project is no longer profitable. Similarly, there may be scope to expand or contract the scope of an investment project, to vary the methods or pace of implementation, or to 'opt out' and terminate a failing project. Some investments increase the range of options available. For example, a farmer who invests in an irrigation system has many more land-use opportunities available as a result. So, if the relative profitability of the crops currently being grown were to fall, there would be better options available to cope with the downturn.

Keeping options open has the merit of providing potentially better opportunities to limit downside risk if things go wrong and to seize opportunities to benefit when things go well – in just the same way as applies to financial options, discussed in Chapter 12, this volume. Agricultural investments typically involve a substantial component of sunk costs, which, by definition, are not recoverable if the investment fails. In other words, the presence of sunk costs implies that investment incurs a loss of the option to recover from adverse events affecting the project by reversing the original decision. Hence there is a need in investment appraisal to consider not only investing versus not investing, depending on whether or not the of project appears to be profitable, but also to consider the option of delaying investment in sunk cost items until more is learned about the uncertain future. Evidently, an investment that closes off important options, such as one with a high component of sunk costs, or one that would require heavy use of credit, limiting future borrowing possibilities, should be required to give a higher expected net return, considered alone, than an investment that preserves or expands the real options available. In public decision making, investments that lead to significant irreversibilities, such as permanent environmental damage or the extinction of some species, clearly entail loss of potentially valuable options and so should be evaluated more critically than simply basing a choice on a measure of NPV alone.

Several ways of incorporating real options into decision analysis have been proposed (e.g. Dixit and Pindyck, 1994). In terms that link back to earlier chapters, real options can be handled by more careful specification of the decision tree for the initial decision problem. The range of possible actions at the first decision node should be extended to include other alternatives than invest or don't, notably the option to postpone investment. The tree can also be extended from branches representing investing or not investing to reflect both possible uncertain events and the scope to respond to those events (i.e. the main future options). The latter will usually be different depending on the initial choice.

Of course, in reality, to follow this prescription may all too quickly make the decision tree into a bushy mess. It will also be impossible to imagine and incorporate every future possible state of the world and every available option to respond to that state. Several different approaches have been developed to try to handle this complexity. These methods are designed to take account of uncertainty about the future evolution of the parameters that determine the value of the investment as well as accounting for a DM's ability to respond to the evolution of these parameters. It is the combined effect of these that makes real options technically challenging.

In early applications, real options were often modelled using a partial differential equation. However, both limitations of this approach and the advanced mathematics involved mean that such methods are more likely to be found in academic papers rather than in use by practitioners. More practical methods include extensions of the decision tree approach (e.g. Gilbert, 2004; Brandão *et al.*, 2005; Alexander and Chen, 2012), and stochastic simulation (as discussed in Chapter 6, this volume). Because of the diversity of methods that have evolved and the mathematical complexity of many of them, the various alternative approaches are not explained here. Interested readers are referred to Dixit and Pindyck (1994) or Schwartz and Trigeorgis (2001) for a general introduction. Examples of agricultural applications include Tegene *et al.* (1999) and Luong and Tauer (2006).

In what follows, we have chosen to focus on more straightforward approaches that embody aspects of real options valuation.

Preliminary screening

Because investment appraisal accounting for risk is difficult and time-consuming, it is obviously to be avoided if possible. Therefore, we suggest that a first step in evaluating some proposed investment is to apply some 'quick and dirty' test to see whether the project deserves more detailed consideration. For example, in a case where there is some large initial investment that is expected to yield future benefits, it makes sense to make preliminary estimates of the expected capital costs and the expected value of annual benefits when the project is up and running. Then, provided the project produces a fairly even flow of expected benefits over time, the estimated benefits can be related to the estimated costs to make a preliminary assessment of how worthwhile it seems likely to be. The *amortization formula* introduced above for the EA might be useful here, now slightly re-written as:

$$A = \frac{C\,i(1+i)^T}{(1+i)^T - 1} \tag{11.3}$$

where A is the minimum required expected annual net return for the investment to be profitable over the periods $t = 1$ to T, C is the capital cost, i is the discount rate and T is the projected life of the investment. A zero salvage value is assumed. For example, if an investment has an expected capital cost of \$100,000, the discount rate is 7% and the projected life of the investment is 8 years, application of the formula shows that the minimum average annual net benefit needs to be at least \$16,750 for the investment to be profitable. A project giving less than this amount could be rejected with no further analysis. Similarly, a project yielding a much higher annual benefit than the break-even annuity might be judged to be so attractive that it can be recommended for implementation with only limited further analysis of feasibility and risk.

In practice, for reasons of risk aversion or in the expectation of some downside risk, it might be wise to look for a rather better return than the calculated break-even annuity before taking the analysis further. And, of course, there could well be other uses for the funds to be invested that might return better than 7%, so that finding an estimated average net benefit that just exceeds \$16,750 is hardly grounds for going ahead with no thought for other possibilities. Some sensitivity analysis of the estimates of capital costs, time horizon and discount rate may allow the DM to take some limited account of the inherent uncertainty and imprecision of this preliminary evaluation.

Testing for financial feasibility

If an investment proposal passes the preliminary assessment of profitability, the question of its financial feasibility might sensibly be addressed next, in cases where this aspect is important. Commonly, large investments involve borrowing funds, so that the main risk may be that, if things go badly, it will not be possible for the loan to be serviced and the bank may foreclose. Therefore, rather than focusing immediately on a more thorough assessment of the overall merit of the investment, it may be more important as a next step to assess the risk of loan default.

For most investments that are made using borrowed funds, the risk of default will usually be greatest during or soon after the inception phase of the investment. It may take 2 or 3 years or more for the benefits from the investment to peak. If the assumptions on which the investment decision was based were too optimistic, this misjudgement may become apparent at this time. However, if things go reasonably well, debt servicing charges may decline as repayments are made, or, if the loan is amortized, the lender may be more willing to reschedule payments after a few years of successful debt servicing. And, in some cases, inflation may erode the value of the loan, making debt servicing easier. For all such reasons, feasibility assessment may be based on a budget that runs for only 2, 3 or 4 years into the future, depending on circumstances. Such a budget is obviously easier to construct than one that runs for the whole life of the investment of 10–20 years or more, partly because the circumstances of the next few years are easier to predict than those for the distant future.

What is required to assess the early risk associated with an investment is a stochastic cash flow budget over the selected danger period, linked to a finance budget that keeps track of the debts. In most cases, these budgets will need to be set up on a whole-farm or whole-business basis since, during the development phase or in bad times, there may be a need to cross-subsidize the investment project from other parts of the business. And, of course, there may be other loans to be serviced or risks affecting other parts of the business that need to be reflected in the budgets. Lien (2003) provides an example of such an analysis.

If an investment has passed the initial screening test and appears to be financially feasible, without leading to too high default-risk on loans, it may then be appropriate to move on to a more careful risk analysis of the investment, using one or other of the more formal appraisal methods outlined next.

Practical approaches to risky investment appraisal

Despite some theoretical qualifications, the pragmatic approach to the investment appraisal task under uncertainty may be argued to be to treat individual investments as marginal changes in the DM's wealth, as measured by the stochastic value of the project NPV. The discount rate used may be the borrowing rate or the opportunity cost of the invested funds, depending on circumstances. If the opportunity cost is used, it is often argued that the chosen opportunity should be of 'equivalent riskiness' to the project under review. However, if a risk analysis of alternative investments is to be conducted comparing the distributions of NPVs, a risk-free discount rate should be used to avoid double accounting for risk. If the project is small enough compared to the wealth of the DM, risk aversion may be ignored, in which case the expected value of NPV is a sufficient basis for choice. If that is too strong an assumption, as when the investment outcomes could have a significant impact on wealth, expected utility can be calculated using the DM's elicited or assumed utility function for current wealth, as explained in Chapter 5. If even the utility function for current wealth is unavailable, the full distribution of NPV may be generated by appropriate stochastic analysis and submitted to the DM as, say, a CDF, for direct appraisal. Alternatively, when seeking to differentiate between alternative investments, some form of stochastic efficiency analysis, such as SERF, may be used (Chapter 7). Typically, the distribution of returns from the investment is estimated using stochastic simulation, as outlined earlier in Chapter 6.

The important point to note about stochastic analysis of investments in terms of the distribution of NPV, even assuming indifference to risk, is that, provided the decision analysis is done appropriately, any important downside risk will be taken into account. As a result, the calculated expected NPV may be quite different from, often less than, the NPV calculated under the typical assumption that everything goes according to plan.

It follows that stochastic analysis of NPV may also be relevant for the appraisal of public investment projects under uncertainty. In these cases the assumption of no risk aversion is more easily justified, for reasons outlined in Chapter 5 and revisited in Chapter 13. Most international agencies prefer to side-step the issue of setting the relevant discount rate for NPV calculations by undertaking the analysis in terms of IRR, with the choice of the cut-off rate between acceptable and unacceptable projects being essentially a political or administrative matter (such as an 'institutional benchmark of 12%'). However, few such appraisals take any specific account of risk except by means of limited sensitivity analyses. Perhaps as a result, post-project audits typically show that *ex ante* rates of return have generally been too optimistic, owing, it might be argued, to neglect of downside risk (Pohl and Mihaljek, 1992). More careful risk analyses could minimize such problems, we suggest.

If the assumptions of stochastic independence between alternatives seem too strong to allow investments to be appraised individually, a first rough-and-ready approach might be made by assessing the appropriate project risk premium via Eqn 5.25 (Chapter 5), repeated here for convenience:

$$PRPx \approx r_r(w_0)C[x]\ \{0.5ZC[x] + \rho C[w_0]\} \qquad (11.4)$$

where PRP_x is the risk premium as a proportion of $E[x]$, $C[.]$ is the coefficient of variation, Z is the relative size of the marginal risky prospect, approximated by $E[x]/E[w_0]$ and ρ is the correlation between w_0 and x. Implementing this formula typically involves making subjective but informed assessments of the various terms. Such assessments usually demand some stochastic appraisal to determine the expected value and standard deviation of project returns, for example. Similarly, some data analysis may be needed to determine project size relative to national income (the latter being capitalized to estimate wealth), the relevant moments of the distribution of national income, and the likely correlation between project returns and the rest of the economy. Recall, however, that the formula is an approximation – though certainly better than ignoring the impact of risk aversion and stochastic dependency.

Where more careful analysis of the impact of stochastic dependency between investments is sought, MP models can be used to tackle the implied multi-period portfolio selection problem. Unfortunately, such models can become large and cumbersome to work with, especially if they include recognition of embedded risk, as discussed in Chapter 9. For this reason, it is common for strong simplifying assumptions to be made in this type of modelling, for example by ignoring embedded risk and associated options and by casting the analysis as an E,V risk programming approximation (see Chapter 9).

Any methods based on NPV imply an assumption that the capital market works well enough so that DMs can iron out most of the variability in net cash flow between time periods, making such inter-period variability relatively unimportant in the assessment. If this assumption is too strong, the usual recourse is to find some plausible and computable form of the inter-temporal utility function of Eqn 11.1. This function is obviously a multi-attribute function of the type discussed in Chapter 10, so the methods described there could, in principle, be applied to the elicitation of a quasi-separable form of Eqn 11.1. It seems likely that a multiplicative form would make sense, at least for a DM operating a family business, since any pattern of cash income flows through time with zero income for consumption in one or more time periods should surely be assigned zero utility. However, perhaps because of difficulties in elicitation, most studies have assumed the additive form:

$$U = \sum_{t=1}^{T} \alpha_t u_t(c_t) \qquad (11.5)$$

where $u_t(c_t)$ is the (usually risk-averse) utility function for period t, and α_t is a weighting factor, usually interpreted as a discount factor measuring the DM's impatience for future consumption. Commonly, the single-period utility functions are assumed to be the same for all time periods so that $u_t(c_t)$ in Eqn 11.5 is replaced with $u(c_t)$.

A utility function such as Eqn 11.5 can properly be applied to the evaluation of a single (more-or-less stochastically independent) investment only if the single-period utility functions are CARA functions (see Chapter 5) and if the project cash flows are small relative to flows from other sources. Otherwise, the inter-temporal utility function can be applied only to total net cash flows from all sources.

Applications of multi-period utility analysis are rather rare, no doubt because of the difficulty in eliciting or plausibly assuming the relevant utility function. Meyer (1976) and several of the papers in Lind (1982), such as those by Stiglitz (1982) and Wilson (1982), offer conceptually satisfactory approaches to addressing these difficulties. Meyer and Meyer (2005) provide a more recent discussion of some of the puzzles in relating utility for consumption to preferences for wealth.

Dynamic Decision Analysis

In the introduction to this chapter we defined a dynamic decision problem as one in which a sequence of decisions must be made through time, these decisions typically being interleaved with the outcomes of uncertain events that impinge on later choices, as well as on the consequences of decisions taken earlier. We can best illustrate the principles and methods of analysis of such problems by means of a simple example. For this purpose it is convenient to develop the harvester replacement example, introduced in Chapter 3.

The farmer who owns the harvester is concerned about the number of breakdowns it will suffer in the coming harvesting season. Each breakdown could be expensive in terms of repair costs and could cause a costly delay in harvesting. Furthermore, the number of breakdowns in the previous season is an indicator of the likely performance of the harvester in the coming season. A harvester that broke down in the previous year is judged to be more likely to prove unreliable again in the coming harvesting season. For the purpose of the illustration, the number of breakdowns is assumed to be either 'Zero' or 'More than zero' (denoted as '> Zero' or '> 0', i.e. one or more breakdowns). Furthermore, the maximum age of the harvester is assumed to be 3 years – after 3 years the machine will be replaced no matter how many breakdowns there were in the previous season – kept short to make the example simple enough for demonstration purposes. The DM has to decide every year whether to keep or to replace the harvester. Note, by the way, that these choices involve real options, as discussed above.

We can use the following notation to represent the state of the harvester at time t: $S_t = (A_t, b_t)$ where A_t denotes the age of the harvester, and b_t the number of breakdowns during the past year. We consider cases where $A_t = 1$, 2 or 3 and $b_t = 0$ or > 0 (this notation will be explained in detail in the next section). Suppose it is immediately before the coming harvest ($t = 0$). The harvester bought for the previous year is now 1 year old and had one breakdown during the last harvest. Hence it is now in state $S_0 = (1, > 0)$. The farmer could now make one of two possible decisions: keep the harvester for at least one more year, or replace it with a new one. In the first case, the farmer will use the 1-year-old machine for the coming harvest. It will become 2 years old after that harvest. Then, in a year from now, it will be in state $S_1 = (2, 0)$ if there were no breakdowns, or in $S_1 = (2, > 0)$ if there was one breakdown or more. The farmer is uncertain about the number of breakdowns this 1-year-old machine will have in the coming harvest. Suppose the probabilities of different outcomes in terms of breakdowns for the existing machine (with one breakdown

Table 11.1. Net costs ($) for the harvester replacement problem.

	Age of harvester (years)		
Number of breakdowns	**0**	**1**	**2**
Zero	20,000	19,000	18,000
> Zero	24,000	23,500	23,000

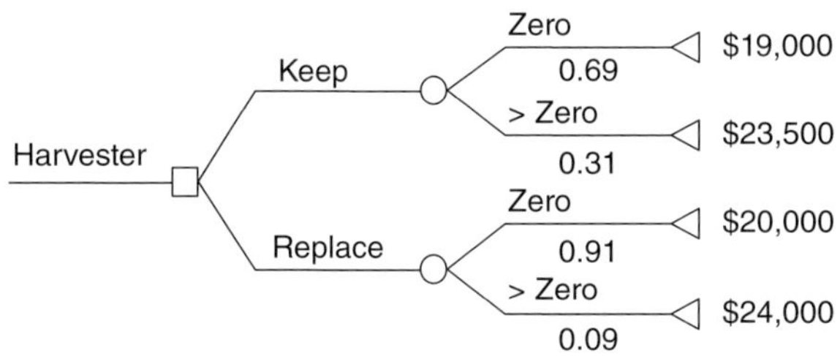

Fig. 11.1. Decision tree of the harvester replacement problem for a machine that is 1 year old with one or more breakdowns last year.

last year) are as follows: *P*(Zero) = 0.69 and *P*(> Zero) = 0.31. Similarly, suppose the farmer believes that the probabilities of breakdown of a new machine, should a replacement be bought, are: *P*(Zero) = 0.91 and *P*(> Zero) = 0.09. Evidently, the farmer thinks a new machine will be more reliable than the existing one.

The relevant net costs of the harvester are given in Table 11.1. The costs include two components. The first includes regular maintenance, depreciation and interest per year. In total, these costs decrease with the age of the machine. The second component comprises the additional costs caused by the breakdowns, including both repair costs and the consequential losses from delays in harvesting caused when the machine is broken down.

All the above assumptions are represented in the decision tree shown in Fig. 11.1.

The expected net costs of the two event nodes can be calculated as $20,395 if the harvester is kept compared with $20,360 if it is replaced with a new one for the coming season. Thus, for the problem as specified, and ignoring any risk aversion, the latter is the optimal decision (i.e. has the lowest net expected costs).

The above decision tree and calculations, however, refer only to one time period (i.e. the coming harvest). It is conceivable that replacing the harvester now may not be optimal taking a longer view. If the farmer wants to optimize the harvester replacement decision not just for 1 year but over the next several years, a different approach is required, as explained below.

Concepts of stages, states and transitions in dynamic decision models

The decision tree in Fig. 11.1 can be expanded to include more time periods, so that it becomes a *dynamic decision model*. Dynamic decision models typically have five main elements that are used to represent the real system over time: *decisions*, *stages*, *states*, *transitions* and *stage returns*. Some of these elements have already been explained in Chapter 2 but will be put into a slightly different context here.

The total number of time periods T over which decisions will be made and consequences experienced is called the *planning horizon*. The planning horizon is divided into T decision moments at times t where $t = 0, 1, \ldots, T - 1$. (We assume no decision is taken at the planning horizon at time T.) The length of the T time intervals between the decision moments depends on the problem at hand. The intervals might be years, months, weeks or even shorter times. A decision moment is generally called a *stage*. Typically, a decision (or set of decisions) a_t is made at each stage. A dynamic decision model consists of a set of stages joined together in a series so that (part of) the output of one stage becomes the input to the next.

At each stage t, the condition or state of the system is described by a set of *state variables* (sometimes simply called just *states*), denoted by the vector S_t. Although state variables may be continuous, analysis is usually made easier by assuming that each state variable can take only a finite number of distinct values that, in combination, describe the state of the system. Examples of states are number of cows, amount of fertilizer available, inventory of potatoes to be sold, the price of products, the temperature in a glasshouse or the age and breakdown history of a machine.

The function that describes how the system changes from one state at stage t to another at the next stage, $t + 1$, is called a *transformation function* (τ_t), and going from one state to another is called a *transition*. In a deterministic dynamic decision model (i.e. under assumed certainty), the transformation function is a function of the current state of the system (S_t) and the decisions made at that stage, a_t, i.e. $S_{t+1} = \tau_t(S_t, a_t)$.

At each stage there is a *stage return* (r_t) that measures the payoff or utility earned at that stage, and that is a function of the current state and the decisions made: $r_t = r_t(S_t, a_t)$ where $r_t(.)$ is the stage return function for stage t. Decisions at any stage may also imply costs or benefits at future stages, but these effects are captured through the state transformation function. There is also usually a terminal value of the system at the time horizon T which is a function of the state of the system at that time: $r_T = r_T(S_T)$.

The main components of a dynamic decision model are illustrated in Fig. 11.2.

The *objective function* of the deterministic T-stage dynamic decision model is some function $f(.)$ of the T-stage returns and the terminal return, over the decision variables $a_0, a_1, a_2, \ldots, a_{T-1}$, which means finding the total return as a function of the initial state S_0. Denoting $R_0(S_0)$ as the total return to the time horizon T, we can write:

$$R_0(S_0) = f\{r_0(S_0, a_0), r_1(S_1, a_1), \ldots, r_{T-1}(S_{T-1}, a_{T-1}), r_T(S_T)\} \tag{11.6}$$

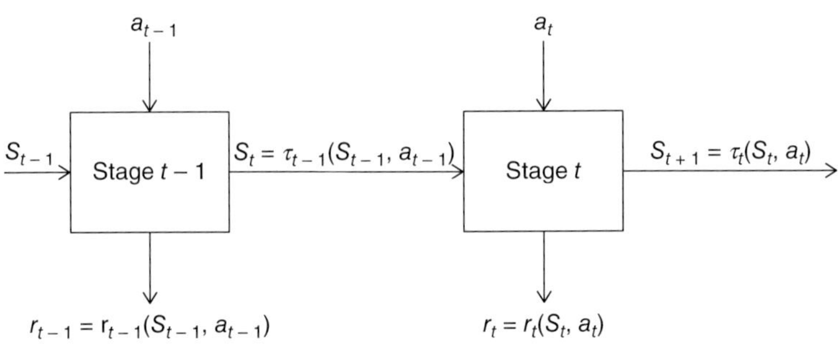

Fig. 11.2. The basic notation for a dynamic decision model.

In a stochastic model, however, either the stage return or the transformation function, or both, are functions of a vector of uncertain quantities y_t with associated probabilities $P_t(y_t)$ (i.e. y_t is a vector of uncertain quantities and $P_t(y_t)$ is the probability of any one of a finite number of combinations of values of those uncertain quantities occurring at stage t). However, for now it will serve to regard y_t as a single uncertain quantity that may take different values $y_{t1}, y_{t2}, \ldots, y_{tk}, \ldots$ at stage t with the defined associated probabilities. Note that the probabilities $P_t(y_t)$ may be conditional on one or both of the state of the system at the start of that stage and the decision taken. The probabilities could therefore be written as $P_t(y_t|S_t, a_t)$, but for convenience the shorter notation shown above is used for the present.

In a general stochastic model, therefore, the stochastic state of the system is given by the transformation function $S_{t+1} = r_t(S_t, a_t, y_t)$ and the stochastic stage return by $r_t = r_t(S_t, a_t, y_t)$. Expected stage return $E[r_t(S_t, a_t, y_t)]$ is the probability-weighted average of r_t, $\Sigma_k P_t(y_{tk}) r_t(S_t, a_t, y_{tk})$. In most applications, the overall expected return from all stages will be some function $g(.)$ of these expected stage returns and of the expected terminal return:

$$E[R_0(S_0)] = g\{E[r_0(S_0, a_0, y_0)], E[r_1(S_1, a_1, y_1)], \ldots, E[r_T(S_T, y_T)]\} \tag{11.7}$$

It can be seen, therefore, that the objective function of a stochastic dynamic decision model is similar to the one under assumed certainty, but now including the uncertain quantities y_t.

Consider now the harvester example. The decision tree in Fig. 11.1 has been redrawn in Fig. 11.3 to show most of these concepts.

The planning horizon T of the tree for now is 1 year, so the model involves only one stage with $t = 0$ leading to the time horizon when $t = 1 = T$. The states S_t for $t = 0$ and 1 are denoted as two-dimensional vectors (A_t, b_t), in which, as above, parameter A_t refers to the age of the harvester at stage t and parameter b_t refers to the number of breakdowns during the previous season. The harvester is initially 1 year old and had one breakdown in the previous season so at the current moment, just before the coming harvest, $S_0 = (1, > 0)$.

If the harvester is kept (a_0 = keep), the age of the harvester will increase by 1 year to 2 just before the next harvest. The number of breakdowns during the coming harvest, y_0 may be zero or more than zero. Hence, at the planning horizon S_1 may be (2, 0) or (2, > 0), as shown in Fig. 11.3.

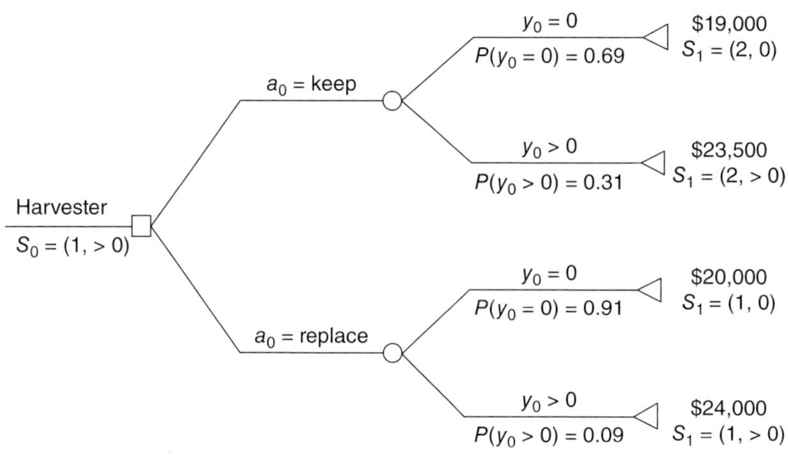

Fig. 11.3. The harvester replacement problem redrawn using dynamic decision model notation.

If the harvester is replaced (a_0 = replace), the farmer will immediately buy a new one to use for the coming harvest, and this new machine also may be in one of two states after the harvest and just before the next, with S_1 = (1, 0) or (1, > 0), depending on the outcome y_0 of the stochastic variable.

Finally given that the current state is (1, > 0) and the farmer decides to keep the machine, the probability of no breakdown, $P_0\ (y_0)$ for $y_0 = 0$, given in Fig. 11.1 is 0.69, and the probability of at least one breakdown is 0.31. It follows that these are the probabilities of reaching states S_1 of (2, 0) and (2, > 0), respectively. If the decision is to replace the machine then, as shown in the figure, the probabilities of going to states (1, 0) and (1, > 0) are 0.91 and 0.09, respectively, corresponding to $P_0(y_0)$ for $y_0 = 0$ and > 0, respectively. The stage returns for this problem were given in Table 11.1 above.

Multi-stage decision trees

The decision trees in Figs 11.1 and 11.3 cover only one stage and therefore are called *one-stage decision trees*. One-stage trees have one decision fork followed by a chance fork after each possible decision and ending with the payoffs.

Our harvester decision tree can be generalized to the tree presented in Fig. 11.4. The transition probabilities, which depend on the current state and the replacement decision, are shown as conditional on the current state of the system and the decisions taken (except that the current state is irrelevant if the decision is taken to replace the harvester). Instead of the payoffs, at the far right of the tree the next state is given.

If our one-stage tree is expanded to cover more stages, we get a *multi-stage decision tree*. Figure 11.5 shows the general outline for the *n*-stage decision tree.

In a multi-stage tree there is a whole sequence of decisions and event forks over time. All stage returns of the whole series of decisions and events are captured in the payoffs for each stage at the far right of the tree. There are as many stage returns as there are possible paths through the tree. Multi-stage decision trees are generally solved by 'averaging out and folding back', as explained in Chapter 6.

The entire multi-stage decision tree for our harvester example becomes a 'bushy mess' if extended over many years, so only two stages of the full multi-stage decision tree are presented in Fig. 11.6. The tree starts with the 1-year-old harvester that had at least one breakdown (i.e. > Zero) in the past harvesting season. If the farmer decides to keep the machine, then in the coming season it can have either zero,

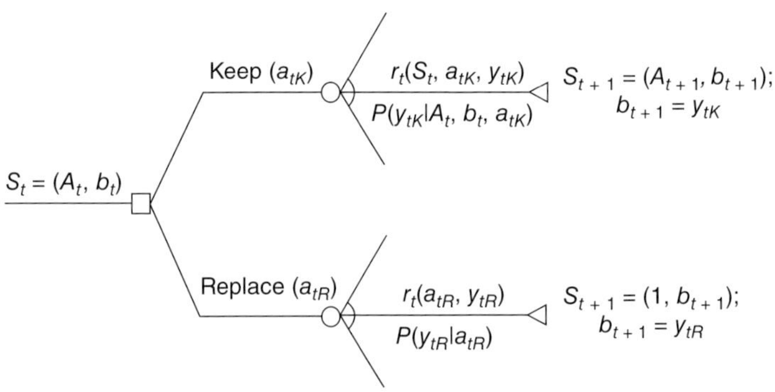

Fig. 11.4 Generalized one-stage decision tree of the harvester replacement problem ($_K$ = keep, $_R$ = replace).

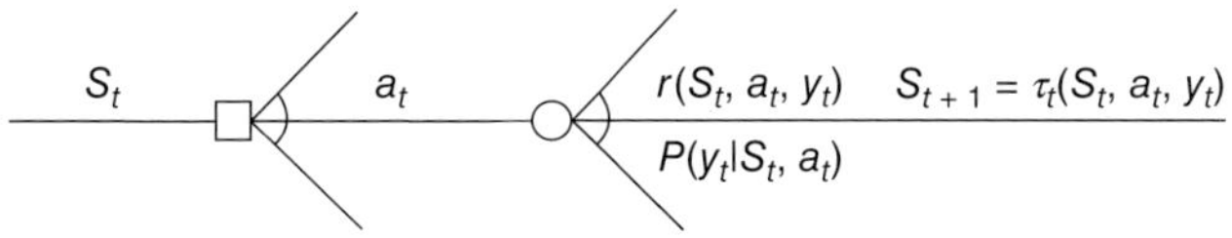

Fig. 11.5. Generalized *n*-stage decision tree.

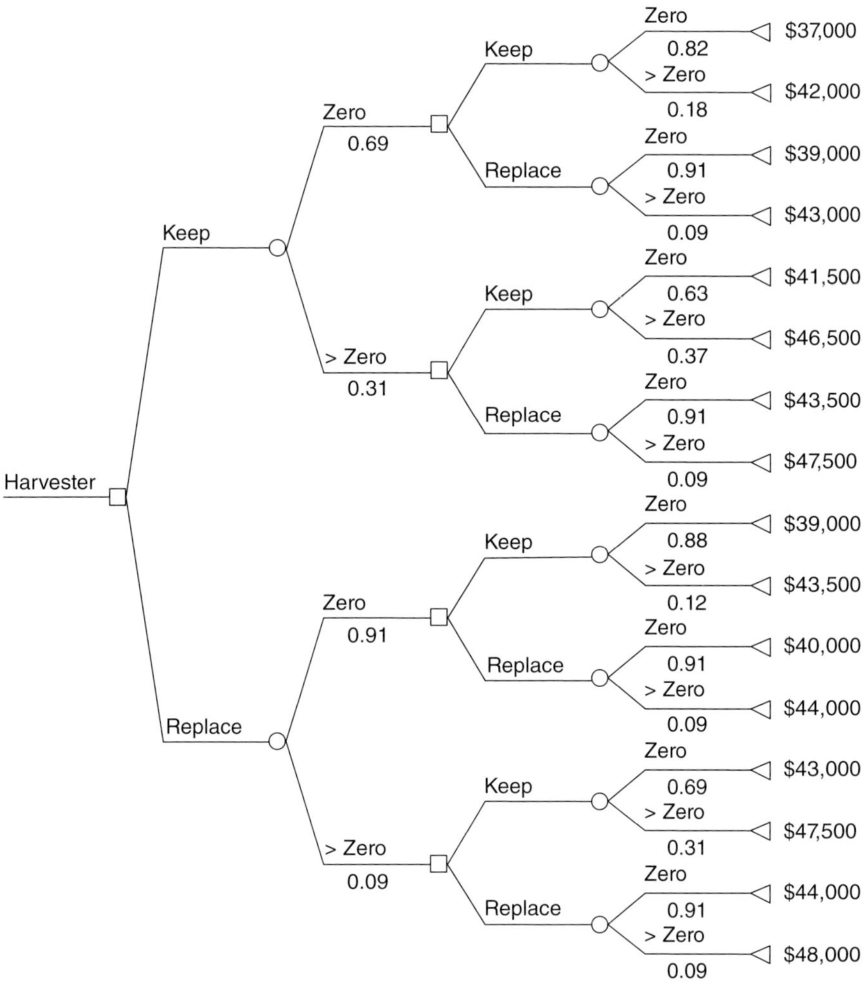

Fig. 11.6. Two-stage decision tree of the harvester replacement problem starting with a harvester of one year old with at least one breakdown during the previous year, i.e. $S_0 = (1, > 0)$.

or more than zero breakdowns. As in Fig. 11.1, the probabilities of occurrence of each breakdown are 0.69 and 0.31, respectively. The other probabilities are given in Table 11.2.

The probability of breakdowns in the future depends only on the age of the machine and the number of breakdowns during the previous harvest. The older the machine gets and the greater the number of breakdowns during the previous season, the greater the probability of one or more breakdowns during the coming harvest (Table 11.2).

Table 11.2. Transition probabilities for the harvester replacement problem.[a]

	Next state following decision a_t					
	a_t = replace		a_t = keep			
Current state[b]	**(1, 0)**	**(1, > 0)**	**(2, 0)**	**(2, > 0)**	**(3, 0)**	**(3, > 0)**
(1, 0)	0.91	0.09	0.88	0.12	..	..
(1, > 0)	0.91	0.09	0.69	0.31	..	..
(2, 0)	0.91	0.09	..	..	0.82	0.18
(2, > 0)	0.91	0.09	..	..	0.63	0.37
(3, 0)	0.91	0.09	..	..	..	..
(3, > 0)	0.91	0.09	..	..	..	..

[a].., Not applicable.
[b]See text for state notation.

The tree represented in Fig. 11.6 was solved using the DATA software package by TreeAge using the 'averaging out and folding back' procedure (Chapter 6), in this case, by selecting the branch with the lowest expected cost. The values at the far right of the tree were calculated directly from Table 11.1 and assuming a zero discount rate. For instance, the upper-right net cost of \$37,000 equals the net cost of keeping the harvester two more stages and observing no breakdowns (so, \$19,000 + \$18,000). The optimal solution is as follows (assuming no salvage value of the harvester after stage 2). At stage 1, the (current) harvester should be kept and, if the number of breakdowns is zero, then it should be kept at stage 2 also; if it had more than zero breakdowns it should also be kept at stage 2. So, in contrast to the optimal solution of the one-stage decision tree, the policy leading to the least expected cost in this particular example is always to keep the machine until it is 3 years old before replacement. This policy has an expected cost of \$39,590, which is lower than the expected costs of immediately replacing the harvester (\$39,974).

Although quite clear and convenient for small problems, multi-stage trees with very many states and stages become impossible to draw in full – although with modern software, quite large trees are certainly feasible. Notice, however, that some sub-trees in Fig. 11.6 are, in fact, identical. After every replace decision, the subsequent branches are basically the same. These repeated patterns provide a clue to the way complex multi-stage problems can be handled, as explained later in the chapter where we explain dynamic programming.

Stochastic Dynamic Modelling Methods

There are several ways that stochastic dynamic decision problems can be modelled. Some, such as stochastic simulation and MP, have already been described in earlier chapters. In this section, two dynamic modelling methods are presented in detail. Both are based on the principles and concepts of the previously discussed multi-stage decision tree. The first method, dynamic programming (DP), is an optimization technique to find the best sequence of decisions over time in a multi-stage decision tree. Deterministic DP is introduced first to serve as a basis for the more complex stochastic DP, which accounts for risk. The second method of Markov analysis is a special form of dynamic probabilistic simulation to determine the behaviour of the modelled system over time.

For didactic purposes, we discuss the methods as applied to discrete events, although application is by no means limited to discrete events. The harvester replacement problem is used to illustrate the principles. We also assume a finite planning horizon, consisting of a finite number of stages. There are techniques available whereby infinite-stage problems can be solved (see, for instance, Kennedy, 1986; Kristensen, 1994). Somewhat counter-intuitively, DP problems with infinite time horizons are often more easily solved than those with finite horizons.

Also for ease of explanation, we assume a near perfect capital market and a linear utility function to allow the objective function of our illustrative examples to be maximization of NPV, or expected NPV in the case of stochastic specifications. The methods presented, however, can also be used with other objective functions discussed earlier in this chapter.

Deterministic dynamic programming (DP)

DP is a mathematical optimization technique for solving multi-stage decision problems, originally developed by Richard Bellman in the 1950s. DP is conceptually close to the way multi-stage decision trees are analysed.

In a DP setting, there are decision moments (i.e. stages) at times 0 to $T - 1$. The objective function for deterministic DP is the maximization of the general formulation for dynamic decision models under assumed certainty (Eqn 11.6):

$$V_0(S_0) = \max_{a0, \ldots, aT-1} f\{r_0(S_0, a_0), r_1(S_1, a_1), \ldots, r_{T-1}(S_{T-1}, a_{T-1}), r_T(S_T)\} \quad (11.8)$$

Note that $R_0(S_0)$ of Eqn 11.6 now appears as $V_0(S_0)$, defined as the maximum (or minimum) value of the objective function. In many cases, simultaneous optimization of this equation over T stages with the associated T decision variables is difficult, even impossible. Fortunately, for many, but not all, multi-stage decision problems, it is possible to decompose Eqn 11.8 into T one-stage optimizations in the form of a *recursive relation* (Bellman, 1957):

$$V_t(S_t) = \max_{at} f_t\{r_t(S_t, a_t), V_{t+1}(S_{t+1})\} \quad \text{for } t = T - 1, \ldots, 0 \quad (11.9)$$

where $V_t(S_t)$ represents the optimal value of the objective function over the remainder of the planning horizon under optimal decisions given that the current state of the system is S_t and with $V_T(S_T) = r_T(S_T)$. In Eqn 11.9 f_t is a decomposable function, and $S_{t+1} = \tau_t(S_t, a_t)$ with the stage transformation function, τ_t, defining the state of the system at time $t + 1$, S_{t+1}, given the previous state, S_t, and the decisions a_t taken at stage t.

While the conditions for *decomposition* are somewhat general (Nemhauser, 1966, p. 35), a very common and useful case is when $V_0(S_0)$ is a discounted sum of the stage returns:

$$V_0(S_0) = \max_{a0, \ldots, aT-1} \Sigma_t\{\delta_t r_t(S_t, a_t) + \delta_T r_T(S_T)\} \quad \text{for } t = 0, 1, 2, \ldots, T - 1 \quad (11.10)$$

where δ_t is the discount factor for returns at time t. Decomposition of this function allows it to be written:

$$V_t(S_t) = \max_{at}[r_t\{S_t, a_t\} + \delta V_{t+1}\{\tau_t(S_t, a_t)\}] \quad \text{for } t = T - 1, \ldots, 0 \quad (11.11)$$

with

$$V_T(S_T) = r_T(S_T) \quad (11.12)$$

$$S_0 = S_0' \tag{11.13}$$

where $r_T(S_T)$ is the final value of the system in state S_T at the end of the planning horizon, S_0' is the initial state at stage 0, and δ is the one-period discount factor.

Optimization of the recursive system generally starts at the end of the planning horizon and moves backwards in time to the initial stage, just as in averaging out and folding back a decision tree. At each stage the optimal decision is determined for all possible states.

More formally, because $V_{t+1}\{\tau_t(S_t, a_t)\} = V_{t+1}(S_{t+1})$ and $V_T(S_T) = r_T(S_T)$, the recursive relation can be solved for $t = T - 1$ to give $V_{T-1}(S_{T-1})$ for all possible values of S_{T-1}. The solution to the whole problem is obtained by repeating the process for all t from $t = T - 1$ to $t = 0$. In other words, the T-stage decision problem can be solved by solving T one-stage problems, almost always with considerable economy in calculation.

The logical principle that justifies such recursive solutions is called *Bellman's principle of optimality* which states that, for a decomposable problem, an optimal decision at any stage can be found provided that all subsequent decisions with regard to the state resulting from that decision are also optimal. More formally (Bellman, 1957, p. 83) stated: 'An optimal policy has the property that, whatever the initial state and decision are, the remaining decisions must constitute an optimal policy with regard to the state resulting from the first decision'. Many people find the application of this principle difficult to understand. The problem seems to be that it is so simple an idea that it is almost incredible that it can be so powerful in allowing otherwise quite intractable optimization problems to be solved. Yet, as we shall illustrate, it works well.

Let's now re-visit our harvester replacement problem and solve it using deterministic DP. For simplicity, we first assume that the discount rate is zero, so δ equals one, that the objective is to minimize the cost of harvesting, and that the planning horizon is 5 years. As we now are dealing with deterministic DP, and in order to illustrate the DP structure for this simplified case, assume for the moment that there are no breakdowns (so, at any stage t: $S_t = (A_t, b_t)$ with $b_t = 0$). All the information for the DP procedure is available, including the stage returns $r_t(S_t, a_t)$. For zero breakdowns, there are only three different stage returns, defined as net costs as presented in Table 11.1 above. At the end of the planning horizon (in this example when $T = 5$), the final value of the system is assumed to be zero for all possible terminal states, i.e. $V_T(S_T) = r_5(S_5) = 0$ for all possible S_5.

A useful procedure for solving DP models is to prepare a series of tables, one for each stage, working back from the final decision stage. Each table has a row for each feasible state, and in that row the cumulative stage returns (in this example costs) from the present stage to the end of the planning horizon, $V_t(S_t)$, are calculated for each feasible decision. The optimal decision for the current stage for each possible state can therefore easily be found, and then the process is repeated until the initial stage has been reached. The solution of the harvester example is worked out in Fig. 11.7. The tables in this figure are more detailed than is strictly required in order to show the links to the models given in the equations above. In a lengthy calculation a much more economical form of presentation could be used.

At the top of the table we find some of the main assumptions, i.e. the final values at the end of the planning horizon, $r_T(S_T)$, all being zero for all states S_T, and the discount rate of zero. In the next part, the costs of the two alternative stage 4 decisions are shown along with the resulting terminal states, from which the total costs from stage 4 states to the terminal states can be calculated. For example, starting in state (1, 0) at stage 4, a decision to keep the harvester means a stage 4 cost of \$19,000 and transforms the state of the system to a terminal state S_5 of (2, 0). From the table of terminal values we know that the value of $V_5(2, 0)$ is zero, so the value of $f_4(1, 0)$ for a_4 = keep is \$19,000 + 0 = \$19,000, as shown. On the other hand, if the 1-year-old machine is replaced at stage 4, the stage return is a cost of \$20,000 and the system

Solving the harvester replacement problem using deterministic DP

$t = 5$	Terminal values for $t = T$
$S_5 =$ (A_5, b_5)	$V_5 =$ $r_5(S_5)$
(1, 0)	0.0
(2, 0)	0.0
(3, 0)	0.0

Discount rate	Discount factor δ
0.0%	1.00

$t = 4$	Costs from stage 4 to terminal state						
$S_4 =$	a_4 = keep			a_4 = replace			V_4
(A_4, b_4)	r_4	S_5	$r_4 + V_5$	r_4	S_5	$r_4 + V_5$	
(1, 0)	19.0	(2, 0)	19.0	20.0	(1, 0)	20.0	19.0
(2, 0)	18.0	(3, 0)	18.0	20.0	(1, 0)	20.0	18.0
(3, 0)	..	..	..	20.0	(1, 0)	20.0	20.0

$t = 3$	Costs from stage 3 to terminal state						
$S_3 =$	a_3 = keep			a_3 = replace			V_3
(A_3, b_3)	r_3	S_4	$r_3 + V_4$	r_3	S_4	$r_3 + V_4$	
(1, 0)	19.0	(2, 0)	37.0	20.0	(1, 0)	39.0	37.0
(2, 0)	18.0	(3, 0)	38.0	20.0	(1, 0)	39.0	38.0
(3, 0)	..	..	..	20.0	(1, 0)	39.0	39.0

$t = 2$	Costs from stage 2 to terminal state						
$S_2 =$	a_2 = keep			a_2 = replace			V_2
(A_2, b_2)	r_2	S_3	$r_2 + V_3$	r_2	S_3	$r_2 + V_3$	
(1, 0)	19.0	(2, 0)	57.0	20.0	(1, 0)	57.0	57.0
(2, 0)	18.0	(3, 0)	57.0	20.0	(1, 0)	57.0	57.0
(3, 0)	..	..	..	20.0	(1, 0)	57.0	57.0

$t = 1$	Costs from stage 1 to terminal state						
$S_1 =$	a_1 = keep			a_1 = replace			V_1
(A_1, b_1)	r_1	S_2	$r_1 + V_2$	r_1	S_2	$r_1 + V_2$	
(1, 0)	19.0	(2, 0)	76.0	20.0	(1, 0)	77.0	76.0
(2, 0)	18.0	(3, 0)	75.0	20.0	(1, 0)	77.0	75.0

$t = 0$	Costs from stage 0 to terminal state						
$S_0 =$	a_0 = keep			a_0 = replace			V_0
(A_0, b_0)	r_0	S_1	$r_0 + V_1$	r_0	S_1	$r_0 + V_1$	
(1, 0)	19.0	(2, 0)	94.0	20.0	(1, 0)	96.0	94.0

Fig. 11.7. Series of tables to solve the harvester replacement problem by deterministic dynamic programming (DP) (net costs in thousand \$). .., means not applicable.

passes from state (1, 0) at stage 4 to S_5 = (1, 0). Since $V_5(1, 0)$ is zero, the value of $f_4(1, 0)$ for a_4 = replace is \$20,000 + 0 = \$20,000, as also shown. So, the least-cost decision is to keep the harvester, and the cumulative costs until the end of the planning horizon starting in state (1, 0) at stage 4 is the minimum value of {20,000, 19,000}, i.e. $V_4(1, 0)$ = \$19,000. Corresponding calculations are shown in Fig. 11.7 for S_4 = (2, 0) and (3, 0). No calculations are shown for keeping a harvester with S_4 = (3, 0) since the assumption was made that a harvester that reaches 3 years of age is always replaced.

Now we go back one more stage for $t = 3$ and repeat the calculations just explained, this time using the values of $V_4(A_4, b_4)$ just determined in evaluating f_3 for each possible starting state. Other tables of

Fig. 11.7 for $t = 2$, 1 and 0 are calculated in the same way, except that, for $t = 1$ and 0, the possible states to consider are reduced based on the knowledge that the machine is in state (1, 0) at the start, $S_0 = (1, 0)$.

When the optimal stage 0 decision has been determined, the whole problem is solved. The least-cost sequence of decisions is found by tracking through the tables, starting at the bottom (stage 0) to the top (stage 4). Starting in state (1, 0) at stage 0, we see that the harvester should be kept, leading to state (2, 0) for stage 1. At this stage too the machine should be kept, leading to $S_2 = (3, 0)$. For $t = 2$ the optimal decision is to replace, and the second harvester is kept for the remaining years in the planning period. The cost of this policy is given by $V_0(1, 0) = \$94,000$.

Now that we have solved the deterministic harvester replacement problem by DP without discounting, it is easy to solve it with discounting. The only difference, as indicated in the Eqn 11.11, is the inclusion of a discount factor in calculating the cumulative stage returns until the end of the planning horizon, $V_{t+1}(S_{t+1})$. At each stage the cumulative return from later stages is multiplied by the one-period discount factor δ. Suppose the discount rate is 5% per year, then $\delta = 1/(1 + 0.05)^1$ or 0.952. Using this discount factor the optimal sequence of decisions is unchanged and the total discounted costs of the optimal path 'keep – keep – replace – keep – keep' is \$85,510.

The computational advantage of backward recursion is that many fewer decision consequences have to be explored than with total enumeration. In other words, at least when using numerical methods, the principle of optimality is a rule that defines an efficient search procedure.

Stochastic dynamic programming (DP)

In the above example we assumed away uncertainty to make the explanation shorter and to illustrate the application Bellman's principle of optimality. In the risky world that is the main concern in this book, stochastic DP will be the appropriate model to optimize multi-stage decision problems. In stochastic DP, state transitions and/or stage returns depend on the current state (S_t) of the system, the decision taken at that stage (a_t), as before, and also on the vector of stochastic variables y_t (defined as before), which are outside the control of the DM. We assume that, in the discrete case, the vector y_t can take different (sets of) values $y_{t1}, y_{t2}, \ldots, y_{tk}, \ldots$ at each stage t, and that the probability of outcome y_{tk} at stage t is $P_t(y_{tk})$. Again, this probability may be conditional on S_t and a_t.

The objective function for stochastic DP is the maximization of the general formulation for dynamic decision models under uncertainty (Eqn 11.7) and has the following general form:

$$E[V_0(S_0)] = \max_{a0,\ldots,aT-1} g\{E[r_0(S_0, a_0, y_0)], E[r_1(S_1, a_1, y_1)], \ldots, E[r_T(S_T, y_T)]\} \qquad (11.14)$$

where again $E[V_0(S_0)]$ is the optimal expected value of $R_0(S_0)$ in Eqn 11.6, $E[r_t(S_t, a_t, y_t)] = \Sigma_k P_t(y_{tk}) r_t(S_t, a_t, y_{tk})$, and $r_t(S_t, a_t, y_{tk})$ is the function describing the immediate expected value of the stage return at stage t resulting from decision a_t in state S_t and the outcome of y_{tk} with probabilities $P_t(y_{tk})$. The final term, $E[r_T(S_T, y_T)]$, represents the expected value of the system at the planning horizon T. So, the objective function of stochastic DP is similar to the one under assumed certainty but now including the uncertain quantities y_t and hence being expressed in terms of expected values.

After decomposition, the recursive relation for the expected discounted value of stage returns in the stochastic case can be formulated as:

$$E[V_t(S_t)] = \max_{at}\{\Sigma_k P_t(y_{tk})[r_t(S_t, a_t, y_{tk}) + \delta V_{t+1}(\tau_t(S_t, a_t, y_{tk}))]\} \quad \text{for } t = T-1, \ldots, 0 \qquad (11.15)$$

with

$$\Sigma_k \, P_t(y_{tk}) = 1 \tag{11.16}$$

$$E[V_T(S_T)] = E[r_T(S_T, y_T)] \tag{11.17}$$

$$S_0 = S_0' \tag{11.18}$$

where $E[V_t(S_t)]$ represents the maximum expected discounted value of the objective function – in terms of cumulative expected stage returns – during the remainder of the planning horizon under optimal decisions given state S_t. As indicated in Eqn 11.15, given that the system is in state S_t at stage t, then, if decision a_t is taken and the outcome of the stochastic variable or variables is y_{tk}, the stage return is $r_t(S_t, a_t, y_{tk})$. Moreover, the system is transformed at stage $t + 1$ into state $S_{t+1} = \tau_t(S_t, a_t, y_{tk})$, so that S_{t+1} also depends on y_t and hence is stochastic.

We can illustrate stochastic DP using the simplified harvester replacement example. The objective is to find the replacement policy with the least expected costs (i.e. assuming no risk aversion). As with deterministic DP, the optimal solution can be found by creating a set of tables, one for each stage. However, because of the stochastic element, these tables become extensive. Therefore, Fig. 11.8 shows only the tables needed to solve one stage of the harvester replacement problem by stochastic DP. The other tables have a similar structure and are solved in a similar fashion.

Following the procedure illustrated in Fig. 11.8, we can solve the harvester replacement problem over a planning horizon of five stages. The approach is similar to that used before but the scope of the calculations is larger. All optimal decisions per state and stage for this particular example are presented in Table 11.3. The optimal decision sequence for our harvester, which is in state (1, > 0) at stage 0, is underlined in each column of the table. In all columns after stage 0, two entries are underlined because the optimal policy has to be specified for the two possible breakdown and repair histories in the previous year.

Solving the harvester replacement problem using stochastic DP, stage 4

$S_4 = (A_4, b_4)$	$P_4(y_4)$		r_4	
$a_4 = K$	$y_4 = 0$	$y_4 > 0$	$y_4 = 0$	$y_4 > 0$
(1, 0)	0.88	0.12	19.00	23.50
(1, > 0)	0.69	0.31	19.00	23.50
(2, 0)	0.82	0.18	18.00	23.00
(2, > 0)	0.63	0.37	18.00	23.00
$a_4 = R$	0.91	0.09	20.00	24.00

K = keep; R = replace

$S_5 = (A_5, b_5)$	V_5
(1, 0)	0.00
(1, > 0)	0.00
(2, 0)	0.00
(2, > 0)	0.00
(3, 0)	0.00
(3, > 0)	0.00

Disc. rate	Disc. factor
5.0%	0.9524

Costs from stage 4 to terminal states, r_4							$E[r_4 + V_5]$		V_4	Optimal a_4
	S_5 given y_4									
$S_4 = (A_4, b_4)$	$a_4 = R$		$a_4 = K$							
	(1, 0)	(1, > 0)	(2, 0)	(2, > 0)	(3, 0)	(3, > 0)	$a_4 = R$	$a_4 = K$		
(1, 0)	20.00	24.00	19.00	23.50	..	..	20.36	19.54	19.54	K
(1, > 0)	20.00	24.00	19.00	23.50	..	..	20.36	20.40	20.36	R
(2, 0)	20.00	24.00	..	..	18.00	23.00	20.36	18.90	18.90	K
(2, > 0)	20.00	24.00	..	..	18.00	23.00	20.36	19.85	19.85	K
(3, 0)	20.00	24.00	..	..	..	..	20.36	..	20.36	R
(3, > 0)	20.00	24.00	..	..	..	..	20.36	..	20.36	R

Fig. 11.8. Part of the tables to solve the harvester replacement problem by stochastic DP (net costs in thousand $). .., means not applicable.

Table 11.3. Optimal decision for the stochastic harvester problem (K = keep harvester; R = replace harvester).

	Optimal decision by state and stage[a]				
State	**Stage 0**	**Stage 1**	**Stage 2**	**Stage 3**	**Stage 4**
(1, 0)	K	K	K	K	K
(1, > 0)	K	R	R	K	R
(2, 0)	K	K	K	K	K
(2, > 0)	R	K	R	R	K
(3, 0)	R	R	R	R	R
(3, > 0)	R	R	R	R	R

[a]The optimal decision for each stage is underlined. After stage 0 two entries are optimal (see text for explanation).

It can be seen from the table that the optimal sequence is to keep the machine (which was 1 year old to start with) until it is 3 years old (i.e. stage 2), regardless of the number of breakdowns experienced, then to replace it at stage 2 with a new one and to keep that machine at stages 3 and 4, again regardless of breakdowns. The present value of expected least cost of the optimal decision sequence over the entire planning horizon is $89,740.

As may be apparent, in stochastic DP we have a method for optimizing a wide range of choice options at multiple times in the future, according to the outcomes of uncertain events. The main outcome of such an analysis, of course, is the identification of the best immediate choice, since later a similar analysis could be performed for the next stage, perhaps updating some assumptions made initially. It should be evident, that stochastic DP is a form of real options analysis, since the solution reached accounts for what might happen in the future and what responses to the outcome of uncertain events are best (see Dixit and Pindyck, 1994, Chapter 4 for more details).

Dynamic probabilistic simulation

Markov processes

Sometimes we are not interested in the optimal solution of a stochastic multi-stage decision problem in itself, but rather in how a given process changes over time. For example, we may want to know such things as how the price of a particular commodity, the spread of a particular crop pest or animal disease, or an agribusiness firm's market share evolves through time. All these examples relating to stochastic processes involve stochastic variables that change over time, as was the case with stochastic DP. In this section we focus on simulating those stochastic processes over time, and in particular on a type of stochastic process known as a *Markov process* or a *Markov chain*.

A Markov process is a special type of stochastic process. If the following equation applies, then the discrete-time stochastic process is a Markov process:

$$P(S_{t+1} = i_{t+1} | S_t = i_t, S_{t-1} = i_{t-1}, \ldots, S_0 = i_0) = P(S_{t+1} = i_{t+1} | S_t = i_t) \quad (11.19)$$

Essentially, the equation says that the probability distribution of the state at stage $t + 1$ depends on the state at stage t, defined by i_t, and does not depend on the states the process passed through on the way to i_t. In this discussion of Markov processes we make the assumption usually made that, for all states i and j and all t, $P(S_{t+1} = j|S_t = i)$ is independent of t. This allows us to state that:

$$P(S_{t+1} = j|S_t = i) = p_{ij} \tag{11.20}$$

In this equation, p_{ij} denotes the probability that the system will be in a state j at stage $t + 1$, given that it was in state i at stage t. As with DP, if the system moves from state i to state j at the next stage, we say that a transition from i to j has occurred. The p_{ij} values are called the *transition probabilities*, usually presented as a transition probability matrix, P. Equation 11.20 states that the probabilities of passing from one state to another in the next stage do not change over time, implying what is referred to as the *stationarity assumption*. The probability that the process (system) is in state i at stage 0 (the beginning of the planning horizon) is called the initial probability distribution q_i. So, $P(S_0 = i) = q_i$. Finally, given that the state at stage t is i, and j indexes all the possible states at $t + 1$, then:

$$\Sigma_j P\{(S_{t+1} = j|P(S_t = i)\} = 1 \quad \text{for all } i \tag{11.21}$$

Since negative probabilities are impossible, all entries in the transition probability matrix must be non-negative and, according to Eqn 11.21, all the entries in each row must sum to one.

The p_{ij} values are sometimes also called the *one-step transition probabilities*, as they represent the probability of going from state i at stage t to state j at stage $t + 1$. One-step probabilities are usually denoted as $p_{ij}(1)$. However, for many Markov process problems, an important question is 'if the Markov process is in state i at time t, what is the probability that n stages later the Markov process will be in state j?' Since the Markov processes we are dealing with are stationary, this probability is independent of t, so that:

$$P(S_{t+n} = j|S_t = i) = P(S_n = j|S_0 = i) = p_{ij}(n) \tag{11.22}$$

where $p_{ij}(n)$ is called the n-step probability. Of course, $p_{ij}(1) = p_{ij}$, and $p_{ij}(2)$ is the transition probability of going in two steps from i to j. In other words, it is the joint probability of first passing from state i to an intermediate state, say k, and then from the intermediate state k to state j for all k, i.e.

$$p_{ij}(2) = \Sigma_k p_{ik} p_{kj} \tag{11.23}$$

Thus $p_{ij}(2)$ is the ij-th element of a matrix that we can denote by P^2. More generally, for $n > 1$:

$$p_{ij}(n) = ij\text{-th element of } P^n \tag{11.24}$$

For many Markov processes, not all, there exists a vector of state probabilities $\pi = (\pi_1, \pi_2, \ldots, \pi_s)$ such that (with s indicating the total number of states):

$$\lim_{n \to \infty} P^n = \begin{bmatrix} \pi_1 & \pi_2 & \cdots & \pi_s \\ \pi_1 & \pi_2 & \cdots & \pi_s \\ \cdot & \cdot & & \cdot \\ \cdot & \cdot & & \cdot \\ \pi_1 & \pi_2 & \cdots & \pi_s \end{bmatrix}$$

Or more compactly:

$$\lim_{n\to\infty} p_{ij}(n) = \pi_j \tag{11.25}$$

This equation means that, after a time, the Markov process settles down in state j with probability π_j, independent of the initial state i. The vector with these probabilities $\pi = (\pi_1, \pi_2, \ldots, \pi_s)$ is called the *steady-state distribution* or *equilibrium distribution* for the Markov process. There are two ways to find the steady-state probability distribution. First, we can multiply P^1 by P^1 to obtain P^2, and then multiply P^2 by P^2 to obtain P^4, and so on, until the probabilities in the matrix do not change with further multiplication. For example, consider the following simple transition matrix:

$$P^1 = \begin{bmatrix} 0.90 & 0.10 \\ 0.20 & 0.80 \end{bmatrix}$$

Then P^2, P^4 and (with P^{16} not shown) P^{256} are as follows:

$$P^2 = \begin{bmatrix} 0.83 & 0.17 \\ 0.34 & 0.66 \end{bmatrix} \quad P^4 = \begin{bmatrix} 0.747 & 0.253 \\ 0.507 & 0.493 \end{bmatrix} \quad \cdots \quad P^{256} = \begin{bmatrix} 0.6667 & 0.3333 \\ 0.6667 & 0.3333 \end{bmatrix}$$

Since the two rows of the matrix for P^{256}, found after four matrix multiplications, are almost identical, we can assume a steady state has been reached.

The second way is to use the properties that, for large n, $P^n = P^{n-1}P^1$ and $P^n = P^{n-1}$ so $\pi = \pi\, P^1$, which can be solved to find π. For the simple example above:

$$\pi_1 = 0.90\,\pi_1 + 0.20\,\pi_2 \tag{11.26}$$

$$\pi_2 = 0.10\,\pi_1 + 0.80\,\pi_2 \tag{11.27}$$

$$\pi_1 + \pi_2 = 1 \text{ (all probabilities add to one)} \tag{11.28}$$

We can solve for π_1 and π_2 using Eqn 11.26 or 11.27 together with 11.28, resulting in $\pi_1 = 2/3$ and $\pi_2 = 1/3$. Hence, after an infinite number of stages, there is a 2/3 probability that the Markov process will be in state 1 and a 1/3 probability that it will be in state 2.

Note that the stochastic DP method and the Markov process method have some features in common. Although the first is an optimization method usually starting at the end of the planning horizon and working backwards, and the second is a simulation method starting at the beginning of the planning horizon and working forwards, the concepts of stages, states and transitions are the same.

Armed with all the ingredients of a Markov process, which ingredients we already used in stochastic DP, we return to the harvester replacement problem. First, we define a decision policy to be evaluated. Our initially chosen policy (policy 1) is to keep the harvester if it had no breakdowns in the past season, and to replace it if it had one breakdown or more or when it is 3 years old. We can now enter all the relevant transition probabilities and stage returns into the Markov analysis option of the DATA software of TreeAge and run a simulation over the stages 0–5 (as with the stochastic DP example). DATA determines the discounted expected net costs. The simulation reveals that the expected value of the cumulative stage returns (expected net costs) of policy 1 are \$89,940.

It is easy to evaluate another replacement policy (policy 2) with the Markov process model, for instance the same policy as before, but now replacing the harvester that has not had breakdowns when it reaches 2 years old, rather than 3 years as under policy 1. The discounted expected cumulative net costs of policy 2 are \$91,091, slightly higher than for policy 1.

DATA also determines the steady-state distribution of both policies. The steady-state distribution π for policy 1 is: π = (0.34, 0.03, 0.30, 0.04, 0.24, 0.05) – in which the order of states is (1, 0), (1, > 0), (2, 0), . . ., (3, > 0) – with an expected stage return of $19,651 per stage. Policy 2 results in the following steady-state distribution π = (0.47, 0.05, 0.42, 0.06, 0.00, 0.00) with expected stage returns of $19,965 per stage. So we can conclude that policy 1 results in slightly lower costs. The steady-state mean age of a harvester under policy 1 can be calculated as 1.92 years, compared with 1.48 years for policy 2. The steady-state probabilities of one or more breakdowns are almost identical for the two policies at 0.12 for policy 1 and 0.11 for 2.

Stochastic simulation

Stochastic simulation was discussed in Chapter 6. The method can be used for stochastic dynamic simulation, as a variant of the methods outlined in that chapter. In such dynamic stochastic simulation, the concepts and definitions of stages, states, transitions, stage returns and transition probabilities are the same as illustrated previously in this chapter. The only difference is that in stochastic simulation the probability distributions are used as a basis for sampling to simulate how a system could evolve over time.

The simulation of the behaviour of a system that is being modelled starts at the beginning of the planning horizon and proceeds stage by stage until the last stage at the end of the planning horizon is reached. For each iteration, every state transition that follows a certain (predefined) decision is determined by a random sample from the appropriate probability distribution. Consequently, the state transition that actually occurs in each iteration of the model is determined by the sampled outcome. Next, the corresponding payoffs are calculated based on a decision specified in advance (as with Markov process simulation), again accounting for the sampled outcome as appropriate. Then the next stage is simulated given the state of the system, and so on, until the end of the planning horizon. Each iteration includes one set of samples to the end of the planning horizon and represents one possible combination of values for the risky events and the resulting states. As explained in Chapter 6, samples are drawn from the underlying probability distributions for many iterations until the distributions of the relevant output variables are stabilized.

Stochastic simulation of the harvester replacement problem was carried out again using the DATA software of TreeAge. As with Markov process simulation, we ran the Monte Carlo model for the two replacement policies described in the previous sub-section. To get a reliable insight into the behaviour of the system over time (in this example, as with DP and Markov process simulation, over five stages), it is necessary to run the model many times and record the distribution of cumulative stage returns. We did only 250 iterations with this simple model, and the CDFs of stage returns resulting from the policies 1 and 2 are presented in Fig. 11.9.

Some descriptive statistics for policy 1 (over stages 0–4) are: the minimum observed discounted cost was $87,330, the maximum discounted cost was $99,434, the mean or expected cost was $89,780, and the standard deviation was $2931. The corresponding statistics for policy 2 were, respectively: $89,103, $97,467, $91,061 and $2344. To compare the policy choices, these descriptive statistics, could be used in an *E*, *V* analysis or the full CDFs could be used in a stochastic efficiency analysis such as SERF, as explained in Chapter 7. Alternatively, expected utility analysis could be used provided the DM's utility function was available.

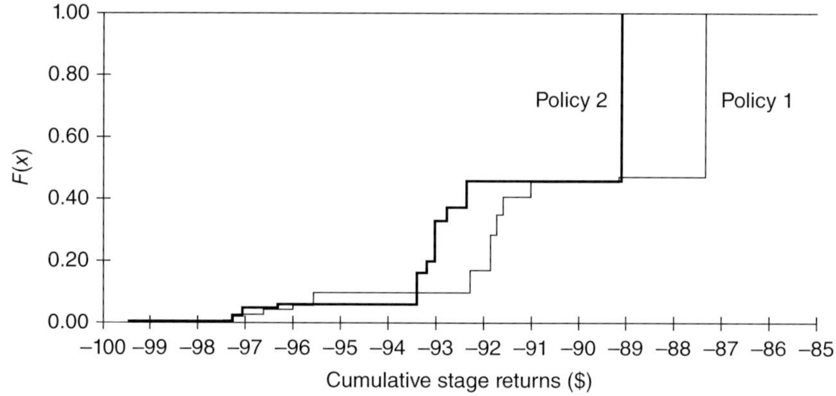

Fig. 11.9. Cumulative distribution functions (CDFs) of two policies resulting from Monte Carlo simulation (250 iterations) for the harvester replacement problem (stage returns in thousand $).

Selected Additional Reading

No list of a few references can do justice to the vast literature on investment appraisal and temporal uncertainty. However, Jean (1970) is a good entry to the theory of finance and investment. Dixit and Pindyck (1994) were important contributors to developments in the understanding of real options. More recently, Chavas and Mullarkey (2002) have developed a general model of private and public choice under temporal uncertainty. The model incorporates the effects of risk preferences and the prospect of future learning and resolves some aspects of options.

The general principles of DP are set out in several texts, such as Bellman (1957) and Nemhauser (1966). More recent treatments are to be found in Winston and Goldberg (2004) and Bather (2000). Kennedy (1986) wrote an influential text on DP in agriculture. Kristensen (1994) and Kristensen and Jørgensen (2000) describe algorithms that reduce the dimensionality problems within DP. Putterman (1994) discuses Markov decision processes and stochastic DP.

References

Alexander, C. and Chen, X. (2012) *A General Approach to Real Option Valuation with Applications to Real Estate Investments.* ICMA Centre Discussion Papers in Finance: DP2012-04, ICMA Centre, University of Reading, Reading, UK.

Bather, J. (2000) *Decision Theory: an Introduction to Dynamic Programming and Sequential Decisions.* Wiley, Chichester, UK.

Bellman, R. (1957) *Dynamic Programming.* Princeton University Press, New Jersey.

Brandão, L.Z., Dyer, J.S. and. Hahn, W.J. (2005) Using binomial decision trees to solve real-option valuation problems. *Decision Analysis* 22, 69–88.

Chavas, J.-P. and Mullarkey, D. (2002) On the valuation of uncertainty in welfare analysis. *American Journal of Agricultural Economics* 84, 23–38.

Dixit, A. and Pindyck, R. (1994) *Investment under Uncertainty*. Princeton University Press, New Jersey.

Gilbert, E. (2004) Investment basics XLIX: an introduction to real options. *Investment Analyst Journal* 60, 49–52.

Jean, W.H. (1970) *The Analytical Theory of Finance: a Study of the Investment Decision Process of the Individual and the Firm*. Holt, Rinehart and Winston, New York.

Kennedy, J.O.S. (1986) *Dynamic Programming: Applications to Agriculture and Natural Resources*. Elsevier, London.

Kristensen, A.R. (1994) A survey of Markov decision programming techniques applied to the animal replacement problem. *European Review of Agricultural Economics* 21, 73–93.

Kristensen, A.R. and Jørgensen, E. (2000) Multi-level hierarchic Markov processes as a framework for herd management support. *Annals of Operations Research* 94, 69–89.

Lien, G. (2003) Assisting whole-farm decision-making through stochastic budgeting. *Agricultural Systems* 76, 399–413.

Lind, R.C. (ed.) (1982) *Discounting for Time and Risk in Energy Policy*. Johns Hopkins University Press, Baltimore, Maryland.

Luong, Q.V. and Tauer, L.W. (2006) A real options analysis of coffee planting in Vietnam. *Agricultural Economics* 35, 49–57.

Meyer, D.J. and Meyer, J. (2005) Relative risk aversion: what do we know? *The Journal of Risk and Uncertainty* 31, 243–262.

Meyer, R.F. (1976) Preferences over time. In: Keeney, R.L. and Raiffa, H. (eds) *Decisions with Multiple Objectives: Preferences and Value Tradeoffs*. Wiley, New York, pp. 473–514.

Nemhauser, G.L. (1966) *Introduction to Dynamic Programming*. Wiley, New York.

Pohl, G. and Mihaljek, D. (1992) Project evaluation and uncertainty in practice: a statistical analysis of rate-of-return divergences of 1,015 World Bank projects. *World Bank Economic Review* 6, 255–277.

Putterman, M.L. (1994) *Markov Decision Processes: Discrete Stochastic Dynamic Programming*. Wiley, New York.

Schwartz, E.S. and Trigeorgis, L. (eds) (2001) *Real Options and Investment under Uncertainty: Classical Readings and Recent Contributions*. MIT Press, Cambridge, Massachusetts.

Stiglitz, J.E. (1982) The rate of discount for benefit–cost analysis and the theory of the second best. In: Lind, R.C. (ed.) (1982) *Discounting for Time and Risk in Energy Policy*. Johns Hopkins University Press, Baltimore, Maryland, pp. 151–204.

Tegene, A., Wiebe, K. and Kuhn, B. (1999) Irreversible investment under uncertainty: conservation easements and the option to develop agricultural land. *Journal of Agricultural Economics* 50, 203–219.

Wilson, R. (1982) Risk measurement of public projects. In: Lind, R.C. (ed.) (1982) *Discounting for Time and Risk in Energy Policy*. Johns Hopkins University Press, Baltimore, Maryland, pp. 205–249.

Winston, W.L. and Goldberg, J.B. (2004) *Operations Research: Applications and Algorithms*. Thomson/Brooks/Cole, Belmont, California.

12 Strategies Decision Makers Can Use to Manage Risk

Introduction

We have emphasized that risk is everywhere and is substantially unavoidable. It follows that management of risk is not something different from management of other aspects of a farm, since every farm management decision has risk implications. There are, however, some types of farm management decisions that bear strongly on the riskiness of farming, and some of these are reviewed in this chapter. The treatment is general because, as we have shown, every decision should be considered in the context of the particular circumstances, notably the beliefs and preferences of the DM. Therefore, specific prescriptions about strategies to manage risk are seldom possible. Instead, we canvass some of the main areas where DMs can act to manage risk and indicate how choices in some of these areas might be analysed.

As outlined in Chapter 1, there are two reasons why risk in agriculture matters: risk aversion and downside risk. Moreover, we have argued that, at least in capitalist agriculture, the latter will often be at least as important as the former since extreme risk aversion by relatively wealthy DMs is irrational and unlikely to exist, at least for important risky choices. In the light of this view, it might seem natural to draw a distinction between management strategies that deal with risk aversion and management strategies that deal with downside risk. That, however, does not work well because effective strategies to manage downside risk will also have benefits in terms of increased utility for risk-averse DMs.

By way of introduction, it is useful to start with a brief overview of some empirical surveys that have sought to elicit from farmers their perceptions of the important risks they face and the main strategies they use to deal with these risks. As might be expected, primary sources of risk farmers typically mention are production, price/market and institutional risks. Human and financial risks are also highly ranked risks. In coping with these risks, farmers often report the use of: (i) financial and debt management; (ii) flexibility; (iii) collecting information/use of consultants/advisers; (iv) prevention of diseases and pests; (v) insurance; and (vi) producing at lowest possible costs or in the most productive and profitable way. Understandably, risk perceptions and management strategies differ significantly, between different types of farms, geographical location, agricultural policy regimes and countries. For more see, for example, Patrick and Musser (1997), Meuwissen *et al.* (2001), Koesling *et al.* (2004), Lien *et al.* (2006) and Greiner *et al.* (2009) and the references therein. We discuss several of these risk-management strategies below.

As the cited literature shows, there can be an imperfect connection between what are identified by farmers as important risks and the risk-management strategies they nominate as most useful. There can be several reasons for this. Farmers, like the rest of us, cannot be expected to act rationally all the time. Indeed, if they were always able to cope effectively with all risks, there would be no need for this book. Yet it may not be irrational for them not to have a close match between their perceived main risks and the risk-coping strategies they use. One reason could be that ways of mitigating some risks are easier and/or cheaper than for other risks. So it might make perfect sense to take measures to deal with the easy ones, then, having reduced their total risk exposure, it may be optimal to bear the remainder. In other words, as we hope should be evident, and as we further emphasize below, coping with risk is about risk efficiency, not about risk reduction or risk minimization. Moreover, risk-management strategies need ideally to be looked at in a whole-farm or whole-business context.

Before considering specific risk-management strategies in more detail, some general comments need to be made. First, it is important to reiterate that the objective of risk management is to maximize utility, not to minimize risk, as is often wrongly presumed. It is particularly dangerous to equate risk minimization with minimizing the variability of returns. This is apparent from the application of stochastic dominance methods described in Chapter 7. Indeed, in many situations the best route to risk efficiency is by finding strategies that improve the expected value of returns, rather than those that reduce the dispersion of returns, as measured, for example, by the variance. The point is illustrated in Fig. 12.1.

Suppose that Fig. 12.1 relates to two farm management strategies with A being diversification and B specialization. In the case illustrated, specialization (B) does indeed offer more volatile returns than A, as indicated by the relative slopes of the two CDFs, but no wise DM would prefer A over B. As drawn, B is first-degree stochastically dominant over A and so, as explained in Chapter 7, is superior for all DMs who prefer more to less return, regardless of their attitudes to risk.

Even in the case of risky alternatives for which there is no stochastic dominance in the way illustrated in Fig. 12.1, an alternative with more volatility may still be preferred by risk-averse DMs, provided the upside gains are sufficiently substantial. Such a case is illustrated in Fig. 12.2, which is a hypothetical illustration of the so-called green revolution technology in comparison with traditional methods previously used in less developed countries. The new technology requires farmers to make investments in increased levels of inputs to gain the rewards of higher potential returns. As a result, returns from the improved methods are inferior to those from the traditional ways if things go wrong. Yet, as experience shows, even risk-averse resource-poor farmers have often been willing to accept this risk, and the greater volatility in returns that may thereby be entailed, for the good prospect of better returns.

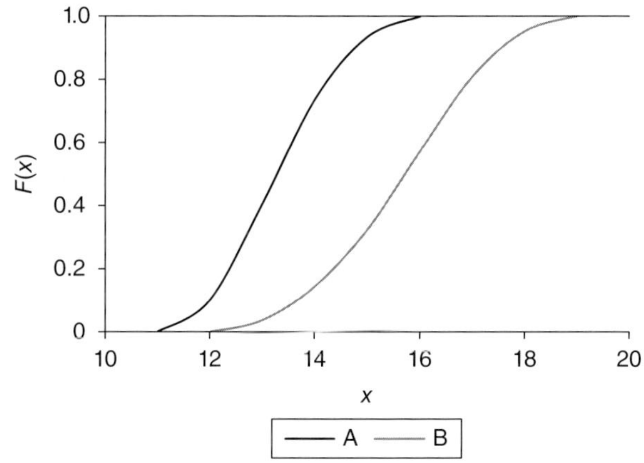

Fig. 12.1. Comparison of cumulative distribution functions (CDFs) for two risk-management strategies.

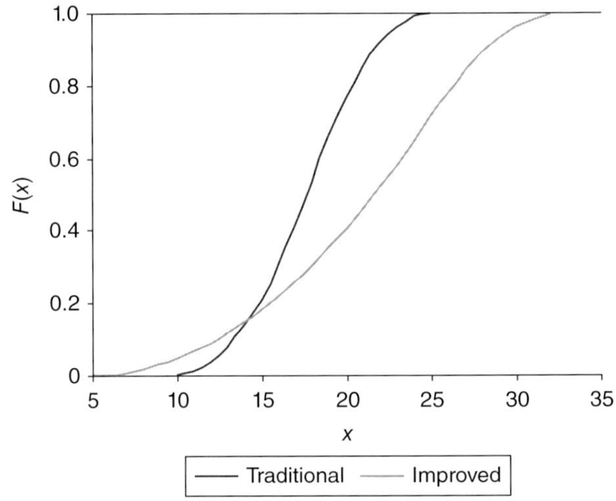

Fig. 12.2. An illustration of the CDFs of improved versus traditional technologies.

It is such logic that leads us to some scepticism about the widespread interest by both farmers and policy makers in developing and using financial instruments for risk sharing in agriculture. Most, though not all risk-sharing strategies will reduce expected returns and so, at best, will appeal only to risk-averse DMs. If the loss in expected returns from such instruments is too great relative to the reduction in volatility, such schemes will not (or should not) be attractive to even moderately risk-averse farmers. Exceptions, of course, are those schemes that are heavily subsidized by governments, so that the expected returns to farmers who participate are positive.

The related point that needs to be made at the start of this discussion of risk-management strategies is that one strategy that should always be included among the alternatives to consider is risk retention, meaning that DMs should be prepared to take some risks if they expect to earn some profit. After all, profit is the reward for bearing risk. If it were possible to insure all risks, a DM who did so would end up with a secure loss! So the choice of risk-management strategies has to be selective, with DMs accepting most risks while taking measures to avoid or abate those risks judged too great to be borne. Perhaps the best conceptual view of this process is to cast risk management in a portfolio selection context with the objective or finding the optimal combination of risk taking, risk abating and risk avoiding options.

In what follows we have adopted the usual classification in distinguishing on-farm risk-management strategies and risk-management strategies to share risk with others.

On-farm Strategies

Compared with risk-sharing strategies, on-farm management strategies can more readily be used to avoid or soften the impact of undesirable events (downside risk). Of course, they can also accommodate measures reflecting the risk aversion of the DM.

Avoiding or reducing exposure to risks

While life without risk would be deadly dull, it is also the case that some unwanted risks do not need to be faced and others can be managed to reduce their impact. Two important management strategies, therefore, are risk avoidance and risk abatement.

Risk avoidance is concerned with reducing or eliminating the possibility of events with unfavourable consequences occurring. With foresight and the adoption of preventive measures, many accidents can be avoided. Most obviously, this applies to the safety of people working on the farm or in an agribusiness. In many countries there are support services available and regulations in place to help avoid accidents. The most common form of fatal accidents on farms, at least in more developed countries, are vehicle accidents, tractors and four-wheel motor bikes being particularly dangerous. Hence steps on farm to encourage safe use of farm vehicles deserve high priority.

Suicide is a significant cause of death in many countries, with the incidence usually being highest among males, especially younger males. Farm people may be more at risk than others, particularly in more remote rural areas where support networks are weaker, signalling a need for families and friends of people working in such environments to provide more emotional support, especially during hard times, and for health professionals and others interacting with farm people to be more alert to the risks and to take action when they spot danger signs.

Some farming risks can be avoided by good preventive maintenance. Farm production can be adversely affected by such actions as careless introduction of new animals potentially carrying a disease on to a farm, or improper use of hazardous chemicals. During a severe drought in Australia, for example, many cattle were fed cottonseed waste and it was only later discovered that the feed was contaminated with a persistent pesticide, making the meat unfit for human consumption.

Regular drenching of sheep for internal parasites will keep these pests under control and prevent a costly epidemic. Likewise, if farm machinery is serviced regularly, chances of breakdowns at critical times are reduced, and farmers who are concerned about downside risk associated with breakdowns can choose to replace machinery before it becomes unreliable.

Risk abatement relates to measures to deal with bad outcomes when they occur and to manage the subsequent recovery. For foreseeable risks, it will often be important to have contingency plans in place as well as the means to implement those plans. For example, it would be wise to have a plan to deal with an outbreak of fire in farm buildings housing valuable animals or machinery. Possible priority actions would include calling for help, evacuating the buildings and fighting the fire. Fighting the fire would be possible only if effective firefighting equipment were in place and operational.

To a large degree, risk abatement in farming is simply good animal and crop husbandry. A good farmer will spot trouble early and will quickly do something about it. Loss of condition in cattle might indicate parasite infestation that can be controlled with appropriate treatment before the economic consequences become serious. Similarly, an insect infestation of a crop that is noticed and dealt with quickly would do less damage than if left untreated until much later. In agribusiness, risk abatement will often be closely related to quality control – a supplier of chicken meat to a retail chain would be wise to have in place monitoring and control procedures to identify any quality problems, particularly any affecting food safety, as well as procedures for speedy identification of the source of the problem so that it can be corrected quickly. Operators of business involved in rearing and slaughtering farm animals need to have good policies in place to assure the welfare of those animals and to prevent deliberate cruelty. Otherwise, they risk serious penalties, even closure of their operations, as happened with the live beef trade from Australia to Indonesia.

A risk abatement strategy should also include plans and related measures for recovery after a disaster. For example, a farmer who has spent years breeding a superior strain of dairy cows could lose everything if the animals were compulsorily slaughtered during a foot-and-mouth disease (FMD) outbreak. At least partial recovery would be possible if stocks of semen from the best bulls had been kept deep frozen. As this example shows, risk abatement strategies are unlikely to be costless, so again some analysis may be needed to estimate how much should be spent to abate which risks.

The above strategies of risk avoidance and risk abatement are both based on the proposition that major risks can be identified and planned for. The truth is, of course, that some disasters come out of the blue. No amount of planning can protect against the unimaginable. The best that can be done in such circumstances is to mitigate the adverse consequences in whatever ways can be found.

In confronting a range of risks characterized by uncertainty, it is often suggested that a *precautionary principle* should be invoked. As originally proposed at the 1992 United Nations Conference on Environment and Development in Rio de Janeiro, the principle was stated as: 'Where there are threats of serious or irreversible damage, lack of full scientific certainty shall not be used as a reason for postponing cost-effective measures to prevent environmental degradation.' Subsequently, a number of variants of the principle have been developed, some of which have been enshrined in legislation. The logical foundation of the principle lies in the theory of real options, as discussed in Chapter 11, and in that respect is consistent with the approach to risky choice taken in this book. However, while the sentiment embodied in the principle is uncontentious, efforts to implement it can encounter problems, especially as several different reformulations and interpretations exist. Worse, the principle can be abused, turning it around to support just the sorts of actions it was intended to avoid. We therefore suggest that a carefully conducted decision analysis will generally lead to the same decision about environmental protection in the face of uncertainty as would be reached using the precautionary principle, yet would be more soundly based, accounting for well-considered probabilities and a careful consideration of associated costs, benefits and community values.

Notwithstanding the above remarks, when information about risk is very incomplete but where it is believed that any action or inaction carries with it the possibility of serious and irreversible consequences, such as bankruptcy or even death, a high degree of caution may well be appropriate. Such a strategy might include:

1. Postponing a decision to change the existing situation until more information is obtained about the possibility of serious negative results from the change.
2. In a situation where continuation or projection of present practices creates the threat of serious negative consequences, strict 'safety standards' may be imposed, at least until more is known.
3. Somewhat in the spirit of the postponement option above, a decision that does not depart too much from the status quo can be taken, deliberately choosing a course that is overly cautious but nevertheless better than indecision or inaction.

The first rule might be called the 'look before you leap' principle, and the second the 'better safe than sorry' principle. The third may not really qualify as a principle, but could perhaps be characterized as 'slow and steady wins the race'.

A danger with precaution is that it may induce DMs to adopt an excessively risk-averse attitude. As noted in Chapter 1, we all take actions with the risk of very bad outcomes, such as death, often for very small reward. Who had not risked crossing a busy street to buy an ice cream or newspaper? Evidently extreme caution should be invoked only when the circumstances of the situation indicate a significant but hard-to-quantify risk of a seriously bad and irreversible outcome. As discussed above, many farming risks are so small as to be relatively insignificant in the context of the survival and overall profitability of the farm business. The same applies even more to agricultural policy making and planning at the national level where risk aversion can be argued to be largely irrelevant because of the opportunity that often exists to spread the adverse consequences of a single decision over a large number of people (Chapter 13, this volume). Precaution is called for, however, when the risks could be catastrophic, such as the possible irreversible environmental and human health consequences of some new and untried farming practices.

Operationally, the precaution implies adding to the list of options usually considered by DMs. Thus, the 'look before you leap' principle implies that, for example, instead of a choice between building a new piggery or not building it, the option is also explicitly considered of delaying any start to construction until more information has been gathered on likely future profits and on alternative designs that may be more successful than the one currently planned. The 'better safe than sorry' principle means that, for example, if there is uncertainty about the human health impacts of a pesticide in regular use on the farm, its use might be restricted or stopped until more conclusive evidence is available. The 'slow and steady' principle suggests that adaptive change may be safer than dramatic change. For example, in switching from conventional to organic farming, a farmer might be wise to start on only part of the farm for the first few years, rather than aiming for immediate full-scale changeover.

Note that precautionary behaviour means that some actual or opportunity costs are incurred by acting cautiously, and these costs must be subjectively weighed against the achieved reduction in the risk of seriously unfavourable consequences.

It might be argued that the very familiarity of the aphorisms used to characterize these precautionary principles shows that they are merely common sense, which indeed they are. As many past mistakes in many realms reveal, however, common sense is too often not applied in decision making. Adding a requirement for precautionary options to the list of choice alternatives in decision analysis may reduce the frequency of such mistakes.

Collecting information

Whether or not risk aversion matters, better decisions in a risky world can always be made – though better outcomes not guaranteed – if more and better relevant information is obtained. It is very important for a farmer to know what is going on, both on the farm itself and in the outside world. Collecting more information can be an important strategy to reduce downside risk. Examples where investments of time and money in collecting information can have substantial payoffs in agriculture include collecting information about more productive technology options and about marketing opportunities and market trends.

Bearing in mind that, according to the view adopted herein, all probabilities are subjective, information gathering can be seen to have two impacts on subjective distributions. First, the dispersion of the distribution will usually be reduced as knowledge is accumulated. For example, a farmer who has no knowledge of a new technology may be thought of as having a prior distribution for the returns from that technology with virtually infinite dispersion. Once the farmer has accumulated some information about the technology, for example by observing it in place on a neighbour's farm or by learning through the media of the results of trials of the technology, the distribution will be modified and become much 'tighter'.

Moreover, if still more relevant information is obtained, for example by the farmer trying out the technology on a pilot scale on his own farm, a still lower variance for the subjective distribution will usually result.

The second impact of information on subjective distributions is likely to be a shift in the location of the distribution. Although in fact, the whole distribution may shift, for convenience we can think of this change as an adjustment to the subjective mean as excessive optimism or pessimism is corrected, based on accumulated information. Again, we can imagine that our farmer who is learning about a new technology may initially be pessimistic about how it will perform on the farm. The farmer may well have 'had fingers burnt' in the past by giving too much credence to the optimistic claims of agricultural scientists, farm advisers or rural sales representatives, and so will heavily discount any new claims (see also the section on bias in Chapter 3). Yet, if the farmer takes the trouble to find out more, perhaps trying out the technology on a small scale if this is possible, the initial scepticism may be modified.

As explained in Chapter 4, this process of modifying subjective prior distributions in the light of accumulating information can be formalized using Bayes' Theorem. Moreover, given that information gathering and processing has costs, the methods of decision analysis, based around the application of Bayesian analysis, can be used to resolve the risky decision of how much information to collect, as was illustrated in Chapter 6.

Note, of course, that Bayesian revision of probabilities can be applied sequentially, as new pieces of information become available. Thus, the posterior probability resulting from the first application of Bayes' Theorem becomes the prior probability that goes into the next revision. After t pieces of information, $I_1, \ldots, I_t$, have been obtained, the t-th application of Bayes' Theorem is:

$$P(S_i|I_1, \ldots, I_t) = \frac{P(S_i|I_1, \ldots, I_{t-1})P(I_t|S_i, I_1, \ldots, I_{t-1})}{\Sigma_i P(S_i|I_1, \ldots, I_{t-1})P(I_t|S_i, I_1, \ldots, I_{t-1})} \quad (12.1)$$

where $P(S_i|I_1, \ldots, I_t)$ is the posterior probability of S_i after t pieces of information, $P(S_i|I_1, \ldots, I_{t-1})$ is the prior probability of S_i after $t-1$ pieces of information, and $P(I_t|S_i, I_1, \ldots, I_{t-1})$ is the likelihood of observing I_t when S_i is the true state and $t-1$ previous pieces of information that have been received.

While not necessarily the most useful form of this equation (Anderson *et al.*, 1977, p. 54), this formulation emphasizes the sequential nature of probability revision in the light of accumulating knowledge. In this form, it represents a formal model of learning that is useful, for example, in explaining the process of changing opinions that a farmer may pass through in moving towards the adoption of a new technology. Yet this formal model contrasts with evidence from studies of behaviour that: (i) people often have a too optimistic view of what they know, and so tend to collect too little information before making risky decisions; and (ii) people commonly under-utilize new information when they get it. Clearly, the careful decision analyst would wish to guard against these sources of error, at least for important decisions, checking intuition by using systematic methods to decide whether new information is worth collecting and to work out how that information should be incorporated into probability assessments (Chapter 6, this volume).

Selecting less risky technologies

It is evident that some farming activities give higher and more stable returns over time than others. For example, intensive livestock production is likely to be more stable, at least in terms of levels of production achieved, than extensive grazing, since the latter is exposed to the effects of weather variability and the former usually is not. Similarly, for some farming activities in some countries, prices are more or less guaranteed by government market interventions, while prices for other agricultural outputs are determined in fluctuating world markets. Risk-averse farmers will obviously consider these aspects in deciding what to produce.

Farmers also often have the possibility of selecting between relatively more or less risky ways of producing the same commodity. For example, investments in irrigation may give more assured levels of crop or pasture production in areas of unreliable rainfall. Similarly, investments in actions to control or prevent outbreaks of pests and disease may be successful in limiting the chance of serious losses.

Methods of whole-farm planning accounting for risk, such as those discussed in Chapter 9, are appropriate for decision analyses of what to produce and how to produce it. However, simpler, less formal approaches may be followed, often within a stochastically efficient framework, as described in Chapter 7. As explained at the start of this chapter, the aim of the analysis is to find the risk-efficient choices, not to minimize the volatility of returns.

Diversification

Similar considerations apply to the analysis of diversification of farming activities to manage risk as for choice of technologies. The idea of diversification is to reduce the dispersion of the overall return by selecting a mixture of activities that have net returns with low or negative correlations. Again, however, the aim should be to find the risk-efficient combinations of activities, not the one that merely minimizes variance. In general, farmers will diversify more with increasing degree of risk aversion. However, more diversification can be increasingly costly if it means forgoing the advantages that specialization confers through better command of superior technologies and closer attention to the special needs of one particular market.

Farm planning models to find the appropriate degree of enterprise diversification have conventionally been cast in the *E,V* or portfolio selection framework and have typically been solved by quadratic risk programming, as described in Chapter 9. However, as also described in that chapter, a better approach may be to formulate the model in terms of direct expected utility maximization, perhaps in the form of utility-efficient (UE) programming. While results have generally been similar, the latter types of models put more weight on bad outcomes and are in any case more consistent with the expected utility hypothesis.

While the potential gains in risk efficiency from on-farm diversification are an empirical matter to be resolved on a case-by-case basis, these gains are often less than may be imagined, for a number of reasons. First, even farm plans to maximize expected return will often be reasonably diversified before risk aversion is considered. Risk concerns aside, mixtures of activities will typically make best use of available resources. Mixed cropping allows more productive and sustainable crop rotations; labour and machinery requirements for a mixed system will be more evenly spread throughout the year, resulting in more efficient use of these resources; and seasonal cash-flow troughs will be filled by having income from diverse sources received at several stages during the farming year, again favouring some degree of system diversification. Moreover, having only two or three enterprises can mean that the majority of the risk-reducing benefits from diversification are captured (although, again, of course, this is an empirical matter).

More generally, the fact that returns from different activities are typically strongly positively correlated limits the gains from diversification on farm. Better opportunities to spread risks may lie in spatial diversification, meaning owning farms in several locations sufficiently widely scattered to reduce positive correlations owing to weather effects. Carrying the logic a step further, farms located in different regions or countries will limit risk impacts of unexpected and unfavourable policy changes by individual governments. Of course, this form of diversification is open only to the largest businesses and would obviously create a whole new range of management problems. Less ambitiously, investments of capital, management and time in off-farm activities may provide an effective risk-spreading avenue that should not be overlooked, even for relatively poor farmers. For example, farm families in less developed countries often diversify income sources, engaging in such non-farm activities as handicraft production, or investing in relocation costs, and education costs, for family members to find wage employment off the farm. Such measures are adopted in the expectation that those who leave will remit part of their earnings to those family members left at home. Investing in children's education is often seen as a good form of diversification to protect the long-term welfare of the farm family, in both more and less developed countries.

Flexibility

Flexibility refers to the ease and economy with which the farming business can adjust to changed circumstances. It is therefore related to real options, as discussed in Chapter 11, the theory of which provides a framework for the analysis of decisions affecting flexibility. A flexible strategy is one that maintains or increases options in ways that help a farmer accommodate risk aversion, manage downside risk and increase expected returns. Greater flexibility means better possibilities to respond to unfavourable events and to benefit from opportunities that arise. Measures that farmers can take to enhance flexibility include asset, product, market, cost and time flexibility.

Asset flexibility means investing in assets that have more than one use. For example, when constructing a farm building for a specific use, it may not cost much more to modify the design so that, if required, it could be readily adapted to an alternative use at modest cost. Similarly, land that can be used for several different types of production is a more flexible investment than land that is limited, by soil type or climate, to only one or two uses. Of course, the most flexible form of asset is cash, and maintaining an adequate level of liquidity in the farm business is an important part of prudent financial management, a topic taken up later in this chapter. Other forms of assets, such as fodder reserves held as a precaution against drought, may also add to flexibility.

Product flexibility exists when an enterprise produces a product that has more than one end use, or when the enterprise yields more than one product. Both attributes may enhance flexibility. For example, coconuts may be used for home consumption, stock feed, copra production, be sold as green nuts for drinking, as whole nuts for direct consumption, or processed into desiccated coconut for sale. In addition, the trunks, fronds, husks, shells and milk of coconuts all have economic uses, and the coconut palms can be stimulated to produce sweet sap to make an alcoholic beverage that can be drunk at home, sold as is, or distilled into a spirit beverage, making coconuts perhaps the world's most flexible crop.

Related to product flexibility is *market flexibility*, whereby a product can be sold in different markets that may not be subject to the same risks. For example, a small domestic market may be subject to greater change than an export market. However, only those producers who are able to meet export quality standards may have access to the more stable export market. Products with high value to weight or value to volume ratios can more easily be shipped to distant locations, which may become an attractive option when market prices there are relatively high.

With *cost flexibility* the idea is to organize production while keeping fixed costs low and incurring higher variable costs as necessary. For example, land or machinery may be leased rather than purchased, labour may be hired on a contract or casual basis rather than in the form of permanent workers. By such means, fixed costs are kept to a minimum and there is greater scope to change levels of resource use or to switch to other types of production should circumstances warrant it.

Finally, *time flexibility* relates to the speed with which adjustments to the farming operations can be made. Activities with short production cycles are obviously more flexible than those with long cycles. Contrast tree crops, which may have a production cycle of several decades, with short-term seasonal crops that may be planted, grown, harvested and sold all within 6 months or so. Time flexibility may also be relevant on a scale that is more limited, but perhaps still important. For example, a crop may be grown using a schedule for the application of inputs of fertilizer and sprays that is adjusted according to developing seasonal conditions rather than being predetermined.

Flexibility can be accommodated in decision analysis by making sure that options are not overlooked. Thus, for example, in a decision tree, stochastic simulation or in any other of the forms of analysis described in earlier chapters, it is important that a reasonably comprehensive range of alternatives is considered. It is equally important to extend the analysis far enough into the future to reflect differences among those alternatives in the ease or difficulty with which adjustments can be made in light of favourable or unfavourable outcomes in the unfolding of uncertainty. As mentioned in Chapter 11, it may be particularly important to include the alternative of waiting until more information is available before a choice with substantial sunk costs is implemented.

Strategies to Share Risk with Others

Farm financing

The way a farm business uses debt can have major implications for risk exposure. A key concept in this regard is *financial leverage*, defined as the use of credit and other fixed-obligation financing relative to the use of equity capital (Robison and Barry, 1987, Chapter 16). From the point of view of the owner, increases in financial leverage magnify the impact of variability of firm returns. If the return on total assets is above the borrowing rate, the rate of return on the owner's equity will be increased. Conversely, if the overall rate of return is less than the borrowing rate, the owner will suffer, in the extreme case receiving a negative rate of return on equity. The effect is illustrated by the simple example given in Table 12.1.

As the table shows, when the debt to equity ratio is zero (no debt), the return on total capital is naturally the same as the return on equity. Also, when the return to total capital is equal to the rate of interest on debt, leverage has no effect on the return to equity. But, when the return on total capital falls to 5% and the debt to equity ratio is 50% or 100%, return to equity is below this rate at 2.5% and zero, respectively. On the other hand, when the overall return rises to 15%, returns on equity rise more, to 17.5% and 20% for debt to equity ratios of 50% and 100%, respectively. The example therefore illustrates the magnification of risk in the total farm return for the equity holder(s).

Table 12.1. Effect of financial leverage in magnifying the impact on equity of variability of returns.

Equity capital ($\$10^3$)	50	50	50
Debt to equity ratio	0%	50%	100%
Debt capital at 10% ($\$10^3$)	0	25	50
Total capital ($\$10^3$)	50	75	100
Rate of return on total capital	10%	10%	10%
Total income ($\$10^3$)	5.0	7.5	10.0
Interest on debt at 10% ($\$10^3$)	0.0	2.5	5.0
Balance to equity owner ($\$10^3$)	5.0	5.0	5.0
Return on equity	10.0%	10.0%	10.0%
Rate of return on total capital	5%	5%	5%
Total income ($\$10^3$)	2.5	3.8	5.0
Interest on debt ($\$10^3$)	0.0	2.5	5.0
Balance to equity owner ($\$10^3$)	2.5	1.3	0.0
Return on equity	5.0%	2.5%	0.0%
Rate of return on total capital	15%	15%	15%
Total income ($\$10^3$)	7.5	11.3	15.0
Interest on debt ($\$10^3$)	0.0	2.5	5.0
Balance to equity owner ($\$10^3$)	7.5	8.8	10.0
Return on equity	15.0%	17.5%	20.0%

The effect of financial leverage in magnifying risk raises the question of the optimal financial structure for a farm business. The answer depends on the risk preferences as reflected in the investor's utility function (assuming that a totally riskless asset does not exist, so that the Separation Theorem does not fully apply – see Chapter 7). Given this information and information on the DM's beliefs about future income levels, it is possible to determine the optimal level of debt for any given interest rate on loans. If the borrowing rate itself is uncertain, this too can be included in the analysis. While some rather elegant analytical methods can be used to solve this problem (e.g. Robison and Barry, 1987, pp. 234–236), it is also relatively straightforward to solve as a stochastic simulation model using a trial-and-error solution procedure, as briefly indicated in a simple example in Chapter 6.

There are, however, some limitations to this relatively simple and static approach to the determination of optimal financial structure. Notably, the dynamics of farm debt are ignored. Given a run of bad years, the farmer may reach the limit on borrowing set by the bank, or may run up more debt than can be serviced, especially if there is a persistent downturn in farm profitability. Such situations can obviously lead to bankruptcy. It is evidently unwise routinely to borrow to the limit of available credit set by lending institutions since holding a credit reserve is an efficient way to provide liquidity to sustain the business through hard times. However, while the direct costs of holding a credit reserve are usually low, the opportunity costs, in terms of the return on the forgone investment, may be considerable. Again, some careful analysis may be called for to work out the best strategy.

The reality of the financial management of a modern commercial farm business is still more complicated than we have indicated. For example, in many countries there is now available to farmers a wide range of financial instruments such as fixed or flexible interest rate loans, loans with more or less flexible repayment conditions, financial and commodity derivatives, and even various arrangements such as investment trusts that in effect enable the farmer to sell a portion of the equity to outside investors. Some farms, of course, are set up as limited liability private or public companies with the equity held by the share owners.

Evaluating the financial structure of a modern commercial farm business is obviously more complex than is suggested by the few issues we have briefly addressed. Decision analysis for such complexity requires a relatively comprehensive and flexible model. For one such model that used stochastic simulation to assess the financial feasibility of development alternatives for a Norwegian dairy farm see Lien (2003). That work drew in part on a similar study by Milham (1992) of the financial feasibility of Australian wheat and sheep farms.

Insurance

There are various types of insurance contracts available to farmers, including: (i) fire and theft cover for assets; (ii) death and disability cover for proprietors; (iii) cover for workers' injuries and for public liability; and (iv) mortality and infertility cover for stud stock. It is usually possible to insure crops commercially for fire and storm damage but comprehensive crop insurance is mostly provided under subsidized government schemes (see also Chapter 13). Insurance companies also commonly offer various types of savings contracts, such as superannuation schemes, that may be called insurance but in fact are not.

The principle of insurance as a risk-sharing device is that, by accepting appropriate *premiums* from a large number of clients, the insurance company is able to pool the risks. Moreover, by use of information on the frequency and level of claims, the company aims to set premiums at levels that will enable it to pay all *indemnities* (compensation for insured losses) from the aggregate contributed premiums and still leave a margin for operating costs and profit. Some of the profit may also come from investing the premiums collected (prior to payment of indemnities) along with accumulated financial reserves, in various more or less secure financial instruments.

From the point of view of the farmer considering buying insurance, the way commercial insurance usually works means that the expected value of insuring is almost always negative. The expected value can only be positive if the probability judged by the farmer of making a successful claim is considerably higher than the probability judged by the company from actuarial information. Such an outcome is unlikely since most policies require the person taking out the insurance to disclose to the company any special circumstance that might increase the risk of the insured event occurring. Failure to do so may invalidate the policy, leading to claims being rejected.

It follows that commercial insurance will usually be attractive only for risk averters, and only then for risks that are sufficiently serious to warrant paying a premium equal to significantly more than the expected loss (actual loss times probability) without insurance. For most farmers, and indeed most people, this means that insuring small, easily borne risks will not be worthwhile. However, it may well be worth insuring large risks that otherwise could threaten the continued existence of the farm business or that could seriously damage the welfare of the owners.

The principles of decision analysis applied to choosing whether or not to insure were illustrated by the simple example, developed and worked through in earlier chapters, of the dairy farmer undecided about buying insurance against losses from FMD.

Actual insurance decisions will often be more complex than this simple example may suggest. For instance, a major threat to the survival of a family farm business is the death or serious disability of one of the principal partners. This is an insurable risk, but it is seldom clear when buying insurance what value should be attached to the insured event. Other policies may include a so-called deductible or excess that the person taking out the insurance agrees to pay if the insured event happens, with the company being liable for only the balance of the loss. Accepting policies with such a deductible can be a sensible way to insure only the bigger risks, as mentioned above. In some forms of insurance, such as vehicle cover, insurance companies may offer a no-claim bonus to discourage small claims for minor damage. The person taking out the policy therefore has to consider whether, if some claimable damage does occur, it is worth making a claim and sacrificing the accumulated bonus.

Some area-based index insurance contracts have been developed to allow farmers to insure against low crop yields (e.g. Skees and Barnett, 1999; Turvey, 2001). Under these contracts, the indemnity is calculated on the average yield in the area where the farm is located, not on the actual farm yield. There are also some products available for which indemnity is based on a weather index, such as rainfall in a defined period, as recorded at a nearby official weather station. The advantages to the insurer of such arrangements are further discussed in Chapter 13. From the point of view of a farmer contemplating buying such index-based insurance, a key issue is the degree of correlation between the on-farm risk and the index on which indemnity payments are based. Unless the correlation is high, such insurance will not be worthwhile.

The methods of decision analysis described in earlier chapters provide a framework for analysing these kinds of questions. Stochastic dependency between the contract and returns from the rest of the farm business often makes it difficult to evaluate correctly the net effect of introducing a new risk-management instrument such as insurance. Consequently, such decisions may be best considered in a whole-farm portfolio selection context. As described in Chapter 9, a utility-efficient programming approach is especially suitable for evaluating the merits of crop insurance in a farm plan since this method, implemented via a state of nature matrix, can capture the effect of insurance on the distribution of returns.

Share contracts

Share contracts and other arrangements between a farmer and somebody else (landlord, labourers, etc.), can include arrangement whereby risks such as those caused by poor yields, low output prices or unexpectedly high input costs can be shared between the parties. Crop or livestock share leases, labour share leases and variable cash share leases are examples of share contracting. Through specification of the share contracts it is possible to make a trade-off between the lessee's and lessor's incentives and responsibilities, accounting for differences between them in the capacity to bear risk (e.g. Petersson and Andersson, 1996; Allen and Lueck, 2003).

Contract marketing

Farmers, particularly those in more developed countries, can often use various marketing arrangements to reduce price and other types of risks for commodities not yet produced or for inputs needed in the future. The most important alternatives, from a risk-management perspective, include cooperative marketing with price pooling, and forward contracts for commodity sales or for input delivery. This category also includes hedging on futures markets and the use of options to reduce price risk. The former two are outlined below in this section, and the latter two are dealt with in the next section.

Price pooling arrangements are usually of limited efficacy for risk management. They operate by a group of farmers collectively buying their inputs or selling their outputs through a cooperative or marketing board. Membership of the group may be voluntary or compulsory. The price pooling arrangements may be operated in various ways but are generally designed to protect the individual from short-term fluctuations in prices by some form of averaging of prices across multiple sales. It may also be claimed that increased market power and economies of size result in lower input prices or higher product prices than could be obtained by the individual. However, if these benefits do indeed exist, they will be at least partially offset by the administrative costs of the scheme.

Forward contracting of sales or purchases is a much more effective and relatively widely used form of risk management for farmers. For simplicity and because it is by far the most important form of such arrangements, we shall concentrate on forward contracting of sales of farm production. Similarly, we shall assume that the contract is for the sale of a crop, although forward contracts may also be written for the sale of animal products. The contract is written, perhaps at planting time or maybe later in the season, between the farmer and the purchaser who agree on a price (or on a basis for determining a price, such as a price scale according to grade or other characteristics of the product). The contract may also stipulate the quality or quantity of produce to be delivered by the farmer, or may relate to the whole production, which will obviously depend on the yield. The price offered is likely to be discounted below the generally expected price for the future delivery date, since the merchant is taking a risk of loss should the market drop between the contract date and the delivery date while a risk-averse farmer may be willing to accept such a discounted price for the security of an assured payment for the product.

Again, the methods of decision analysis described in earlier chapters can be used to evaluate contract marketing opportunities. Once again, the decision should ideally be cast in a whole-farm or whole-business context because of likely stochastic dependencies among prices. It would also be wise to look at a wide range of alternatives such as different forms of contracts with different proportions of the production sold on contract, contracts versus hedging or options trading on the derivatives market (see below), and crop revenue insurance (if available).

Note that a farmer who contracts to deliver a specified quantity of the commodity, regardless of the yield attained, is facing an extra risk. If the yield is lower than contracted, the farmer may have to purchase on the open market to meet the contracted delivery requirements, or may face some other penalty under the contract. Moreover, if the contracted crop has yielded below expectations, the same may be true for other crops grown, making the cost of the shortfall in contracted supply more serious in a year when overall farm income is depressed.

Trading in commodity derivatives

Derivatives trading can be used to reduce price risks for both future inputs and future outputs. The most important examples are hedging on the commodities futures market and options trading.

Hedging on the futures market is rather similar to forward selling on contract but with a number of differences to be explained. One important difference is that futures contracts are standardized, widely traded contracts, so prices are more competitively determined than for a specific contract between a single farmer and a single merchant. That might mean that the farmer can get a better deal by hedging on the futures market than by selling on contract.

In principle, the seller of a futures contract makes a commitment to deliver a contracted quantity of a defined grade of the traded commodity at a particular date. The buyer makes a commitment to take delivery of that amount of the commodity on that date at the contracted price. In practice, delivery of the commodity normally does not take place. For example, a farmer can hedge to reduce future price risk by selling a futures contract for a commodity that matches as closely as possible the product the farmer expects to have available to sell later. However, the farmer will normally *close out* the position by buying back a futures contract, effectively cancelling it, at around the time the real commodity is sold on the spot market. In other words, a futures market is mainly a speculative market in contracts, not a market in the commodity itself.

Table 12.2 illustrates how an Australian wool producer could use a futures hedge to reduce price risk.

The wool producer plans to sell the wool in September. In May the producer decides to hedge on the futures market and so sells a futures contract for an amount of wool approximately equal to total production. We assume, for convenience, that the October futures price at that moment is 1000 ¢/kg. By selling such a contract the farmer is agreeing to supply the specified quantity of wool of the specified quality in October at the contract price.

In the example, the current cash price for similar wool is only 930 ¢/kg and the difference between this current price and the futures price is called the *basis*. In this case the basis is –70 ¢/kg, making hedging

Table 12.2. Possible outcomes for a farmer who hedges on wool futures.

Prices and transactions	Price rises	Price falls
Prices:	¢/kg	¢/kg
May		
Current cash price	930	930
Futures price for October	1000	1000
Current basis	–70	–70
September		
Current cash price	1116	744
Futures price for October	1122	746
Current basis	–6	–2
Farmer's transactions:		
May		
Sell October futures contract	1000	1000
September		
Sell wool	1116	744
Buy back October futures contract	–1122	–746
Net price received	+994	+998

look attractive to the farmer. The basis is attributed to location, quality and timing discrepancies between commodities traded in the cash market and those deliverable on futures. Because the basis will vary over time in an unpredictable way, it is a source of risk that cannot be eliminated by the farmer. However, as the contract date draws nearer, the basis will narrow, approaching the actual market price. If this were not so, speculators could make a sure profit by simultaneously trading in wool and futures (e.g. Paroush and Wolf, 1989; Pennings and Meulenberg, 1997).

By September, when the sheep have been shorn and the farmer is ready to sell the wool, price levels are likely to have changed. We show two cases in the table – a rise of 20% and a fall of 20% in the spot price of May. In either event, the farmer will sell the wool on the ordinary auction market and receive the current September price. The producer will also buy back the October futures contract to close out that position, thereby formally cancelling the obligation to deliver wool under that contract. By this date, the basis will be much less, as shown. The result is that, regardless of the actual September wool price, the farmer receives close to the futures contract price of 1000 ¢/kg. There is a small loss owing to the non-zero basis assumed for October, and the farmer would also have to pay commissions and other charges on the various transactions, not accounted for here.

The decision on whether to hedge with futures hinges principally on the farmer's expectations about the cash price when the commodity will be sold relative to the futures contract price for that period. Risk aversion apart, hedging will be attractive only if the more or less certain futures contract price is above the DM's expected value of the subjective distribution of future cash price by an amount more than sufficient to cover the costs of the transactions. A risk averter would be prepared to accept a small expected loss from hedging for the greater security of price so obtained. If a decision is taken to hedge, it would be usual to hedge an amount approximately equal to projected actual sales, provided this quantity is known with reasonable certainty. However, as with forward contract selling, it is possible to hedge only a portion of expected sales. It is also possible to speculate on futures markets, but that is another story!

The other risk-management strategy that deserves to be mentioned here is *options trading* to reduce price risk. An option is a contract giving the buyer the right, but not the obligation, to buy or sell a defined commodity (often named the underlying asset) at a specific price on or before a certain date. The idea behind options trading is straightforward and relates readily to many everyday situations. For example, a farmer may want to purchase a plot of land for farm expansion but will not have the required funds available for another 6 months. After negotiation with the owner, a deal is made whereby the farmer gets an option to buy the land in 6 months at a price of $100,000. For this option the farmer has to pay the owner $1500. Now, consider two theoretical situations that might arise after the deal is struck:

1. It is revealed that the plot of land is included in the expansion plans of a nearby city. As a result, the market value of the plot increases to $300,000. Because the owner sold the option to the farmer, the owner is obligated to sell the plot of land for $100,000. In the end, the farmer's profit is $198,500 ($300,000 – $100,000 – $1500).

2. The farmer discovers that the quality of the soil is much worse than initially thought. In fact, the farmer thinks that the plot is almost worthless. Fortunately for the farmer, because of the nature of the option contract, there is no obligation to go through with the sale. Of course, the farmer still has to pay the $1500 price of the option.

This simple example demonstrates three key features of options. First, purchase of an option confers on the purchaser the right, but not the obligation, to acquire something. The purchaser can choose to let the expiration date of the option go by, at which point the option is worthless. If this happens, the purchase price of the option cannot be recovered. But this is the only payment required, not the entire value of the actual asset. Second, an option is merely a contract that deals with an underlying asset. For this reason, options are called derivatives. That is, an option derives its value from something else. In our example, the plot of land is the underlying asset. Usually, the underlying asset is a number of shares in a company or a specific amount of a defined commodity. Third, the seller of the option will require the buyer to pay a *premium* reflecting the value of the option. In this example, the premium is $1500.

Most agricultural options have futures contracts as the underlying asset. An option on the futures exchange is a contract that gives the option buyer the right, but not the obligation, to buy or sell a futures contract at an agreed *strike price* at or before the exercise date of the option. An option to buy is known as a *call option* and an option to sell is a *put option*. For every call there must be a corresponding put for a contract to exist. European style options can be exercised only on the exercise date itself, whereas American style options, which are the most usual in agriculture, can be exercised at any time up to and including the exercise date.

A would-be seller of an option makes an offer at a specified strike price and premium. The offer is made through a broker on the futures exchange. The strike price (also named exercise price) is the price at which the option buyer (or the holder) can later buy or sell a futures contract. The non-refundable premium is paid by the buyer to the seller (or the writer) for the right to trade the futures contract that is guaranteed by the option. Readers interested in how option prices are formed are referred to Lore and Borodovsky (2000) or Hull (2011).

There are two main reasons for an investor to use options: to hedge and to speculate. Only the hedging is discussed here using the same example of the sheep farmer who is planning to sell wool in September and in May is seeking to protect against a fall in price. We now suppose that the farmer wants to keep open the opportunity to benefit from a rise in the price of wool. Instead of selling a futures contract in May, as in Table 12.2, the farmer could buy a put option (European style) to sell a wool futures contract in September. The transactions are illustrated in Table 12.3.

The table shows that, in May, the farmer pays the non-returnable premium of 80 ¢/kg for the put option. Then, if the price rises, the farmer allows the option to expire, and ends up receiving the higher spot price (minus the option premium) when selling the wool. However, if the spot price falls, the farmer exercises the option by selling a futures contract at the advantageous price of 1000 ¢/kg, and immediately closes out that contract by buying an offsetting contract at the lower current futures price. The result is a net return equal to the strike price for the option less the premium and less the small basis on closing out the futures contracts. The farmer would also have to pay brokers' fees (not shown here) on the futures traded.

Buying a put option has the advantage over hedging with futures themselves, since, as illustrated, the farmer can still benefit if the price rises. Of course, the premium payable is likely to reflect the value of this opportunity. As a result, if the price falls, the net return from using an option is usually lower than that from buying a futures contract, as illustrated for this hypothetical example by comparing Tables 12.2 and 12.3.

The power of options trading lies in their versatility. They enable traders to adapt their position to any situation that might arise. With an option contract the buyer can 'lock in' what appears to be a favourable price at the moment without being obliged to accept that price should the market move to make it not a good deal to trade. Put contracts give the buyer of the option the right to 'sell dear' and so are attractive when the price of the underlying asset is expected to fall. Buying a put option is equivalent to buying insurance against a price fall. Call options give the buyer of the call contract the right to 'buy

Table 12.3. Possible outcomes for a farmer who uses wool futures put options.

Prices and transactions	Price rises	Price falls
Prices:	¢/kg	¢/kg
May		
Current cash price	930	930
Futures price for October	1000	1000
Current basis	–70	–70
Put option premium	–80	–80
September		
Current spot price	1116	744
Futures price for October	1122	746
Current basis	–6	–2
Farmer's transactions:		
May		
Buy put option for October wool futures sell contract	–80	–80
September		
Exercise futures put option	..	1000
Close out futures contract	..	–746
Sell wool	1116	744
Net price received	1036	918

cheap' and so are attractive when the price of the underlying asset is expected to rise. Buying a call option is equivalent to insuring against a price rise. Call options might be used by an intensive livestock producer who wants to insure against the risk of a rise in grain prices.

Agricultural economists have devoted considerable amounts of time to attempts to analyse futures and options markets systematically and to show how risk-averse farmers 'should' use such markets. To sample this literature, see, for example, Tomek and Peterson (2001) and Garcia and Leuthold (2004). Yet the reality is that rather few farmers actually use futures and options hedging, probably in part because of lack of knowledge of how the markets work. And in many countries agricultural policies stabilize domestic prices and protect farmers' income, which reduces the attractiveness of marked-based price derivatives. Moreover, there are some limitations to hedging on futures and/or options as a means of risk management. For a number of different reasons, not all risk can be eliminated. First, as we have seen, the basis is not certain. Prices for the grade of wool sold by the farmer may move somewhat differently from prices for the grade specified in the futures or option contract, creating a source of risk. Second, since the farmer does not know in advance the volume of wool that will be produced, there is a risk that the actual volume will not match the volume contracted on the futures or options market. Especially in the event of a shortfall, the revenue earned on the spot market may not cover the cost of closing out the futures contract. Third, the derivatives discussed in this section only reduce price risk. There may be other sources of risk that are more important to the farmer, such as production and yield risk. Hence, there may be a need for derivative contracts where the underlying assets are yields for different commodities within different regions (Tomek and Peterson,

2001). Fourth, only uniform commodities can be traded. So farmers producing (or buying) specialized products adapted to particular market demands of a retailer or a group of consumers cannot use the futures and option markets to manage price risks. Finally, the farmer must be able to finance the futures and/or options transactions. If this entails borrowing, some increased risk is implied.

Most published evaluations of derivatives as risk-sharing devices in agriculture have been done on a partial basis whereas, for reasons already explained, it would usually be best to conduct the analysis on a whole-farm or whole-business basis. Again, the effect in a whole-farm plan context of the introduction of derivatives as additional activities can be analysed in a utility-efficient programming approach, as discussed in Chapter 9.

Concluding Comment

Managing risk in farming through use of many of the principles outlined above is a challenging task – indeed one with many rather artistic dimensions, given the liberal extent of judgement involved in dealing with so many possibilities. Some lend themselves to further depth of insight and greater confidence in the chosen strategies if subjected to decision analysis, as has been indicated in our short review of a topic that is so central to the purpose of this book. Just how formal and extensive such analysis needs to be is still somewhat of an open question, and only a well-monitored set of detailed analyses will make us better informed. We leave more detailed analysis to the readers as they put to work the principles of decision analysis to assist in coping with risk in agriculture.

Selected Additional Reading

Most of the strategies considered in this chapter are treated more comprehensively and analytically by Robison and Barry (1987), whose book is recommended for further reading. Harwood *et al.* (1999) consider strategies farmers can use to manage risk in the USA, while Anderson (2003) gives an overview of strategies and arrangements for dealing with risk in rural development, mainly in less developed countries.

For readers wanting to learn more about the use of derivatives as risk-management tools, the books by Williams and Schroder (1999), Geman (2005) and Hull (2011) are recommended. While Williams and Schroder focus on risk management in agricultural markets, Geman and Hull deal with derivative markets in general.

References

Allen, D.W. and Lueck, D. (2003) *The Nature of the Farm – Contracts, Risk and Organization in Agriculture.* MIT Press, Cambridge, Massachusetts.

Anderson, J.R. (2003) Risk in rural development: challenges for managers and policy makers. *Agricultural Systems* 75, 161–197.

Anderson, J.R., Dillon, J.L. and Hardaker, J.B. (1977) *Agricultural Decision Analysis*. Iowa State University Press, Ames, Iowa.

Garcia, P. and Leuthold, R.M. (2004) A selected review of agricultural commodity futures and options markets. *European Review of Agricultural Economics* 31, 235–272.

Geman, H. (2005) *Commodities and Commodity Derivatives – Modelling and Pricing of Agriculturals, Metals and Energy*. Wiley, Chichester, UK.

Greiner, R., Miller, O. and Patterson, L. (2009) Motivations, risk perceptions and adoption of conservation practices by farmers. *Agricultural Systems* 99, 86–104.

Harwood, J., Heifner, R., Coble, K., Perry, J. and Agapi, S. (1999) *Managing Risk in Farming: Concepts, Research, and Analysis*. Agricultural Economic Report No. 774. Market and Trade Economics Division and Resource Economics Division, Economic Research Service, United States Department of Agriculture, Washington DC.

Hull, J.C. (2011) *Options, Futures, and Other Derivatives*, 8th edn. Prentice-Hall, Upper Saddle River, New Jersey.

Koesling, M., Ebbesvik, M., Lien, G., Flaten, O., Valle, P.S. and Arntzen, H. (2004) Risk and risk management in organic and conventional cash crop farming in Norway. *Food Economics* 1, 195–206.

Lien, G. (2003) Assisting whole-farm decision making through stochastic budgeting. *Agricultural Systems* 76, 399–413.

Lien, G., Flaten, O., Jervell, A.M., Ebbesvik, M., Koesling, M. and Valle, P.S. (2006) Management and risk characteristics of part-time and full-time farmers in Norway. *Review of Agricultural Economics* 28, 111–131.

Lore, M. and Borodovsky, L. (eds) (2000) *The Professional's Handbook of Financial Risk Management*. Butterworth-Heinemann, Oxford.

Meuwissen, M.P.M., Huirne, R.B.M. and Hardaker, J.B. (2001) Risk and risk management: an empirical analysis of Dutch livestock farmers. *Livestock Production Science* 69, 43–53.

Milham, N. (1992) The financial structure and risk management of wool growing farms: a dynamic stochastic budgeting approach. MEc dissertation, University of New England, Armidale, New South Wales, Australia.

Paroush, J. and Wolf, A. (1989) Production and hedging decisions in the presence of basic risk. *Journal of Futures Markets* 14, 545–573.

Patrick, G.F. and Musser, W.N. (1997) Sources of and responses to risk: factor analyses of large-scale US cornbelt farmers. In: Huirne, R.B.M., Hardaker, J.B. and Dijkhuizen, A.A. (eds) *Risk Management Strategies in Agriculture*, Volume 7. Wageningen Agricultural University, Wageningen, The Netherlands.

Pennings, J.M.E. and Meulenberg, M.T.G. (1997) Hedging efficiency: a futures exchange management approach. *Journal of Futures Markets* 17, 599–615.

Petersson, J. and Andersson, H. (1996) The benefits of share contracts: some European results. *Journal of Agricultural Economics* 47, 158–171.

Robison, L.J. and Barry, P.J. (1987) *The Competitive Firm's Response to Risk*. Macmillan, New York.

Skees, J.R. and Barnett, B.J. (1999) Conceptual and practical considerations for sharing catastrophic/systemic risks. *Review of Agricultural Economics* 21, 424–441.

Tomek, W.G. and Peterson, H.H. (2001) Risk management in agricultural markets: a review. *Journal of Futures Markets* 21, 953–985.

Turvey, C.G. (2001) Weather derivatives and specific event risk in agriculture. *Review of Agricultural Economics* 23, 333–351.

Williams, J. and Schroder, W. (1999) *Agricultural Price Risk Management: the Principles of Commodity Trading*. Oxford University Press, Melbourne.

13 Risk Considerations in Agricultural Policy Making

Introduction

Our illustrations in earlier chapters demonstrate that the type and severity of risks confronting farmers vary greatly with the farming system and with the climatic, policy and institutional setting. This is the case in both more developed countries (MDCs) and less developed countries (LDCs). Nevertheless, agricultural risks are prevalent throughout the world and, arguably, have increased over time, as is suggested by the food, fuel and finance crises that have beset the world since 2007. Moreover, climate change appears to be creating more risk for agriculture in many locations. These prevalent and prospective agricultural risks have naturally attracted the attention of many governments – groups of DMs who have so far received little focus in our discussion. In this chapter we address analysis of risk management from this rather different point of view.

In our treatment we deal first with government interventions that have risk implications. Governments should realize that they are an important source of risk, as explained in earlier chapters, in particular when interventions negatively affect the asset base of farms. Potentially successful interventions are not those that merely reduce variance or volatility, but those that increase risk efficiency and resilience (to shocks, such as occasions of severely reduced access to food in LDCs, or extreme weather conditions). In many cases, this means increasing the expected value rather than decreasing the variance. In regard to specific instruments whereby farmers can share risk with others, we argue below that only in the case of market failure is there any reason for government involvement. Market failure is most severe in the case of so-called '*in-between risks*' or *catastrophic risks*. As explained later, in-between risks are risks that, by their nature, cannot be insured or hedged. Catastrophic risks are risks with low probabilities of occurrence but severe consequences. In this chapter we address issues in developing policies to manage these difficult risks as well as the management of some emerging risks, such as extreme weather, food-price spikes, food safety, epidemic pests and animal diseases, and environmental risks.

Government Intervention and Farm-level Risk

Background

Governments tend to want to intervene when they perceive that market mechanisms will not deliver economically efficient, socially desirable, environmentally sustainable or politically expedient outcomes. Such situations can be broadly conceived as forms of market failure. Unfortunately, however, it is not necessarily the case that all market failures can be fixed by government intervention. First, economic, social, environmental and political goals are often conflicting. Consequently, measures to improve the attainment of one goal will often have unintended and possibly seriously deleterious consequences in terms of other goals. So, for example, subsidies designed to support farmers' incomes may induce serious economic distortions and may lead to undesirable intensification of farming systems, thereby possibly harming the environment. In this chapter, we leave political considerations to the politicians and focus mainly on economic efficiency with some consideration given to social welfare and the environment.

Second, while markets certainly do fail in some situations, it is by no means always the case that government intervention will make things better. Sometimes, the information needed to design an effective policy may be too difficult and expensive to collect, or may even be unobtainable. Then the administrative costs of intervention may be high and unjustifiable relative to the likely benefits. Moreover, market failures have to be balanced against administrative failures. Some policy measures, especially those of a 'command and control' nature, may be too difficult to implement effectively. Even a well-designed policy may fail if the public servants responsible for its implementation are incompetent or corrupt, inappropriately trained and experienced, or inadequately supervised and resourced.

Experience shows that policy intervention in agriculture is a tricky business, to the extent that decisions about such interventions fall clearly into the category of risky choice. Evidently, policy analysts could well use some of the previously described methods of decision analysis to better account for the uncertainty of (often multi-attributed) outcomes that proposed interventions are intended to achieve.

Any government intervention that affects farmers' costs and returns will also affect their risk efficiency. As illustrated in Fig. 12.1 in Chapter 12, it is a mistake to think of policies relating to risk in agriculture solely in terms of measures to reduce volatility of production, prices or incomes. Of course, such measures are certainly important and are addressed later in this chapter. On the other hand, as is also clear from Fig. 12.1, any intervention that has the effect of shifting the CDF of a farmer's uncertain income to the right will be risk efficient in the sense that it will be utility increasing for any farmer who prefers more income to less, regardless of attitude to risk. Even when the intervention does not produce a stochastically dominant improvement in farmers' returns in the sense of Fig. 12.1, the result may still be risk efficient for moderately risk-averse farmers, as discussed in relation to Fig. 12.2. Moreover, the converse is also true. An intervention that reduces the volatility of farmers' returns at the cost of a leftward shift in the CDF may well decrease risk efficiency – a point that we believe is not universally well understood.

Decision analysis and policy making

Dealing with risk in public choice, in principle, merely involves application of the ideas set out in agricultural management contexts in earlier chapters. Such decisions always embody risks. Recognizing these systematically sets the scene for their better management by policy analysts and DMs. Documenting deliberated mechanisms for protecting against specific downside risky outcomes, and for dealing with them if and when they emerge, means that decision analysts have probably done their job properly. Decision analysts working on policy issues also need to alert implementers to their obligations to limit exposure to risk, to monitor progress and prospects, and to respond in an appropriate and timely manner when things go wrong. Our discussion of the virtues of flexibility (Chapter 12) in project design is naturally highly relevant to this theme. Needless to say, due accounting for downside risks should enter any *ex ante* assessment of the expected returns from a given choice in order not to send wrong signals to those making decisions about the acceptability of available options.

The need for a proper *ex ante* appraisal of such public choices raises the issues of what probabilities and what utility function should be used in such analyses. We have presented some ideas on how the formation of 'public' probabilities for risky policy analysis might be addressed in Chapter 3. In regard to choice of utility function, we consider first the case where all the consequences of the policy choice are adequately measured in money units via some form of cost–benefit analysis leading to a probability distribution of a measure of worth to society such as net present value. What degree of risk aversion is applicable to the appraisal of such a distribution?

Arrow and Lind (1970), in a seminal analysis of public investment under uncertainty, argued that society is usually able to pool its risks effectively by means of the large numbers of people sharing the risks, so that society as a whole is, to all intents and purposes, neutral towards risk. Cases that these authors say are unusual or exceptional do, however, occur. They are concentrated on project appraisal in LDCs and deserve brief comment here. The classic analysis is that of Little and Mirrlees (1974, p. 316), who outlined when something other than the maximization of expected net present value would be appropriate. Briefly, when a public project is large relative to national income, or highly correlated with such income, or a particular disadvantaged group is involved, there is a strong case for explicit accounting for riskiness of alternative actions, or equivalently, for explicitly taking account of social risk aversion through use of an appropriate social utility function. Simplified methods of accounting for such risk adjustments have been suggested by Anderson (1989) and are described in Chapter 5, this volume (see Eqn 5.25).

The issues are somewhat less clear when the nature of the public choice requires the evaluation of multi-attributed consequences. Whose preferences are then to be used to assess attribute utility functions and weights? There is no simple answer to these questions. Some analysts have been able to gain access to the DMs themselves (or their senior advisers) to undertake the needed assessments. Others have used sample surveys of members of the affected groups or representative panels thereof. These latter approaches raise hard theoretical problems about the legitimacy of combining utility functions in the face of Arrow's so-called 'Impossibility Theorem' which asserts that such interpersonal utility comparisons are invalid (Arrow, 1963). Perhaps it is not surprising, therefore, that some analysts have sought to dodge these tricky questions by recourse to some arbitrary rating methods. Similarly, some economists have sought to assign money values to intangibles to extend conventional cost–benefit analysis. Unfortunately, all such methods imply the adoption of an underlying utility function, so the problem has been implicitly hidden rather than really solved. We do not expect a solution to this puzzle to be found in the near future.

Interventions to improve farm productivity and incomes

Policies that governments can and often do adopt to achieve desirable shifts in farm incomes and risk efficiency include measures to improve farm productivity and measures to improve farmers' domestic terms of trade.

Because of the public-good nature of much productivity-improving research in agriculture, it is likely that there will be too little investment in research if it is left entirely to the private sector. For this reason, most governments fund agricultural research programmes and many also fund extension programmes to help pass on research findings to farmers.

Policies to improve farmers' human capital may also be expected to improve farm productivity and risk management. Human capital improvement is usually addressed by measures to upgrade rural education and health services. Of course, such interventions have broader objectives and benefits than merely improving farmers' risk management.

Policies that improve farmers' domestic terms of trade can be effective in increasing farm incomes. The most obvious way these risk-reducing benefits can be attained is by governments investing in improved rural infrastructure, such as better farm-to-market roads. Such public goods will not be provided by the private sector, so this is a clear responsibility of governments. Governments can also reform marketing arrangements to promote efficiency improvements through greater competition.

Finally, some public investments in agriculture can be both productivity increasing and income stabilizing. An example can be the provision of irrigation water via public schemes.

Interventions that affect farmers' property rights

A major source of risk for many farmers is insecurity of property rights. In LDCs this risk typically relates to insecurity of access to land, with many farm families being tenants or share croppers who may find themselves dispossessed of their access to land at any time, perhaps with no recourse to legal redress. Particularly in MDCs, governments may act unpredictably to change property rights. Thus, growing concern among the general public may lead governments to act to curtail farmers' rights to use land in particular ways. In Australia, in order to try to protect biodiversity and endangered species, measures have been introduced to prevent farmers from clearing their land without special permission. In general, the affected farmers are not compensated for the loss of what was previously a property right. Evidently, governments, particularly those in LDCs, can reduce farmers' risk exposure by codifying and enforcing reasonable property rights, while those in MDCs might at least think carefully about measures introduced for environmental or other reasons that may seriously curtail pre-existing property rights of farmers to an extent that may threaten the viability of their operations.

Interventions for risk prevention and abatement

Just as farmers need to review measures to prevent or reduce the chances of bad events happening, so do governments, in those areas where they have main responsibility. Usually it is government responsibility to try to maintain quarantine and other border protection measures to prevent the importation of potentially disastrous pests, diseases and weeds. The reason is that such problems entail significant market failure in the form of externalities – those who by their negligence introduce or spread the problem seldom bear the cost of their actions. Partly for the same reason, it is also usually government responsibility to deal with and try to minimize the impacts of outbreaks of such exotic infestations when they occur. That means having plans in place to limit any outbreaks that happen, and making sure that there are resources available to implement those plans speedily and effectively.

The operation of such programmes often raises difficult questions about who pays for the costs of controlling an epidemic. If the levels of subsidy to affected farmers are inappropriately set, they create undesirable incentives. For example, if the compensation to a farmer whose animals are compulsorily slaughtered is set too high, there may be an incentive not to take all precautions to avoid an outbreak among animals on that farm, especially if farm profitability is being harmed by restrictions such as movement bans. On the other hand, setting compensation too low may incite some farmers to hide the existence of animals with disease symptoms in the hope of avoiding having all stock slaughtered.

There are other areas where government actions to avoid and abate risks affecting agriculture (and often other sectors) are important. Governments can do little to prevent extreme weather events but they can plan to reduce their impacts by, for example, making appropriate investments in such public goods as levees to contain swollen rivers or providing fire breaks in forested areas to limit the spread of wildfires. Warnings of impending disasters, such as floods, can help to avoid or limit their impacts, as can government disaster management systems that swing into operation quickly and effectively when such disasters strike.

Interventions to deal with the casualties of risk

The nature of risk means that sometimes things go wrong, perhaps very wrong, so that individual farmers or, more often, groups of farmers, experience serious losses. For family farms, this can mean significant economic distress. Living standards may have to be cut and, in the extreme, even the basic necessities may not be affordable. While family, friends or charities may help, there may also be a need for government help.

Disaster relief

Most governments will try to help if there is a disaster leading to human suffering. There can be little argument about the need for such assistance. The problems begin when the definition of a disaster is broadened to encompass more modest falls in incomes, perhaps even merely rather bad outcomes lying well within the typical dispersion of outcomes. Such 'disaster relief' policy, or sometimes the lack of it, represents a significant opportunity for analysis of potential public intervention (e.g. Anderson and Woodrow, 1989). There has been a tendency for emotion and public outcry to drive a process that leads governments to intervene in ways that, with the wisdom of hindsight, are demonstrably ineffective and distorting of individual incentives to plan more carefully for what in many situations are inevitable occasional bad outcomes. Such planning would naturally include prudent management of finances and selective purchase of risk-sharing contracts such as insurance, as discussed in Chapter 12. Hence the routine provision of disaster relief will have predictable negative consequences for broad participation in formal insurance markets. If governments rush to bail people out of the effects of otherwise-insurable natural disaster risks whenever there is clamour to do so, development of commercial insurance markets will be fatally compromised.

Poverty relief and social welfare

In most MDCs, safety nets exist to support those in society who strike hard times. For example, it is common for income support payments to be made to families on low incomes. In principle, the same support payments might be available to help farm families whose incomes are low owing to falls in production or prices. The availability of such payments would reduce, perhaps eliminate, the need for expensive special schemes of farm income support, such as subsidized farm revenue insurance.

A problem with this solution is that social welfare payments are often subject to both income and asset eligibility tests. Farmers, especially owner occupiers, are typically asset rich but income poor. Under some social security schemes, they might be required to sell the farm and live off the proceeds before they could be entitled to support. That makes little sense if the fall in income is temporary and if the farm can continue to generate a reasonable income once the current crisis is over. The problem can be solved by appropriate revision of eligibility rules. Equity considerations suggest that the same revised rules should apply to all self-employed people in similar situations, not just farmers.

In LDCs, the payment of income supplements can seldom be afforded, but it may still be possible for governments to implement measures to deal with short-term poverty via such means as food-for-work schemes and other safety-net schemes such as schemes to transfer cash to targeted recipients, perhaps conditional on them qualifying in some socially determined way such as maintaining attendance of children in school. In cases of serious downturn in farm production, international agencies may come in with famine relief, although disillusion with many past interventions has led to a shift towards making resilience a focus.

Resilience, according to the Organisation for Economic Co-operation and Development (OECD, in some 2013 advice to its Members), is most often defined as the ability of individuals, communities and states and their institutions to absorb and recover from shocks, while positively adapting and transforming their structures and means for living in the face of long-term changes and uncertainty. In a variety of development agencies, resilience has emerged not so much as a new conceptual construct but rather as an organizing framework that facilitates integration of humanitarian and development efforts. As an organizing framework, there seems scope for refinement into practical interventions in the diverse risky situations of the developing world. Efforts to mitigate and adapt to climate change constitute one important example. Using resilience as an organizing framework may facilitate more effective integration of both adaptation to risk and risk mitigation into broader development efforts. Work on strengthening informal and formal collective action, including work on governance, can also become better integrated into broader development efforts (Constas *et al.*, 2014).

Programmes designed to limit or prevent human suffering should, in principle, be well justified as well as meritorious. If implemented effectively, they should ideally somewhat reduce the need for other forms of government intervention in agricultural risk management.

Market Failure for Risk

In the ideal world sometimes assumed by economists, a full range of 'contingency' markets would exist that would enable economic agents to neutralize risks. In such a world, a risk-averse DM would be able to 'offload' as much risk as desired, no doubt sacrificing some reduction in expected income in consequence. In reality, not all such contingency markets exist and some of those that do are not fully effective in enabling farmers and others to share risks.

Individuals such as farmers would be able to smooth consumption flows to a large degree if the financial system were perfect. As discussed in Chapter 11, if farmers had access to a perfect finance system so that they could save in good times and borrow when times were hard, they could afford to make decisions on the basis of maximizing expected returns. Of course, no perfect finance market exists. In MDCs, where finance markets are relatively well developed, saving is easy enough, but credit may not always be available when needed. Even when available, it is offered at a higher interest rate than is available on deposits. Moreover, a prolonged spell of hard times, such as a long drought, could leave a farmer with an unsustainable level of debt. In LDCs the position is often worse. Many farmers have limited access to formal credit and may face usurious interest rates on informal credit. While improvements in farmers' access to financial services may ameliorate the situation, there may still be a need for other risk-sharing mechanisms.

Especially in MDCs, many mechanisms exist for risk sharing, but they are neither costless nor as widely available as might be thought desirable. Key questions are the extent to which there is market failure for risk in agriculture, how serious that failure is, and what can sensibly be done about it. The answers to these questions provide a basis for evaluating the need for, and merit of, government intervention in making risk markets more complete.

In thinking about markets for risk it is important always to recall that not all risks are the same. There is a spectrum of risks that can afflict farmers. At one end of this spectrum there are *independent risks* that are not appreciably correlated across farms. For example, the risk of theft of farm property usually strikes only one or two farms in an area at one time. Independent risks of this type are usually insurable, subject to some conditions that we shall examine below. At the other end of the spectrum are risks associated with falls in commodity prices, interest rate hikes or changes in exchange rates that have highly correlated impacts on many farmers at the same time. For example, a commodity price fall will affect all farmers who produce that commodity, just as all farmers in debt will be affected if interest rates rise. These kinds of risk can be called *covariate risks* because they affect most farmers operating in a particular market system. Risks of this kind are not usually insurable because insurers are not able to pool them. On the other hand, at least some covariate risks can be managed through derivative markets.

Many of the risks that confront farmers and others in agriculture lie on the spectrum of types of risks between covariate risks at one extreme and independent risks at the other. Following Skees and Barnett (1999), we call these *in-between risks.* They are neither independent nor highly correlated. Yet they can, on occasions, lead to high losses for insurers. For this and other reasons, discussed below, such risks can seldom be insured at affordable costs to farmers. On the other hand, since these risks are only moderately correlated, they generally cannot be managed through (traditional) derivative markets. It is therefore for these kinds of risk that there has usually been greatest pressure for government intervention. We examine these risks further below.

Derivative markets for agriculture

As far as covariate risk is concerned, it can be expected that these markets will evolve and develop as financial and risk markets deepen. It is of some concern, however, that, to date, rather few futures and related derivative products exist in agriculture. Exchanges that trade agricultural derivative products are located primarily in the USA, but also exist in Australia and Europe. Some agricultural derivatives trade on very thin markets. For others, trading was so limited that the derivatives were discontinued. Such outcomes seem at variance with the supposed need for farmers to be able to hedge risks. The reasons why so few farmers use these products to hedge risk were discussed in Chapter 12. They include the possibility that farmers are not as risk averse as is widely presumed.

Insurance markets for agriculture

In order to understand why there may be market failure in insurance markets for agriculture, we first need to consider the requirements for a risk to be insurable.

Conditions for risks to be insurable

The following are the ideal conditions for a risk to be insurable (adapted from Rejda, 2003):

1. A large number of homogeneous insured clients facing independent risks – necessary for the insurer to be able to pool the risk.
2. Accidental and unintentional losses – insurance can be problematic if losses are influenced by the management of the insured.
3. Determinable and measurable losses – the amount of loss and the extent to which it was caused by an insured event need to be unambiguous for proper loss assessment.
4. No catastrophic losses – the losses must be sufficiently independent and individually constrained so that there is an acceptably low risk of total losses so large as to threaten the solvency of the insurer.
5. Calculable chance of loss – necessary for the insurer to be able to rate the risk to set a premium, which may be problematic for low frequency, catastrophic loss events.
6. Economically feasible premium – if the premiums are too high, as would be required for high frequency but non-catastrophic events, clients will find it more profitable to retain the risk and absorb it as part of normal operating expenses.

When all these conditions are reasonably well met, insurance products are likely to be available, albeit at premiums that will be somewhat above the expected value of the risk by a margin for the insurer's normal administrative costs and profit. In so far as some conditions are not met, the cost of insurance will be higher, or only partial, or no cover at all may be available.

Reasons for failure of insurance markets for agriculture

Asymmetric information

A main source of failure in risk and related markets arises because of *informational asymmetry* between parties to potential contracts. Such asymmetry occurs when one of the parties has more or better information about a risky outcome than the other. As a result, it can be difficult or impossible for the parties to strike an effective contract. For instance, the market for many factors of production (such as farm finance or rented land) is somewhat imperfect because farm operators invariably know the operating environment and its risks better than potential providers of such services.

In decision making under uncertainty, informational asymmetry causes two main problems (Milgrom and Roberts, 1992): *adverse selection* and *moral hazard*.

Adverse selection is a problem of pre-contractual opportunism related to unobserved or hidden characteristics of a good or service traded, as in the productive services example. In the insurance industry adverse selection is the tendency of those who face higher risks of experiencing an insurable loss to buy insurance cover to a greater extent than those with average or lower expectations of loss. Adverse selection therefore represents a breach of the first condition listed above for risks to be insurable. Yet insurance may still be feasible if less effective. Faced with substantial adverse selection the insurer must set a higher premium (or face a loss) if informational asymmetry makes it impossible to identify the clients with higher risk and to rate policies differentially.

To manage adverse selection, many insurance contracts include a duty of disclosure clause making the contract invalid if the insured party fails to inform the insurer about any adverse circumstances affecting the insured risk. Insurers can also discriminate against clients who make frequent claims by loading premiums, but such measures may not completely solve the problem.

The second cause of market failure owing to informational asymmetry, moral hazard, refers to unobserved or hidden actions by one of the parties to a contract to the detriment of the other party. Moral hazard is a form of post-contractual opportunism. In the insurance industry the term is used to describe the tendency of people with insurance to change their behaviour in ways that lead to larger or more frequent claims against the insurer. For example, a farmer with crop insurance may choose to neglect a poor crop, knowing that the insurer will pay for any shortfall in yield below the insured level. Such behaviour is a breach of the second condition set out above for risks to be insurable. Moral hazard impairs the ability of parties to make mutually beneficial agreements and so limits the effectiveness of markets for risk. Overcoming moral hazard problems by monitoring behaviour is often impossible or at least too costly. Problems may sometimes be reduced by including an appropriate incentive in the contract. For example, an insurance contract may be written with a substantial deductible, meaning that the insured party has to bear the first part of any loss.

Moral hazard and adverse selection problems are not confined to insurance products. They can also be important in other actual or potential markets for risk. It can be argued, for example, that credit institutions, which offer loans at high interest rates, may induce adverse selection of clients, since those who do not plan to repay may be more likely to borrow. High interest rates may also induce moral hazard if borrowers then start to take more risks to try to meet the loan costs. Some ill-conceived government interventions in markets also suffer from problems of informational asymmetry. For example, as noted above, the routine provision of disaster payments to farmers in hard times is likely to encourage greater risk taking, leading to more frequent 'disasters'.

Although both adverse selection and moral hazard are often present for many kinds of risks, it may still be possible to develop a feasible insurance scheme using tools such as deductibles and premium differentiation. For example, optional motor vehicle insurance can be vulnerable to both adverse selection and moral hazard, yet insurance companies have been able to offer policies by being able to rate clients differentially to deal with adverse selection and by the use of deductibles and the like to control moral hazard.

Catastrophic risks

The fourth condition (i.e. no catastrophic losses) usually means that such risks as floods, droughts, hurricanes, all-risk crop yield losses and livestock epidemics have, until recently, seldom been commercially insurable. The reason is that insurers need to hold much larger reserves, or buy expensive reinsurance.

Insurers also confront a problem in rating catastrophic risks because such events are, by their nature, both rare and highly variable in scope and impact. Indeed, the next catastrophe may well be far worse than anything experienced in recorded history. Faced with such imperfect knowledge, insurers are generally obliged to inflate premiums considerably, in case a really bad outcome eventuates.

Despite the problems, the insurance market for catastrophic risks has been growing as a consequence of developments in capital markets, such as the increasing 'securitization' of reinsurance. We discuss the scope for such further growth for catastrophic risks in agriculture below.

In-between risks

In-between risks violate at least some of the ideal conditions for insurability given above. Many in-between risks embody aspects of adverse selection and moral hazard. Crop yield or crop revenue insurance are a good examples. Farmers whose farm conditions, such a soil type, aspect, etc. make them more vulnerable to crop failure than their neighbours are more likely to purchase such insurance. Yet the insurer may not be able to garner the information about such fine differences between farms, or it would be very expensive to do so. Moreover, there is an obvious moral hazard aspect in that yields tend to depend to a considerable extent on crop management. Further, if many farmers in a region were to purchase such insurance and the season was particularly adverse, the insurer could be confronted with catastrophic indemnity payments.

The problems with many in-between risks in agriculture may not prevent the development of products for farmers to share these risks with others, but the violations increase the cost of providing such products and reduce the supply. The result may be failed markets since the socially optimal amount of risk sharing cannot occur.

Other risks that have some of the characteristics of in-between risks for which anxieties about market failure are gaining new prominence include environmental risks, food safety and associated public health issues. However, for many such risks it seems likely that the best approach to remedy perceived problems will not lie in developing new insurance products, since, in many cases, these are likely to be too costly to be economic. Rather, solutions are more likely to lie in better management of the risks themselves.

The Possible Role of Government in the Market for Risks in Agriculture

Understandably, the development of derivative products is proceeding more quickly in those countries where, and for those products for which, government interference is least. For example, there can be no effective futures market for farm products that have prices subject to government manipulation through minimum price schemes. So, if governments want to see farmers offered good commercial opportunities to hedge commodity price risks, they should get out of the price formation process. Governments can encourage the development of derivative products for agriculture by ensuring that the appropriate legal and regulatory frameworks are in place to promote fair trade in such products. Then it is more likely that entrepreneurs will be encouraged to develop and launch new and innovative products to face the test of market acceptability.

Similarly, there would seem to be little need for governments to concern themselves about the market for insurable risks, except again to ensure that appropriate market rules are in place and are enforced. Such supervision should extend to requiring companies to follow sound prudential practices to make sure that funds are available to pay indemnities. When governments step in with more heavy-handed regulation, the results are often unfortunate. Political pressures mean that premiums are typically set too low and indemnity payments are made too generous, so that, sooner or later, a gap in funding opens up and governments may then have to pay for the shortfalls their actions have created.

Only in the case of in-between risks does there seem to be a reasonable case to be made for government to participate actively. The central policy problem in regard to these kinds of risks is that the markets for transferring them, if they exist at all, tend to clear at less than socially optimal quantities of risk sharing. Such failure raises the questions of whether and how governments might intervene to improve the situation. Unfortunately, however, the very reasons that make these risks hard to share through market arrangements also make it hard for policy makers to find ways to protect people (including farmers) from losses suffered owing to in-between risks.

One possible form of intervention that may have some merit is for governments to take a share of the reinsurance of in-between risks. Such reinsurance could be at commercial or subsidized rates, although experience discussed later in the chapter shows that subsidized agricultural insurance has had a rather bleak history. However, there are some arguments in favour of a limited government subsidy for this type of reinsurance, summarized below.

1. Many governments already provide disaster relief. Providing assistance through reinsurance may be more efficient because disaster relief tends to be ad hoc and often involves problems regarding who gets paid and when. For example, it is rare for relief to be paid if only a few farmers suffer severe losses, whereas cover could be individual, or at least local, under an insurance scheme. There are also considerable administrative costs incurred in setting up special agencies to deliver disaster relief. By providing reinsurance, governments can use the experience and capacity of insurance companies in handling large numbers of claims as well as in dealing with moral hazard and adverse selection problems.
2. In some parts of the world, governments already provide continuous payments to farmers (e.g. through price support). Providing assistance through reinsurance seems likely to be more effective, because governments provide assistance when and only when farmers' incomes are low and income enhancement is therefore most needed. The cost to the exchequer (or to consumers) is also much less that way.
3. Having the government financially involved may address a moral hazard problem in government behaviour: many catastrophes (e.g. losses from floods) can be either prevented or magnified by government policies, or lack thereof. Having governments financially responsible for some losses might be an incentive for them to put in place appropriate hazard management measures.
4. Financial involvement of a government in reinsurance may reduce political pressure to provide distorting and often capricious ad hoc disaster relief.
5. Governments can potentially provide reinsurance more economically than can commercial reinsurers. Governments have advantages because of their deep credit capacity and their unique position as the largest social entity in a country. These advantages enable them to spread risks broadly.

While more research is warranted to get a better understanding of the economic, social and practical advantages and disadvantages of government involvement to cover in-between risks, two critical factors in dealing appropriately with in-between risks can be listed. First, government participation, such as providing reinsurance, should be carefully designed with respect to exposure to adverse selection and moral hazard. Otherwise, both farmers and commercial insurers will seek to off-load their losses onto the government. Second, transaction costs, including monitoring and administrative costs, must be kept low if the insurance products developed are to be attractive to potential buyers.

While in principle such public–private partnership for insurance might appear to be rather straightforward, we shall see in the next section that practical results to date have generally been discouraging.

Experience with Public Risk-management Instruments

Given the diversity of their nature, and their proliferation over time and geopolitical boundaries, the experience of risk-intervention policies is, not surprisingly, diverse and varied. A significant focus of the profession of agricultural economics has been on dealing with the instability inherent in the sector, both in MDCs and LDCs. The generalizations presented below represent a synthesis of some relevant studies in the field.

The emphasis here on public intervention reflects the situation that has prevailed in agriculture in most countries. Yet this emphasis on intervention is in stark contrast to the fundamental importance and success of the private sector in overall risk management around the world, such as underpins international trade in agricultural and other products. For instance, the insurance industry, which supports such trade and deals with a diversity of inherent political and other risks, is vital to commerce in general and to the globalization of agriculture. This industry is almost completely private. The importance of the private sector in risk markets is expanding and spreading with globalization. On the other hand, of course, increasing globalization does not yet mean that producers all around the globe have ready access to insurance or other markets, so the many cases in rural areas where such access is absent must be considered.

Price stabilization

Price uncertainty in traded agricultural commodities has long been seen as a significant problem for everyone concerned, from small-scale producer to statutory marketing authority. Price stabilization is the major 'traditional' intervention in the agricultural sector. Various mechanisms have been used to pursue such stabilization objectives, with varying degrees of success and many failures. Probably the most common, and certainly the most significant in terms of the production inefficiencies and distortions to world trade they have caused, have been various forms of guaranteed price schemes for farmers. Because it seems that at last some serious efforts are being made to eliminate or at least scale back these schemes, we shall not dwell on them here. However, the proposed phasing out of such support under World Trade Organization (WTO) agreements has focused interest on other measures that might be used to stabilize farmers' prices. In the past, buffer stocks, buffer funds, variable tariffs and the like have been among the most popular alternatives to guaranteed prices.

The theoretical justification for price stabilization measures has been explored in detail in a number of studies (e.g. Newbery and Stiglitz, 1981). An important frequent finding, however, is that the welfare gains that are possible from price stabilization are relatively small. Moreover, the practical implementation of stabilization schemes raises many thorny problems to be overcome by programme administrators. These include the difficult-to-assess supply responsiveness to induced stability.

Among the important reasons for taking a cautious approach to farm price stabilization schemes is the tendency for political forces to intrude into the management of schemes virtuously put in place, and to modify the rules (e.g. concerning parameters such as trigger prices) in ways that benefit particular groups and inevitably bankrupt the scheme itself, or cause it to be such a drain on the public purse that it becomes impossible to sustain.

Government insurance schemes

Insuring farmers' yields, revenues or incomes has long attracted the attention of governments. As discussed above, these are in-between risks for which covariances of losses are fairly high, aggregate losses often large, and for which there are big problems of adverse selection and moral hazard to be confronted. Naturally, few commercial insurers have ventured into this potential minefield, but the problems have not dissuaded many governments from getting involved. Many schemes have been tried with few successes.

Crop yield insurance

For reasons of asymmetric information, full crop yield insurance has seldom been provided by commercial insurers, although specific risk cover is often available for such defined events as fire and hail damage. There are some recent exceptions to this generalization, but it is still largely true, in both MDCs and LDCs, that such full yield insurance schemes as exist are provided or supported by governments. The motivation for such programmes often originates in political concern about catastrophic risks such as drought, or the desire to reduce the incidence of loan defaults to banks (e.g. see the critical analysis of Hazell *et al.*, 2001).

With few exceptions, the financial performance of public crop insurers has been ruinous (Hazell, 1992). To be financially viable without government subsidies, an insurer needs to keep the average value of annual outgoings – indemnities plus administration costs – below the average value of the premiums collected from farmers. In practice, many of the larger all-risks crop-insurance programmes pay out $2 or more for every dollar of premium they collect from farmers, with the difference being paid by governments. Even at these high levels of subsidy, many farmers are still reluctant to purchase insurance. As a result, some public crop-insurance programmes have been made compulsory, either for all farmers growing specified crops (e.g. Japan), or for those who borrow from agricultural banks (e.g. Mexico). On the face of it, such compulsory schemes cannot be utility increasing for farmers reluctantly forced to join.

The primary reason for the high cost of public crop-insurance schemes is that they invariably attempt to insure risks that, as noted above, are prone to severe informational asymmetry problems in terms of both adverse selection and moral hazard.

Another overwhelming factor is the moral hazard problem that arises in the contract between government and private insurer if the government establishes a pattern of guaranteeing the financial viability of an insurance provider. If commercial insurers know that any losses will automatically be covered by the government, they have little incentive to pursue sound insurance practices when setting premiums and assessing losses. In fact, they may even find it profitable to collude with farmers in filing exaggerated or falsified claims.

Yet another common reason for failure has been that governments undermine public insurers for political reasons. In Mexico, the total indemnities paid have borne a strong statistical relationship with the electoral cycle, increasing sharply immediately before and during election years, and falling off thereafter. In the USA, the government has repeatedly undermined the national crop insurer by providing direct assistance to producers in 'disaster' areas. Why should farmers purchase crop insurance against major calamities (including drought) if they know that farm lobbies can usually apply the necessary political pressure to obtain direct assistance for them in times of need at no financial cost?

Income and revenue insurance

For farmers, insuring their whole-farm income is likely to be more attractive (i.e. closer to optimizing the welfare of the farm family) than insuring separate components of the income, such as the revenue for a particular commodity. From a commercial insurer's point of view, however, insuring whole-farm incomes is not attractive at all. It includes aspects such as farm operating costs and inventories, which are strongly influenced and easily manipulated by an insured farmer. As a result, severe problems of moral hazard are likely, and adverse selection will also be a problem because farmers usually know more about their future income prospects than do insurers.

Even insuring the revenue (price × yield) for a particular commodity is not a simple task. For instance, in order to rate such a policy an insurer would need information on the stochastic dependency between prices and yields. If the correlation is negative (i.e. lower yields result in higher prices, and vice versa) revenue insurance should be less expensive than insurance for yields only. The correlation may change over time, for example as a result of market liberalization, requiring rate adjustments. Also, insurers need to make forecasts of the price levels at harvest or sale time, yet the volatile nature of agricultural prices means that simple projection using historic information will be unreliable.

These and other problems have not stopped some governments from introducing subsidized farm revenue insurance schemes. The motivation for the introduction of revenue insurance comes from the need for governments to desist from paying subsidies based on farm production under recent WTO agreements. Disaster relief is exempt from these provisions (subject to certain restrictions) and several governments have seized on this loophole to be able to continue subsidizing farmers as a means of retaining their political support.

The best documented of these interventions are the schemes in the USA. After being introduced experimentally in 1996, the US Federal gross revenue insurance scheme quickly grew to a national programme covering an increasingly wide range of products. The level of subsidy support for the scheme has also been growing rapidly.

Skees (1999), in a critique of the schemes at that stage, noted many problems. First, subsidized insurance does not take the risk out of farming. The availability of subsidized revenue insurance is likely to encourage farmers to take more risk in other aspects of their business operations. Second, the value of the subsidized protection, like other farming subsidies, quickly becomes capitalized into land values, making land owners richer, but doing nothing for tenant farmers or new entrants. Third, Skees documented some unfortunate unintended consequences of the schemes. He found that those farmers with the highest risk and those in the higher risk regions gained most from the subsidies. This effect has induced a shift in production of major crops away from the most stable and productive areas to more marginal areas. Moreover, because the subsidies are not 'decoupled' from production, they induce a positive supply response that has a negative effect on market prices. The price declines hurt the producers who are most productive yet for whom insurance is unattractive. The distribution of the benefits among farmers is also inequitable. These subsidies, paid under the guise of revenue insurance, like all such subsidies that are linked to the level of output, go mainly to the better-off farmers who produce most of the insured production, the very ones who least need subsidized support. Perhaps most importantly, Skees concluded that, but for the existence of subsidized insurance, there is no reason why commercial multi-peril crop yield and revenue insurance products could not be developed, as well as a range of other products that would help farmers better manage their risks without the distorting effects of the present schemes.

Evidently, governments might be better advised to look at how they might facilitate the evolution of new and effective insurance products, rather than stepping in with ill-conceived subsidized agricultural insurance schemes that may do more harm than good.

Where governments are encouraging innovation rather than impeding it by provision of competing subsidized products, some commercial solutions to sharing farming risks are emerging. Some of these developments are discussed in the next section.

Continuing Issues in Risk Management

Index-based crop insurance and derivatives

In agriculture, analysts have long had the idea of eliminating moral hazard and adverse selection problems that have bedevilled crop insurance so persistently by insuring, instead of an individual's crop and its performance, some more objectively measured index that is less subject to the unplanned-for influence of the insured. One such index that has been proposed is crop yield assessed over a local area so as to avoid the moral hazards of insuring yields on an individual farm or field basis. However, recently there has been a growing interest in products based on weather indexes.

The original idea of index products for transferring risks was published in the middle of the last century (Halcrow, 1949). With the deepening of financial and insurance markets in the 1990s, it got renewed attention. With index insurance products, payment of an indemnity depends on an objective index based, for example, on observations of rainfall or temperature. To avoid problems of informational asymmetry, the index should be independent and reliable – thus beyond control of both insured and insurer. If the index falls below (or rises above) an agreed threshold value, then indemnities are paid by the insurance company. Because there is a (single) objective index, which is easy to measure at low cost, it is usually relatively easy to calculate the probability that indemnities are due. Index products have the potential, therefore, to be cost-efficient and easy to administer – although climate change, especially change that increases the frequency of extreme weather events, adds an extra degree of difficulty.

It is essential that there is a high correlation between the index (such as amount of rainfall at a specified recording station and in a specified time interval) and the losses suffered by the insured farmers. To the extent that the index is not perfectly correlated with the insured property (for instance, crop yield), the insured is left with a basis risk. The lower the basis risk the higher the effectiveness and the efficiency of the risk transfer. The disadvantage of the basis risk, which will always be present though varying in size, may be more than compensated for by the cost advantages of an index product in terms of lower premium and administration costs.

This idea has been vigorously pursued in various ways, and has been under experimental implementation in a few countries (Greatrex *et al.*, 2015). The general idea is that specifically defined perils are insured, such as failure to reach a defined fraction of normal rainfall at agreed recording stations. Insurance policies consist of standard contracts for each unit at a fixed price for a defined region and there are no limits to the number of units an individual can purchase. Even with these simplifications relative to conventional crop-insurance contracts presently used in agriculture, there are implementation issues yet to be ironed out, so it is still premature to declare that such index insurance instruments will necessarily meet the test of the market.

There are also emerging derivative products tied to similar indexes. So, there are reports of what are in effect weather futures being traded that are also based on some objective weather index. Some deals in innovative instruments are commercial in confidence, so it is not easy to assess the current extent of market developments, but it is likely that many more innovative products will be launched in coming years. Some may pass the test of market acceptability and some will no doubt fail. Those that survive will surely broaden the scope for farmers and agribusinesses to share risks in more cost-effective ways.

Food safety and public health

Risks relating to food safety and public health have recently become of increasing concern. These risks are usually caused by hazards in food such as microbial (e.g. aflatoxins and *Salmonella*), chemical (e.g. dioxin, antibiotics, herbicides or insecticides) or physical (e.g. pieces of glass or metal) contamination. The emergence of new pathogens, changes in the food system, and increased trade in food products have all led to more attention being given to food safety issues. A number of countries have updated their food safety regulations, such as the European Union (EU) Food Safety Law of 2002. The food industry is increasing its efforts to certify food safety and reassure consumers. Consumers' confidence in the safety of food has been affected by a number of major food-related crises, such as bovine spongiform encephalopathy (BSE) in beef, dioxin in beverages, melamine in Chinese milk products, formalin in Bangladeshi fish and vegetables and *Escherichia coli* in hamburgers. Although many experts believe that food has never been as safe as it is now, that is not the perception of consumers. The risk of food contamination is an 'imposed' risk for consumers – something over which they have no control. Hence, as discussed by Gardener (2008), they are likely to over-estimate the importance of the risk.

While food may indeed be safer than ever, the risks of widespread problems have also increased. Because food ingredients are increasingly traded worldwide, and it is impossible to check all these ingredients for potential contamination, there is always a risk that something harmful will end up in the food chain. In contrast to earlier years, affected consumers more often try to sue food suppliers for harm they suffer caused by unsafe food. The increased scale of food production and processing means that many consumers can be affected by a single contamination event, raising the prospect of potentially crippling class actions against suppliers of unsafe food. Analysis of the BSE problem showed that, in the worst case, almost all consumers in the UK could have been infected by eating beef or beef products. Similarly, if there were to be an undetected dioxin contamination on a farm producing milk, the processing of this milk into pasteurized milk and milk products (yoghurt, cheese, etc.) could endanger a whole population.

The scale of food safety issues was illustrated by a survey by Buzby *et al.* (1998), which revealed that between 6.5 and 33 million people/year in the USA fall ill from microbial pathogens in their food and, of these, up to 9000 die. In addition, between 2% and 3% of people falling ill following *Salmonella* infections develop secondary illnesses or complications such as arthritis. According to Buzby *et al.*, in the summer of 1997, 25 million lb (about 11,440 t) of hamburger meat was recalled because of potential *E. coli* problems.

The basic problem with food safety is that consumers do not know the level of food-borne-illness risk – contaminants, especially pathogens, cannot usually be identified by visual inspection of the product. Because food suppliers usually have more and better information about the ingredients used and about production and handling, there is informational asymmetry between buyer (consumer) and seller (producer or supplier), implying market failure. The result can be unsatisfactory levels of contamination of food supplied with associated excessive risks to human health. Evidently, in such circumstances public health and social welfare are affected, indicating that intervention by governments may be appropriate.

Most governments do intervene via food safety regulation, and also often through criminal law, so that people who supply unsafe food may face heavy fines or jail sentences. Moreover, as food supply chains nowadays get better organized, using techniques such as electronic tracing and tracking, at least in MDCs, it is growing more likely that blame for a food-borne illness can be traced back to those responsible. Food producers may therefore face large and perhaps catastrophic damages claims. Most therefore seek to protect themselves against such risks by the purchase of product liability insurance that indemnifies them against such claims (though not against criminal penalties). Unfortunately, such insurance cover may create moral hazard problems, to the possible disadvantage not only of the insurer but also of consumers. Partial solutions to this problem, at least for insurers, may be sought through deductibles and co-payments.

Choosing the right mix of measures to improve the safety of food products is itself a risky decision problem. Most such measures are costly and the benefits are usually uncertain. Both for a private business concerned with profitability and financial sustainability, and for a government, concerned about economic and social costs and benefits, the required analyses are challenging. The methods of decision analysis described in this book should surely be relevant to such choices, yet apparently rather little work has been done along these lines. Instead, much of the public debate on these issues seems to be about 'making food safe', which is impossible if taken literally. Even aiming for a near zero risk of any contamination across all products would be impossibly expensive. Therefore it is important to focus on the main risks and to find ways to reduce them that are as effective as possible in terms of the actual and opportunity costs incurred. On the other hand, proper analysis will confront some tricky issues about priorities, valuation of human health and life, and willingness of consumers to pay for products that may be safer but not perceptibly so when offered for sale.

Epidemic pest and animal diseases

Outbreaks of epidemic animal diseases such as swine fever (affecting pigs), foot-and-mouth disease (affecting all cloven-hoofed animals including cows, pigs, sheep and goats), avian influenza (affecting poultry and humans) and BSE (affecting cattle and humans) have shown that such diseases can have devastating social impacts. These impacts derive from the costs (in all forms) of:

1. the typical (e.g. EU) stamping-out strategy of killing and destroying infected, suspected and contact stock (resulting in loss of assets and animal welfare concerns);
2. the limitations on exports (resulting in lost export markets and extremely low domestic producer prices); and
3. limitations on the movement of animals and people in certain regions (resulting in inconvenience, lost incomes from tourism, etc.).

In terms of animal welfare, economics, personal anguish and societal outcry, the 'losses' can be large indeed. Society has grown increasingly concerned about these issues to the point that many people now think that such losses can no longer be borne. Therefore, it seems inevitable that more attention will be given to preventing such outbreaks and to managing them better when they occur. In the past, some farmers, officials and governments have been seen to be too slow to react to new outbreaks.

Epidemic pests in crops can also have large effects on economics and trade, although the ethical factor that is present in animal disease control (killing and destroying 'healthy' animals) is not usually significant in crop epidemic control.

In some countries, in particular those that have experienced outbreaks of contagious diseases, there is an increasing interest in reviewing the current regulations and practices, including the option to apply emergency vaccination, where available. The choice of preventive and control strategies for a contagious animal disease is a complicated one that should be based on the best available information and a sound trade-off of all aspects – evidently a case requiring decision analysis with multiple objectives (Chapter 10, this volume). Moreover, the issue is inevitably one fraught with uncertainty, so a proper risk analysis needs to be undertaken in reaching a decision about the best control strategies to use.

In formulating policies to try to prevent and control epidemic pests in crops, a similar approach is needed. Epidemiological objectives need to be balanced with economic (damage to affected crops, export consequences) and environmental (including use of chemical control agents) objectives. Again, choosing the best strategy is a risky choice, so here too a multi-objective decision analysis is indicated, at both policy and individual farm levels.

Growing concerns with environmental risks in agriculture

One emerging aspect of agriculture that is riddled with uncertainties is 'the environment'. Most of the agricultural risk literature has been concerned with the production and economic risks facing farmers. Since agricultural activities may often have uncertain but significant impacts on natural resources and the environment through negative and positive externalities, resource depletion, and reduction of environmental amenities, these risks should also be taken into account in farm-level and society-level planning. There is a growing perception among policy makers and the population at large that some of the environmental threats created by modern farming methods may be unacceptable.

The complexities that confuse this topic are intense, and range from the biological, through the physical and chemical, to the social and economic. The phenomena of concern are diverse and include, for instance: (i) loss of biological diversity at the genetic, species and ecosystem levels; (ii) threats to critical ecosystem goods and services; (iii) agricultural contributions to global warming; and (iv) agricultural responses to the resulting changing but yet uncertain agronomic circumstances. Solutions being advanced include integrated pest management, organic farming, and various approaches to sustainable agriculture and sustainability. Risk plays many parts in influencing environmental outcomes of agricultural producers' decisions. Through changes induced in, say, cropping practices or livestock farming intensity, risk policy itself may have environmental consequences.

In the past, debates about environmental policies have been bedevilled by the polarization of positions by 'greenies' and 'developers'. A decision-analytic framework suggests that such differences can be dissected into different beliefs about the (risky) consequences of alternative actions, and different preferences for those consequences. People of good will should be able at least to narrow differences in beliefs about consequences by gathering and sharing more information. When good will is lacking, however, there is a tendency for both sides to make exaggerated claims that are not consistent with the evidence. Resolution of such problems is not aided when so-called experts retained by the opposing sides present as 'facts' conclusions they have reached about fundamentally uncertain phenomena. The role of policy analysis in this process, therefore, should be to strike a balance between the opposing views being advanced and to make the best possible assessment of the probability distributions of the uncertain events and consequences of concern and ultimately reach the best decision possible.

It is less clear how differences in preferences for consequences can be resolved. The polar viewpoints can be characterized as 'ecocentric' and 'anthropocentric', meaning focused primarily on ecological values and on human values, respectively. The view that all living organisms are sacred, for example, is a legitimate position for someone to take, but problems loom when believers seek to impose their views on the rest of society, including many people with different values. It is, of course, a matter for politicians to resolve such irreconcilable views, accounting, at least in a democracy, for the impact of the stance they take on the support they will receive at the next election.

Concluding Comment

Risk management for most farmers is a challenging task that relates to almost all decisions they make, ranging from the everyday to once-in-a-lifetime investment decisions. As argued above, policy makers and governments often have a big impact on farm-level risk management, both directly, by providing (or not providing) certain instruments, and indirectly, such as by creating an enabling environment for competitive farming. Institutional risk has been categorized in Chapter 1 as one of the main sources of risk for farming. This additional risk within the rural sector is a result of policy interventions that have uncertain outcomes, or which are subject to frequent and unpredictable changes in their design and implementation. Evidently, policy makers and governments have a special responsibility with respect to farmers' risk management. They should not always take it for granted that their decisions or projects are technically sound and that their interventions will be effectively implemented. But since risks pervading agriculture are not going away, policy makers must continue to be alert to the challenges of farm communities to manage their risks successfully, and to identify interventions that are actually helpful and defensible.

Selected Additional Reading

General overviews of the roles of government in promoting farm productivity growth have been provided by many authors including Lee and Barrett (2001), Pardey (2001) and Runge *et al.* (2003). Ways of reducing farmers' transaction costs are discussed, for example, by Gabre-Madhin *et al.* (2003).

A general overview of risk-management instruments available to farmers in the USA is given by Harwood *et al.* (1999), while Barry (1984) deals with many of these from a more theoretical standpoint. Fafchamps (2003) covers similar ground but from a development perspective.

Policy issues relating to risk in US agriculture are discussed by, for example, Glauber and Collins (2001). Several contemporary views are reported in the publication edited by Buschena and Taylor (2003).

For reading on derivatives in agriculture we recommend Carter (1999), Harwood *et al.* (1999), Williams (2001), Garcia and Leuthold (2004) and Geman (2005).

The OECD has published a number of reports on the policy aspect of risk in agriculture. These are not cited individually but interested readers can see what is available and in most cases download the documents at: http://www.oecd.org/tad/agricultural-policies/risk-management-agriculture.htm (accessed 2 June 2014).

An overview of the economics of crop insurance has been provided by Hueth and Furtan (1994) with more recent events covered in some several publications by Skees and his collaborators (e.g. Skees and Barnett, 1999; Skees, 2000; Skees *et al.*, 2001). Advantages of using market-based instruments for risk management have been argued by many, including Varangis *et al.* (2002). Analyses of the risks associated with sharp spikes in food prices have been assembled by Barrett (2013) and Chavas *et al.* (2015).

References

Anderson, J.R. (1989) Reconsiderations on risk deductions in public project appraisal. *Australian Journal of Agricultural Economics* 28, 1–14.

Anderson, M.B. and Woodrow, P.J. (1989) *Rising from the Ashes: Development Strategies in Times of Disaster*. Westview, Boulder and UNESCO, Paris.

Arrow, K.J. (1963) *Social Choice and Individual Values*, 2nd edn. Wiley, New York.

Arrow, K.J. and Lind, R.C. (1970) Uncertainty and evaluation of public investment decisions. *American Economic Review* 60, 364–378. Comments by Mishan, E.J., McKean, R.N., Moore, J.H. and Wellington, D. Reply by Arrow, K.J. and Lind, R.C. *American Economic Review* 62, 161–172.

Barrett, C.B. (ed.) (2013) *Food Security and Sociopolitical Stability*. Oxford University Press, Oxford.

Barry, P.J. (ed.) (1984) *Risk Management in Agriculture*. Iowa State University Press, Ames, Iowa.

Buschena, D.E. and Taylor, C.R. (eds) (2003) Advances in risk impacting agriculture and the environment. *Agricultural Systems* 75, Nos. 2–3.

Buzby, J.C., Fox, J.A., Ready, R.C. and Crutchfield, S.R. (1998) Measuring consumer benefits of food safety risk reductions. *Journal of Agricultural and Applied Economics* 30, 69–82.

Carter, C.A. (1999) Commodity futures markets: a survey. *Australian Journal of Agricultural and Resource Economics* 43, 209–247.

Chavas, J.-P., Hummels, D. and Wright, B. (eds) (2015) *The Economics of Food Price Volatility*. University of Chicago Press, Chicago, Illinois.

Constas, M., Frankenberger, T. and Hoddinott, J. (2014) *Resilience Measurement Principles*. Food Security Information Network Technical Series No. 1. Food and Agriculture Organization of the United Nations (FAO) and World Food Programme, Rome.

Fafchamps, M. (2003) *Rural Poverty, Risk, and Development*. Edward Elgar, Cheltenham, UK.

Gabre-Madhin, E., Barrett, C.B. and Dorosh, P.A. (2003) *Technological Change and Price Effects in Agriculture: Conceptual and Comparative Perspectives.* Discussion Paper 62. International Food Policy Research Institute, Washington, DC.

Garcia, P. and Leuthold, R.M. (2004) A selected review of agricultural commodity futures and options markets. *European Review of Agricultural Economics* 31, 235–272.

Gardener, D. (2008) *Risk: the Science and Politics of Fear.* Scribe, Melbourne, Australia.

Geman, H. (2005) *Commodities and Commodity Derivatives – Modelling and Pricing of Agriculturals, Metals and Energy.* Wiley, Chichester, UK.

Glauber, J.W. and Collins, K.J. (2001) Risk management and the role of federal government. In: Just, R.E. and Pope, R.P. (eds) (2001) *A Comprehensive Assessment of the Role of Risk in U.S. Agriculture.* Kluwer, Boston, Massachusetts, pp. 469–488.

Greatrex., H., Hansen, J., Garvin, S., Diro, R., Le Guen, M., Blakeley, S., Rao, K. and Osgood, D. (2015) *Scaling Up Index Insurance for Smallholder Farmers: Recent Evidence and Insights.* CCAFS Report No. 14, CGIAR, Copenhagen, Denmark.

Halcrow, H.G. (1949) Actuarial structures for crop insurance. *Journal of Farm Economics* 31, 418–443.

Harwood, J., Heifner, R., Coble, K., Perry, J. and Agapi, S. (1999) *Managing Risk in Farming: Concepts, Research, and Analysis.* Agricultural Economic Report No. 774. Market and Trade Economics Division and Resource Economics Division, Economic Research Service, United States Department of Agriculture, Washington, DC. Available at: http://www.agriskmanagementforum.org/sites/agriskmanagementforum.org/files/Documents/Managing%20Risk%20in%20Farming.pdf (accessed 2 June 2014).

Hazell, P. (1992) The appropriate role of agricultural insurance in developing countries. *Journal of International Development* 4, 567–581.

Hazell, P., Oram, P. and Chaherli, N. (2001) *Managing Droughts in the Low-Rainfall Areas of the Middle East and North Africa.* Environment and Production Technology Division (EPTD) Discussion Paper 78, International Food Policy Research Institute (IFPRI), Washington, DC. Available at: http://ebrary.ifpri.org/cdm/singleitem/collection/p15738coll2/id/48017/rec/6 (accessed 2 June 2014).

Hueth, D.L. and Furtan, W.H. (eds) (1994) *Economics of Agricultural Crop Insurance: Theory and Evidence.* Kluwer, Boston, Massachusetts.

Lee, D.R and Barrett, C.B. (eds) (2001) *Tradeoffs or Synergies? Agricultural Intensification, Economic Development and the Environment.* CAB International, Wallingford, UK.

Little, I.M.D. and Mirrlees, J.A. (1974) *Project Appraisal and Planning for Developing Countries.* Heinemann, London.

Milgrom, P. and Roberts, J. (1992) *Economics, Organization and Management.* Prentice-Hall, Englewood Cliffs, New Jersey.

Newbery, D.G.M. and Stiglitz, J.E. (1981) *The Theory of Commodity Price Stabilization: a Study of the Economics of Risk.* Clarendon Press, Oxford.

Pardey, P.G. (ed.) (2001) *The Future of Food: Biotechnology Markets and Policies in an International Setting.* International Food Policy Research Institute, Washington, DC.

Rejda, G.E. (2003) *Principles of Risk Management and Insurance,* 8th edn. Addison-Wesley, Boston, Massachusetts.

Runge, C.F., Senauer, B., Pardey, P.G. and Rosegrant, M.W. (2003) *Ending Hunger in Our Lifetime: Food Security and Globalization.* Johns Hopkins University Press, Baltimore, Maryland.

Skees, J.R. (1999) Agricultural risk management or income enhancement. *Regulation* 22, 35–43.

Skees, J.R. (2000) A role for capital markets in natural disasters: a piece of the food security puzzle. *Food Policy* 25, 365–378.

Skees, J.R. and Barnett, B.J. (1999) Conceptual and practical considerations for sharing catastrophic/systemic risks. *Review of Agricultural Economics* 21, 424–441.

Skees, J., Gober, S., Varangis, P., Lester, R. and Kalavakonda, V. (2001) Developing rainfall-based index insurance in Morocco. Policy Research Working Paper, No. 2577. World Bank, Washington, DC.

Varangis, P., Larson, D. and Anderson, J.R. (2002) *Agricultural Markets and Risks: Management of the Latter, Not the Former*. Policy Research WPS 2793. World Bank, Washington, DC.

Williams, J.C. (2001) Commodity futures and options. In: Gardner, B. and Rausser, G. (eds) *Handbook in Agricultural Economics*, Volume 1A. Elsevier Science, Amsterdam, pp. 745–816.

Appendix: Selected Software for Decision Analysis

In developing the examples in this book, we made use of the following software:

- @Risk and RiskOptimizer from Palisade Corporation. Available at: http://www.palisade.com (accessed 2 June 2014).
- DATA, later replaced with TreeAge Pro from TreeAge Software, Inc. Available at: http://www.treeage.com (accessed 2 June 2014).
- GAMS from GAMS Development Corporation. Available at: http://www.gams.com (accessed 2 June 2014).
- Logical Decisions from Logical Decisions. Available at: http://www.logicaldecisions.com (accessed 2 June 2014).
- Microsoft Excel, a component of Microsoft Office from Microsoft Corporation. Available at: http://www.microsoft.com/en-au/default.aspx (accessed 2 June 2014).
- ModelRisk from Vose Software. Available at: http://www.vosesoftware.com (accessed 2 June 2014).
- Solver for Excel from FrontlineSolvers. Available at: http://www.solver.com (accessed 2 June 2014).
- WhatsBest! from Lindo Systems Inc. Available at: http://www.lindo.com (accessed 2 June 2014).

Index

Page numbers in **bold** refer to illustrations and tables.